6th Global Conference on Materials Science and Engineering (CMSE 2017)

IOP Conference Series: Materials Science and Engineering
Volume 283

Beijing, China
24 - 27 October 2017

Editor:

Alexander Khotsianovsky

ISBN: 978-1-5108-5500-7
ISSN: 1757-8981

Printed from e-media with permission by:

Curran Associates, Inc.
57 Morehouse Lane
Red Hook, NY 12571

Some format issues inherent in the e-media version may also appear in this print version.

Printed by Curran Associates, Inc. (2018)

For permission requests, please contact the Institute of Physics
at the address below.

Institute of Physics
Dirac House, Temple Back
Bristol BS1 6BE UK

Phone: 44 1 17 929 7481
Fax: 44 1 17 920 0979

techtracking@iop.org

Additional copies of this publication are available from:

Curran Associates, Inc.
57 Morehouse Lane
Red Hook, NY 12571 USA
Phone: 845-758-0400
Fax: 845-758-2633
Email: curran@proceedings.com
Web: www.proceedings.com

Table of contents

Volume 283

6th Global Conference on Materials Science and Engineering

24–27 October 2017, Beijing, China

Accepted papers received: 14 December 2017
Published online: 2 January 2018

Preface

Papers

IOP Conf. Series: Materials Science and Engineering **283** (2017) 011001 doi:10.1088/1757-899X/283/1/011001

Preface

The 6th Global Conference on Materials Science and Engineering (CMSE2017) was held from October 24th to October 27th in Beijing, China. This conference was organized by Bosheng Academic Conference Solutions Co., Ltd (China).

This Conference was aimed to bring together leading academic scientists, researchers and scholars to exchange and share their experience and research results on all aspects of Materials Science and Engineering, and discuss the practical challenges encountered and the solutions adopted.

CMSE2017 attracted over 130 participants from 26 countries, namely Armenia, Canada, Chile, Czech Republic, China, Egypt, France, Greece, India, Italy, Japan, Lithuania, Malaysia, Mexico, Poland, Portugal, Republic of Korea, Romania, Russia, Singapore, Slovak Republic, Spain, Turkey, UK, Ukraine, and USA. The participants (scholars from universities and research institutes, experts from international companies, and engineers from enterprises) shared their oral and poster presentations, as well as six keynote and sixty-eight invited speeches on 7 topics: Nanomaterials, Energy materials and Semiconductors, Mechanics and Mechanical technologies, Structural materials, Characterization & Testing, Advanced Material Technologies, Surface and Coatings Technology. Eight best oral presenters were awarded to attend the conference in 2018 with a free registration fee.

This volume contains 31 papers in the above domains, which were been peer-reviewed by international experts from the Conference Technical Committee and considered innovative and instrumental to further scientific research in the field of Materials Science and Engineering.

Finally, it is my pleasure to thank all participants for their contributions to CMSE2017 and the Organizing Committee, especially the Conference Secretary Ms. Li Ling, for the impressive results of the conference.

Guest Editor:
Dr. Alexander Khotsianovsky,
Managing Editor of Strength of Materials

Peer review statement

All papers published in this volume of *IOP Conference Series: Materials Science and Engineering* have been peer reviewed through processes administered by the proceedings Editors. Reviews were conducted by expert referees to the professional and scientific standards expected of a proceedings journal published by IOP Publishing.

Hot pressing of nanocrystalline tantalum using high frequency induction heating and pulse plasma sintering

J Jakubowcz[1*], G Adamek[1], M Sopata[1], J K Kope[1] and P Siwak[2]

[1]Institute of Materials Science and Engineering, Poznan University of Technology, Poznan, Poland
[2]Institute of Mechanical Technology, Poznan University of Technology, Poznan, Poland
*E-mail: jaroslaw.jakubowicz@put.poznan.pl

Abstract. The paper presents the results of nanocrystalline powder tantalum consolidation using hot pressing. The authors used two different heating techniques during hot pressing: high-frequency induction heating (HFIH) and pulse plasma sintering (PPS). A comparison of the structure, microstructure, mechanical properties and corrosion resistance of the bulk nanocrystalline tantalum obtained in both techniques was performed. The nanocrystalline powder was made to start from the microcrystalline one using the high-energy ball milling process. The nanocrystalline powder was hot-pressed at 1000 ° C, whereas, for comparison, the microcrystalline powder was hot pressed up to 1500 ° C for proper consolidation. The authors found that during hot pressing, the powder partially reacts with the graphite die covered by boron nitride, which facilitated punches and powder displacement in the die during densification. Tantalum carbide and boride in the nanocrystalline material was found, which can improve the mechanical properties. The hardness of the HFIH and PPS nanocrystalline tantalum was as high as 625 and 615 HV, respectively. The microstructure was more uniform in the PPS nanomaterial. The corrosion resistance in both cases deteriorated, in comparison to the microcrystalline material, while the PPS material corrosion resistance was slightly better than that of the HFIH one.

1. Introduction

Many components used in advanced industry require high-tech materials that are resistant to high temperature, corrosive media or radiation. These components are usually made using, inter alia, refractory elements. However, refractory metals, alloys and composites based on Ta, W, Nb, oxides and nitrides [1-5] are technologically problematic materials that require high temperature processing, i.e. high sintering temperature. These materials are usually made using powder metallurgy [6,7]. In terms of powder metallurgy, the conventional process is usually based on the following stages: (1) preparation of a powder of required composition, size and shape, (2) high pressure compaction for green compact formation and (3) sintering for green compact consolidation [8]. To provide good mechanical properties of sintered material, i.e. strength and low porosity, high pressure and high temperature are required, which is unfavorable in terms of the grain growth at elevated temperature and high requirements for the pressing/sintering equipment. The conventional powder metallurgy process can be successfully used for microcrystalline powders. For nanocrystalline powders, however, other processing routes are required such as hot pressing [8], in which the stages of pressing and sintering are carried out simultaneously. Thus, the process has many advantages including lower pressure, temperature and shorter sintering time. Finally, the hot-pressed material has a greater density, lower porosity and smaller grain size, which ensure better mechanical properties. The hot pressing process has many possible variants of heating. One of them is high frequency induction heating (HFIH), resistance heating (RH), spark plasma sintering (SPS) or pulse plasma sintering (PPS) [1,9-11]. The PPS is a relatively new technique [11,12] developed for the consolidation of high melting temperature compounds [13]. The PPS technique for fast powder heating during pressing, utilizes the pulses of high electric current generated by discharging high voltage capacitors. This allows acceleration of the heating to the steady temperature and shortening the time of sintering. The process is useful in the obtainment of high-density and fine-grained materials [13]. Tantalum can be used in different technical elements, which works in hard conditions [14] and it is very resistant in acidic environments. High corrosion resistance is a consequence of a passive Ta_2O_5 oxide film [15]. Tantalum is an unfavourably expensive element used in very hard operating conditions: in aviation and in the space industry, for chemical, energy-related and biomedical applications. Beside the high cost, which limits its widespread use, one of the downsides of tantalum is its high melting point and difficulties during processing. The powder metallurgy applied to tantalum is a good choice for the production of bulk elements, however, high sintering temperature leads to an increase in the size of the grains. It is commonly known that, according to the Hall-Petch equation, the smaller grain size leads to material strengthening [16]. Therefore, small grains of nanometer size, before the consolidation stage, are much desired to limit the final grain size in the consolidated materials after the sintering process. The formation of nanostructure tantalum can be promising in terms of its strengthening. Decrease of the grain diameter leads to larger volume of the boundaries between them. The nanostructure tantalum powder can be successfully consolidated to form bulk parts of high density, close to the theoretical value. The application of hot pressing limits the grain growth at elevated temperatures (which is lower in hot pressing compared to a conventional sintering process) and should provide nanocrystalline or ultrafine tantalum of greater strength in comparison to microcrystalline counterparts.

The first study of the nanocrystalline tantalum sintering was carried out in 1999 by Yoo *et al* [17]. They found that large surface area of the nanoparticles improves the densification process in comparison to the microcrystalline powder. At high strain rates, tantalum nanograins exhibit ductile deformation behavior without significant fracture and the tantalum nanograins tend to get softer [17]. Ductility and flow stress increase significantly as the initial particle size is reduced below 100 nm. Thus, the nanocrystalline tantalum not only has better mechanical properties but is more susceptible to process in comparison to microcrystalline powder.

The nanocrystalline tantalum was investigated by other researchers [18,19], however, the material still needs further investigation. In this work, we compare properties of HFIH and PPS hot-pressed nanocrystalline tantalum. This is a preliminary study for further preparation of nanocrystalline tantalum based alloys. The new methods of nanomaterials formation can provide new data and properties useful in extending theirs application. Different techniques were successfully used for tantalum nanopowders consolidation [8,10]. The work attempted to determine whether the applied method of hot pressing affects the properties of tantalum.

2. Experimental data

The nanocrystalline tantalum was produced in the process of high energy ball milling (HEBM). The milled powders were next hot pressed. The starting Ta powder (AlfaAesar, Germany) has a size <44 μm. The impurities do not exceed 0.03%. The powder was stored and handled in Unilab glove box (MBraun, Germany). The HEBM process was run using SPEX8000M mill (SpexSamplePrep, USA). Milling time was set experimentally at 48 h. Room temperature (RT) was a standard condition of the process. Powders after HEBM were hot-pressed in vacuum ($<10^{-2}$ Pa). The consolidation was done using graphite equipment (die as well as upper and lower punches), which were covered by BN layer (HeBoCoat, Germany). The BN layer improves punches moving and powders consolidation. The

IOP Conf. Series: Materials Science and Engineering **283** (2017) 012001 doi:10.1088/1757-899X/283/1/012001

authors used two different heating modes during consolidation: high-frequency induction heating (HFIH) and pulse plasma sintering (PPS). The consolidation temperature was 1500°C and 1000°C for micro- and nanocrystalline powders, respectively. The time of sintering at a constant temperature was 5 s. The lower temperature of hot pressing was choosing for nanocrystalline material, which has better sinterability and to avoid its excessive grain growth [17,20]. The pressure exerted on the powder had a value of 50 MPa. In the PPS process the pulses of high electric current going through the die and the powder. The current pulses in the module are generated by discharging of a 250 F capacitor, charged to a voltage of maximum 8 kV. The pulse duration during sintering process is automatically controlled due to a temperature measurement by a pyrometer direct on the upper punch. Voltage, pulse frequency, force, temperature, and real-time vacuum were automatically controlled during the sintering procedure.

The structure was checked using Empyrean XRD (Panalytical, The Netherlands) with a crystallographic database ICDD-JCPDS. The investigations were performed using CuK radiation (45 mV, 40 mA). The crystallite size was calculated by the Scherrer's equation.

The microstructure was recorded by SEM (Tescan, Czech Republic), operated at 20 kV in the secondary electrons mode as well as by AFM (Quesant, USA). The AFM works in the wavemode (SSS-NCLR premounted Nanosensors probe, 145 kHZ probe frequency, and scan speed 3 lines/s, total 1024 lines for image formation). For the microstructure evaluation the hot-pressed samples were grinded, polished and chemically etched in a H_2SO_4 + HNO_3 + HF mixture.

The mechanical properties were measured using Picodentor HM500 (Fischer, USA) nanoindentation tester. The HV – Vickers hardness and E_{IT} – indentation modulus were measured at loading. The indentations were made using diamond indenter at the force of 300 mN acting for 20 s.

The corrosion resistance was measured on hot-pressed materials using Solartron 1285 Potentiostat (Solartron Analytical, England). The potentiodynamic mode was applied, starting at -0.5 V and finishing at 3 V vs OCP. The scan speed was 0.5 mV/s. The bulk tantalum was the working electrode, graphite was the counter electrode and Ag/AgCl/Cl^- was the reference electrode. The corrosion resistance was measured in the Ringer's solution containing the following ion concentrations: 147.2 mmol l^{-1} Na^+, 4.0 mmol l^{-1} K^+, 2.2 mmol l^{-1} Ca^{2+}, 155.7 mmol l^{-1} Cl^-.

3. Results and discussion

The microcrystalline tantalum powder shows a cubic-type Im-3m structure (reference code 01-089-4901). The material is single-phase (figure 1(a)). During high-energy ball milling for 48 h (figure 1(b)) a significant decrease of the crystallite size to 20 nm occurred. The reduction of intensity of peaks and their broadening points formation of nanostructure or even partial amorphization (figure 1(b)). The ball-powder-ball hits results in the introduction of material stress, formation of dislocations and subgrains. These finally lead to cold strengthening, material crushing and nanostructure formation. During hot pressing, a significant relaxation of stress occurs and the provided energy results in grain growth (figures 1(c) and 1(d)). Because of high reactivity of the amorphous/nanocrystalline metallic material, high temperature and residual gases under vacuum, the powder partially reacts with the gases as well as the graphite die (figures 1(c) and 1(d)) and forms intermetallic phases. However, given their properties, these Ta-C, Ta-O or Ta-B type compounds can have a positive effect on the final properties of the bulk nanocrystalline tantalum. The precipitations can strengthening of the tantalum as well as inhibit grain growth at elevated temperature [5,7,8]. In the case of the PPS material, the XRD peaks for additional phases are smaller, indicating that the PPS process does not significantly support the reaction of Ta with the B, C or O elements as is in the case of the HFIH material. The microcrystalline hot-pressed material (figure 1(e)) shows no additional phases, for both HFIH and PPS hot-pressed samples (due to the same spectrum for both PPS and HFIH materials, for comparison, figure 1(e) shows the HFIH sample only).

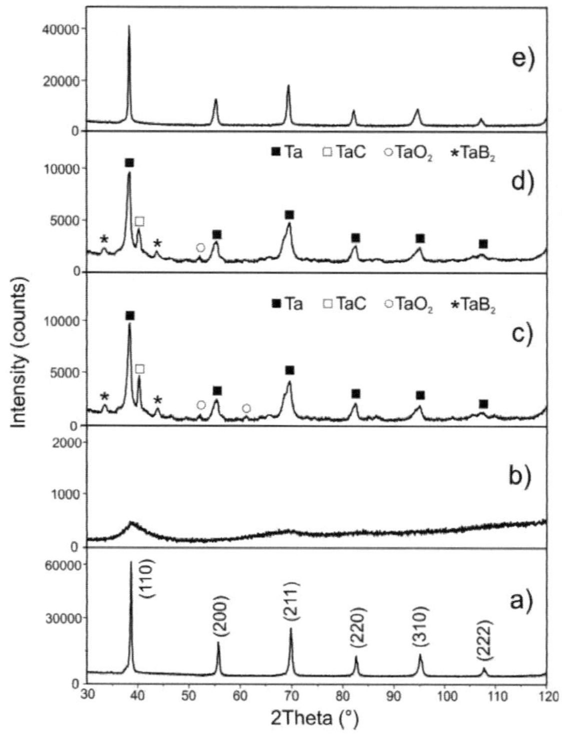

Figure 1. X-ray diffraction data for Ta: microcrystalline powder (a), nanocrystalline powder after mechanical milling (b), nanocrystalline hot-pressed using HFIH (c) nanocrystalline hot-pressed using PPS (d) and microcrystalline hot-pressed using HFIH (e).

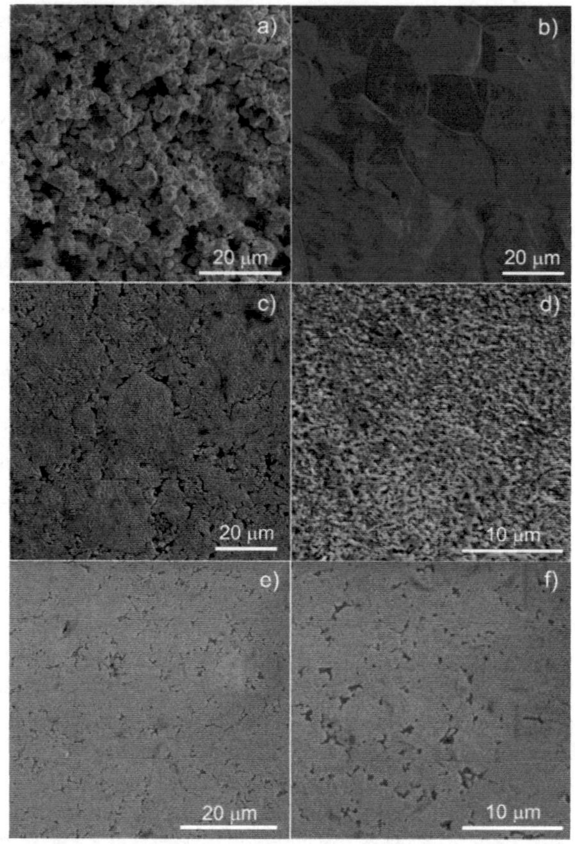

Figure 2. SEM image of tantalum: nanocrystalline powder after mechanical milling (a), microcrystalline hot-pressed (b) nanocrystalline hot-pressed using HFIH (c, d – different magnification) and nanocrystalline hot-pressed using PPS (e,f – different magnification).

In the HEBM process, the tantalum powder undergoes a grain size reduction, however, the powder during milling undergoes not only crushing but strong plastic deformation and cold welding (the powder particles are hit and trapped between the balls and the wall of the milling vial), which results in the formation of powder agglomerates of relatively large size (figure 2(a)). The agglomerates are composed of smaller particles, hence the SEM image (figure 2(a)) shows agglomerate particles (composed of nanograins) rather than nanoparticles. The XRD shows (peaks broadening and decreasing their intensity) that the material after intense milling transforms into nanocrystalline. As the reference material, the authors chose hot-pressed microcrystalline 325-mesh powder. The obtained consolidated microcrystalline sample has large grain size of the average value of 15 μm and low porosity (figure 2(b)). After mechanical milling, the nanocrystalline powders were hot-pressed using HFIH (figure 2(c) and 2(d)) and PPS (figures 2(e) and 2(f)). The hot-pressed nanomaterials have a slightly larger porosity following the pressing of the agglomerates, however, due to limited time of the hot pressing and a relatively low temperature, the metastable nanocrystalline material has insufficient energy for significant grain growth. After the HFIH slight porosity forms around the hot-pressed agglomerates (figure 2(c)). The higher magnification shows ultrafine structure (figure 2(d)). The PPS process results in a lower porosity and more uniform microstructure. The density of hot-pressed microcrystalline tantalum was 15.6 g/cm^3, whereas for HFIH and PPS hot-pressed nanocrystalline tantalum was 14.9 g/cm^3 and 15.3 g/cm^3, respectively. These densities are 95%, 91% and 93% of the theoretical value, respectively.

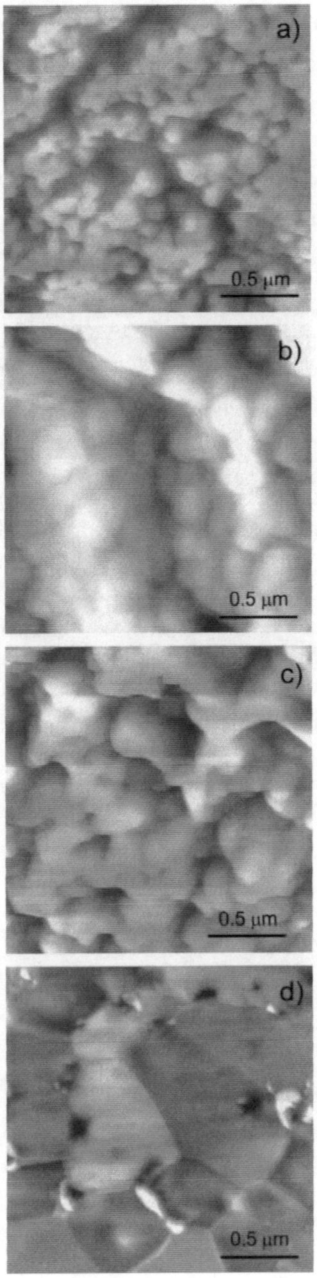

Figure 3. AFM images of tantalum: nanocrystalline powder after mechanical milling (a), nanocrystalline hot-pressed using HFIH (b), nanocrystalline hot-pressed using PPS (c) and microcrystalline hot-pressed (d).

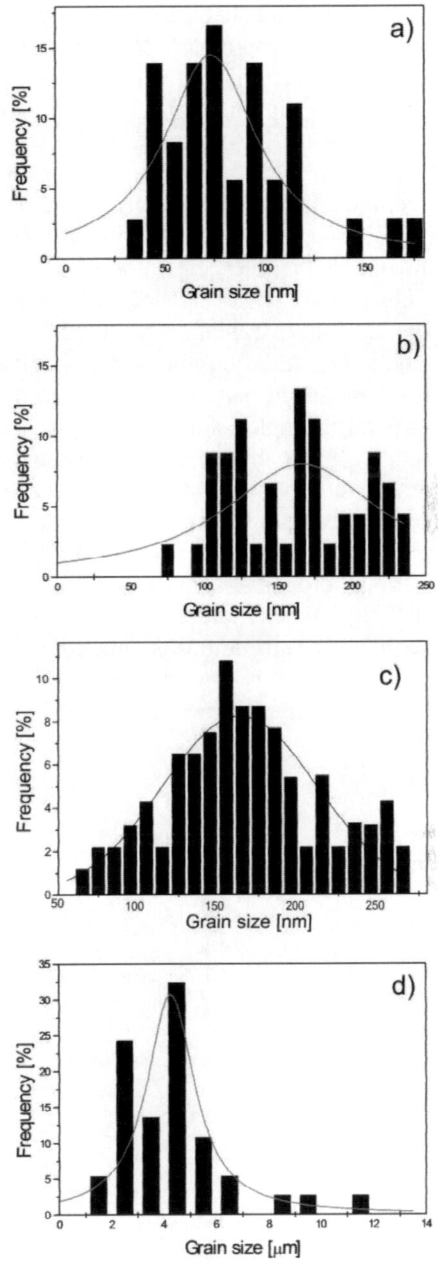

Figure 4. AFM grain size histogram of tantalum in the nanocrystalline powder after mechanical milling (a), nanocrystalline hot-pressed using HFIH (b), nanocrystalline hot-pressed using PPS (c) and microcrystalline hot-pressed (d).

The nanocrystalline materials were investigated using AFM. For comparison, the microcrystalline was also checked. Taking into account the AFM images (figures 3(a)-3(d)), grain size histograms were calculated (figures 4(a)-4(d)). For grain size analysis, we take the AFM pictures of 12 m^2 summarized area. The analysis was done on the pictures recorded on 3 different places to avoid sample inhomogeneity. The mechanically milled tantalum powders are shown in figure 3(a). The presented image shows agglomerated particles of relatively high homogeneity. However, the measured grain size distribution has a relatively broad range from approx. 30 to above 180 nm. The calculated average grain size is 70 nm (figure 4(a)). The hot pressing using HFIH results in the grain growth (figure 3(b)), but the grains are not larger than 230 nm (figure 4(b)). The calculated average grain size is 170 nm. The PPS process results in a smaller grain growth during the consolidation of the powders (figure 3(c)) and the average grain size is 165 nm (figure 4(c)). The microcrystalline sinter has large grains and, due to limited scan size, a relatively low surface area with only few grains in the frame was scanned (figure 3(d)). For the microcrystalline hot-pressed tantalum, the average grain size was calculated at 4.5 μm (figure 4(d)), however, the SEM image shows the average value of approx. 15 μm (figure 2(b)). The difference can be explained with the fact that AFM considered a lower amount of grains for the analysis and the smaller ones are better visibly compared to the SEM image.

The mechanical properties were measured using a nanoindentation tester (figure 5 and table 1). Figure 5 shows example load-displacement curves recorded for nanocrystalline tantalum. The results have a relatively broad spectrum of the curves that do not overlap one another, indicating some inhomogeneity and porosity of the hot-pressed material. The hardness of the nanocrystalline materials is twice as high compared to the microcrystalline and significantly higher than commercially available tantalum (100 HV for annealed and 200 HV for cold-worked). In both, HFIH and PPS nanomaterials, the mechanical properties are comparable. The high hardness in nanocrystalline tantalum can be attributed to the nanostructure as well as hard precipitations formed at elevated temperatures after contact with boron nitride-coated graphite die. The indentation modulus is comparable for all hot-pressed samples and is slightly lower in comparison to bulk commercial tantalum (186 GPa). Generally the porosity deteriorates the modulus.

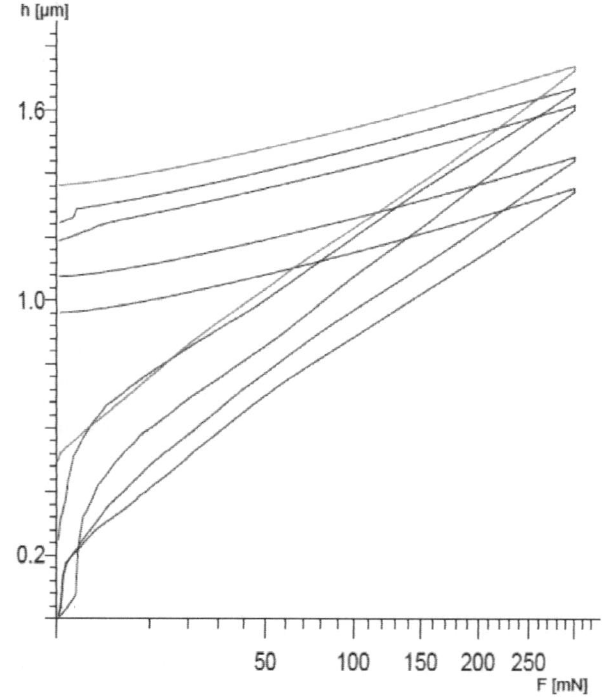

Figure 5. Example load-displacement curves recorded for nanocrystalline PPS tantalum.

Table 1. Mechanical properties of the hot-pressed microcrystalline and nanocrystalline tantalum

Parameter	Microcrystalline (HFIH)	Nanocrystalline (HFIH)	Nanocrystalline (PPS)
HV [HV]	346.4±29.8	625.2±132.1	615.2±149.5
E_{IT} [GPa]	180.3±5.8	180.4±22.8	182.2±23.1

The corrosion resistance was measured for microcrystalline and nanocrystalline hot-pressed materials. It is commonly accepted that the microcrystalline and single-phase structure favour better corrosion resistance in comparison to nanostructure of large volume grain boundaries (smaller grains) and multiphase materials [15,21]. It is, thus expected that nanomaterials have worse corrosion resistance in comparison to microcrystalline materials. The OCP (open circuit potential) and polarization curves are shown in figures 6A and 6B, respectively. OCP for the microcrystalline Ta (a) shows a significantly more negative value in comparison to the nanocrystalline Ta (b, c). The more positive value of OCP suggests a formation of a more stable passive layer. An increase in OCP to a more positive value with time indicates that the material undergoes overall corrosion (a), whereas constant OCP or its slight decrease denotes a stable passive layer (b, c).

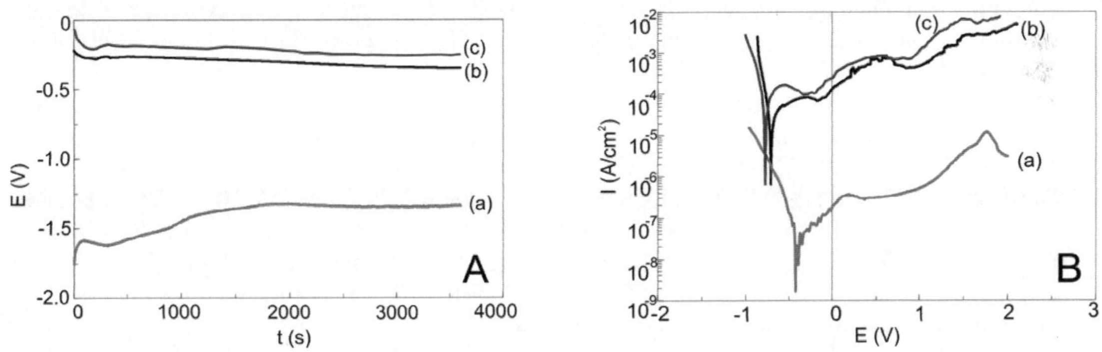

Figure 6. OCP (A) and polarization (B) curves for microcrystalline (a) and nanocrystalline HFIH (b) and PPS (c) tantalum.

The polarization curves shown in B point to better corrosion resistance of the microcrystalline Ta (a) in comparison to the nanocrystalline Ta (b, c). The microcrystalline Ta has a corrosion current density of $I_{corr} = 3.108 \cdot 10^{-8}$ A/cm^2, whereas for the hot-pressed nanocrystalline HFIH and PPS Ta, the $I_{corr} = 3.625 \cdot 10^{-5}$ A/cm^2 and $1.198 \cdot 10^{-5}$ A/cm^2, respectively. The lowest I_{corr} denotes better corrosion resistance. For the nanocrystalline Ta, the I_{corr} has a significantly higher I_{corr}. The high I_{coor} is related to greater volume of the grain boundaries and a smaller grain size. The current in the passive range is significantly higher for the case of nanocrystalline material.

4. Conclusions

The paper has shown the results of nanocrystalline tantalum consolidation using different heating techniques during hot pressing. To prevent excessive grain growth at elevated temperature, the nanocrystalline powders, made using HEBM process were hot-pressed applying induction heating and pulse plasma sintering. Both methods give comparable results, i.e. small average grain size in the range of 165-170 nm and good mechanical properties with the hardness of approx. 620 HV, twice higher compare to the hot-pressed microcrystalline Ta. PPS tantalum exhibits a slightly better corrosion resistance in comparison to HFIH tantalum, however, the microcrystalline tantalum has a best corrosion resistance.

Acknowledgments
The work has been financed by National Science Centre, Poland under project id: DEC-2015/19/B/ST5/02595.

References

[1] Browning P N, Alagic S, Carroll B, Kulkarni A, Matson L and Singh J 2017 Room and ultrahigh temperature mechanical properties of field assisted sintered tantalum alloys *Mater. Sci. Eng.* **680** 141-51
[2] Buckman R W 2000 New applications for tantalum and tantalum alloys *J. Min. Met. Mat. Soc.* **52** 40-1
[3] Yong Q, Lin L and Li Ch 1998 Characterization of nanocrystalline tantalum nitride formed by solid-gas reaction during mechanical alloying *J. Alloys Comp.* **269** 238-40
[4] Zhou J-Q, Akhtar S K, Cai R and Chen L 2006 Comperative study on constructive modeling of tantalum and tantalum tungsten alloy *Int. J. Iron Steel Res.* **13** 68-74
[5] Cardonne S M, Kumar P, Michaluk C A and Schwartz H D 1995 Tantalum and its alloys *Int. J. Refract. Met. Hard Mater.* **13** 187-94
[6] Kofstad P 1963 Studies of the oxidation of tantalum at 1000°-1300°C *J. Less Common Met.* **5** 158-70
[7] Angerer P, Neubauer E, Yu L G and Khor K A 2007 Textur and structure evolution of tantalum powder samples during spark-plasma-sintering (SPS) and conventional hot-pressing *Int. J. Refract. Met. Hard Mater.* **25** 280-5
[8] Kim Y, Kim E P, Noh J W, Lee S H, Kwon Y S and Oh I S 2015 Fabrication and mechanical properties of powder metallurgy tantalum prepared by hot isostatic pressing *Int. J. Refract. Metals Hard Mater.* **48** 211-6
[9] Borodianska H, Demirskyi D, Sakka Y, Badica P and Vasylkiv O 2012 Grain boundary diffusion driven spark plasma sintering of nanocrystalline zirconia *Ceramic Inter.* **38** 4385-9
[10] Dias M, Guerreiro F, Correia J B, Galatanu A, Rosinski M, Monge M A, Munoz A, Alves E and Carvalho P A 2015 Consolidation of W-Ta composites: Hot isostatic pressing and spark and pulse plasma sintering *Fusion Eng. Design* **98-99** 1950-5
[11] Rosinki M, Ciupinska L, Grzonka J, Michalski A and Kurzydlowski K J 2012 Synthesis and characterization of the diamond/copper composites produced by the pulse plasma sintering (PPS) method *Diamond Related Materials* **27-28** 29-35
[12] Atkinson V H and Davies S 2000 Fundamental aspects of hot isostatic pressing: An overview *Metall. Mater. Trans.* **31** 2981-3000
[13] Bischof M, Mayer S, Leitner H, Clemens H, Staron P, Geiger E, Voiticek A and Knabl W 2006 On the development of grain growth resistant tantalum alloys *Int. J. Refract. Met. Hard Mater.* **2** 437-44
[14] Schuster B E, Ligda J P, Pan Z L and Wei Q 2011 Nanocrystalline refractory metals for extreme condition applications *Refract. Met. Alloys* **63** 27-31
[15] Robin A and Rosa J L 2000 Corrosion behavior of niobium, tantalum and their alloys in hot hydrochloric and phosphoric acid solutions *Int. J. Refract. Met. Hard Mater.* **18** 13-21
[16] Tang Y, Bringa E M and Meyers M A 2013 Inverse Hall-Petch relationship in nanocrystalline tantalum *Mater. Sci. Eng. A* **580** 414-26
[17] Yoo S H, Sudarshan T S, Sethuram K, Subhash G and Dowding R J 1999 Consolidation and high strain rate mechanical behavior of nanocrystalline tantalum powder *Nanostruct. Mater.* **12** 23-8
[18] Zhang Y S, Zhang L C, Niu H Z, Bai X F, Yu S, Ma X Q and Yu Z T 2014 Deformation twinning and localized amorphization in nanocrystalline tantalum induced by sliding friction *Mater. Lett.* **127** 4-7
[19] Brar L K, Singla G, Kaur N and Pandey O P 2015 Thermal stability and structural properties of Ta nanopowder synthesized via simultaneous reduction of Ta_2O_5 by hydrogen and carbon *J.*

*Therm. Anal. Calo*r. **119** 175-82
[20] Fang Z, Maheshwari P, Wang X, Sohn H Y, Griffo A and Riley R 2005 An experimental study of the sintering of nanocrystalline WC–Co powders, *Int. J. Refract. Met. Hard Mater.* **23** 249-57
[21] Bermudez M D, Carrion F J, Martinez-Nicolas G and Lopez R 2005 Erosion-corrosion of stainless steels, titanium tantalum and zirconium *Wear* **258** 693-700

IOP Conf. Series: Materials Science and Engineering **283** (2017) 012002 doi:10.1088/1757-899X/283/1/012002

Amphiphilic polymer based on fluoroalkyl and PEG side chains for fouling release coating

W W Cong[1*], K Wang[1], X Y Yu[1], H Q Zhang[1], Z Lv[1] and T J Gui[1]

[1]State Key Laboratory of Marine Coatings, Marine Chemical Research Institute Co. Ltd., Qingdao 266071, China
*E-mail: 15153284902@163.com

Abstract. Under static conditions, fouling release coating could not express good release property to marine organisms. Amphiphilic polymer with mixture of fluorinated monomer and short side group of polyethylene glycol (PEG) was synthesized. And also we studied the ability of amphiphilic polymer to influence the surface properties and how it controlled the adhesion of marine organisms to coated surfaces. By incorporating fluorinated monomer and PEG side chain into the polymer, the effect of incorporating both polar and non-polar groups on fouling-release coating could be studied. The dry surface was characterized by three-dimensional digital microscopy and scanning electron microscopy (SEM), and the morphology of the amphiphilic fouling release coating showed just like flaky petal. The amphiphilic polymer in fouling release coating tended to reconstruct in water, and the ability was examined by static contact angle, which was smaller than the PDMS (polydimethylsiloxane) fouling release coating. Also surface energy was calculated by three solvents, and surface energy of amphiphilic fouling release coating was higher than that of the PDMS fouling release coating. To understand more about its fouling release property, seawater exposure method was adopted in gulf of Qingdao port. Fewer diatoms *Navicula* were found in biofilm after using amphiphilic fouling release coating. In general, coating containing both PEG and fluorinated side chain possessed certain fouling release property.

1. Introduction

During operation, marine biofouling, defined as microbial slimes [1,2], plants [3] and invertebrates [4] onto submerged materials bring huge security risks and economic loss to shipping, aquaculture and oceaneering. Accumulation of marine organisms accelerate corrosion [5] of ships, also they increase in roughness and frictional drag [6] leading to substantial powering penalties. Marine organisms block culture cages and cooling water pipes, also they could weaken detection performance of sonar and action of naval vessels. So it was estimated that hundreds of billions of dollars was used to conquer marine organisms in the world annually [7], also the invasion by exotic marine species is one of the most irreversible environmental problems facing this habitat. The exotic marine species have produced a great deal of negative impacts on the marine ecosystem [8]. They compete or mate with native species, introduce pathogens and toxic algae that will directly or indirectly damage the marine ecosystem. And to date, the most effective method of marine biofouling control has been the use of biocidal antifouling paints [9], but most of biocides are forbidden by severe environmental protection laws and regulations [10,11]. Thus, in recent years, there is considerable interest in developing environmental-friendly antifouling/fouling release coatings [12]. For instance, polydimethyl siloxane (PDMS) fouling release coating (usually referred to as FR coating) is biocides-free, not contaminated the marine environment and easy cleansed of marine organisms, which has been considered as a potential substitute of biocidal antifouling paints. But this material is easily stained at port or with a low speed of ship; also it could not suffer the adhesion of slime with diatom and bacteria [13], while the evolution process of slime is essential to formation of mature communities. And the performance deficiency limited application of this type of coating.

Much work [14,15] has been done to improve properties (such as resistance to slime) of FR coating on resting or at low speed ship. As we know, fluorinated polymers express outstanding capability in low friction and low surface energy. And in the formation of the layer, fluoroalkyl side chains enrich in the surface of coating, which reduces its surface energy and enhances detachment to marine organisms [16,17]. And also it is reported that coating with PEG (polyethylene glycol) segments offer good inhibition to protein [18,19]. Hydrophilic polymer chains can bond abundant water molecule and form a hydration layer on surface. Protein, biological glue and so on should first overcome energy barrier of hydration layer and then touch to polymer matrix shown in figure 1 (The blue style offers hydrophilicity, while the red style offers hydrophobicity).

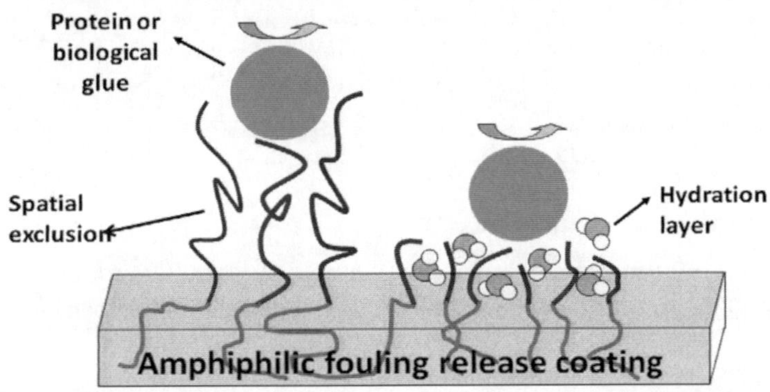

Figure 1. Mechanism of amphiphilic fouling release coating to biofilm.

As we know, some marine organisms prefer to adsorb on hydrophobic surface, but others such as diatom like hydrophilic interface. This paper focuses on amphiphilic polymer, and the amphiphilic polymer combines fluoroalkyl chains as hydrophobic segments and PEG side chains as hydrophilic segments. Fluoroalkyl chains and PEG segments migrate toward to the surface of coating at different speed in seawater, which reconstitute its interface. And this amphiphilic polymer can express blended property to marine organisms. In this paper, amphiphilic polymer with fluoroalkyl and PEG side chains was prepared and its properties were tested by Fourier transform infrared spectroscopy and gel permeation chromatography. FR coating constructed by amphiphilic polymer was obtained and its FR-property was checked at floating raft of Qingdao.

2. Experimental method and materials

2.1. Raw materials

Xylene and n-butyl alcohol were technical grade and were purchased from Qingdao Yongan Chemical Reagent Co., Ltd. (China). Azobisisobutyronitrile (AIBN) was laboratory grade and was obtained from Tianjin Damao Chemical Reagent Co., Ltd. (China). [3-(Methacryloyloxy) propyl] trimethoxysilane (A174) was technical grade and was bought from Momentive. Poly (ethylene glycol) acrylate (PEA6, $CH_2CHCO\ (OCH_2CH_2)_XOH$, average Mn≈336, x≈6) was technical grade and was got from Shanghai Polydex Industry Co., Ltd. (China). 1H, 1H,7H-Dodecafluoroheptyl methacrylate (G04) was technical grade and was purchased from XEOGIA (China). Silanol-terminated polydimethylsiloxane was technical grade and was purchased from Dow Corning. Defoamer was technical grade and was obtained from BYK. Thixotropic agent R972 was technical grade and was obtained from Evonik Degussa GmbH. Iron red was technical grade and was bought from yipin pigment in shanghai. Dibutyltin dilaurate was laboratory grade and was obtained from BASF. PDMS fouling release coating was obtained from State Key Laboratory of Marine Coatings.

2.2. Synthesis of amphiphilic polymer

Xylene and n-butyl alcohol (w/w 1:1) were poured into a four-neck round bottom flask with a thermometer, a Teflon stirrer and a water condenser pipe. The flask was expelled air by bubbling nitrogen and heated to 85℃. Then required quantities of A174, G04, PEA6, AIBN, xylene andn-butyl alcohol were mixed and dropped into the flask in 1 h, and the temperature was kept for 2 h after the addition. Then, some AIBN, xylene and n-butyl alcohol were introduced into the flask and the system temperature was maintained for 2 h. After that, reddish brown mixture was obtained. And the reaction scheme for the preparation of amphiphilic polymer was shown in figure 2.

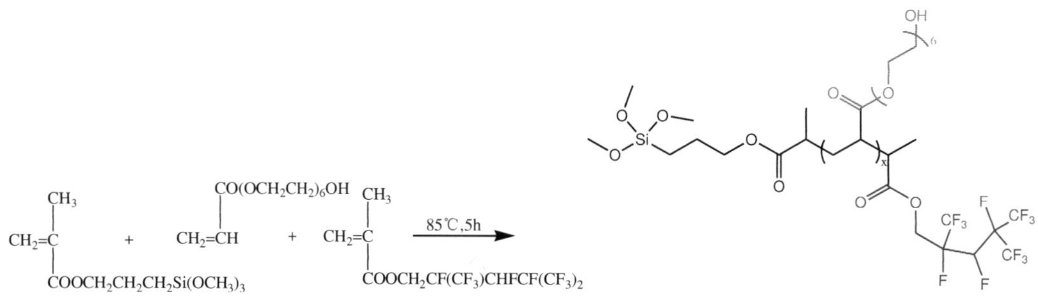

Figure 2. Synthetic route of amphiphilic polymer.
The red style: hydrophilic segment, the blue style: hydrophobic segment

2.3. Preparation of amphiphilic fouling release coating

To obtain amphiphilic fouling release coating, three-component products were used. Part A required silanol-terminated polydimethylsiloxane, defoamer, thixotropic agent R972, xylene, iron red; all the substances were mixed and grind to 40 μm or below. Part B used amphiphilic polymer as the crosslinking agent also with some xylene, which made the fouling release coating amphipathic, while part C applied dibutyltin dilaurate and some xylene as a catalyst system. Amphiphilic fouling release coating was a mixture of the three-component products (w/w/w 10:2:1) and dried naturally for 24 h.

2.4. Fourier transform infrared spectroscopy (FTIR) and gel permeation chromatography (GPC) of Amphiphilic polymer

Reddish brown mixture was removed of solvents through rotary evaporator as much as possible, and then it was on the vacuum drying condition to be purified to get amphiphilic polymer concentrate. After that, amphiphilic polymer concentrate was painted over the KBr tablet evenly and carried out on AVATAR-380 Spectrophotometer (Thermo Nicolet Company) in the wavelength region 500-4000 cm^{-1} at room temperature.

The number and average molecular weights of amphiphilic polymer, Mn and Mw, respectively, were determined by size exclusion chromatography on Waters 1515 gel permeation chromatograph (Waters Company, USA). Amphiphilic polymer was extracted by rotary evaporator, tetrahydrofuran was served as solvent. The differential refractive index detector was used at 40°C; the flow was slowly increased to 1 mL/min at a rate of 0.01 mL/min.

2.5. Contact angle and surface energy tests

Amphiphilic fouling release coating was pasted on pretreated tin plates with dimensions of 50 mm×120 mm×0.3 mm, and dried naturally. Water contact angle measurements [20] were conducted by the sessile drop method on an OCA 20 contact angle system (Dataphysics Instruments GmbH, Germany) at ambient temperature. After placing the panel with amphiphilic fouling release coating on the horizontal workbench, a 3-μL droplet of distilled water was put on the surface using a syringe. Digital images of the droplet silhouette were captured with a charge-coupled device (CCD) camera and the contact angles were evaluated using the measuring angle method. Every database was obtained within 30 s after the droplet contacting with the film surface. For each sample, the average of three measurements was calculated. Surface energy [21] of amphiphilic fouling release coating was calculated through distilled water, ethylene glycol and cyclohexane.

2.6. Surface morphology of amphiphilic fouling release coating

Scanning electron microscope (SEM) images [22] were recorded with an electronic microscope HITACHI TM-1000 operating at 15 kv, solid state backscattered electron detector (BSE) was adopted for its detector. Amphiphilic fouling release coating (one layer, 70-100 μm) was pasted by air spraying and dried naturally on tin plates. Also surface morphology and advanced 3D digital observation were obtained on HIROX KH-7700 digital microscope by the way of progressive scan.

2.7. Seawater exposure methods

Low carbon steel panels were applied to substrates with the dimension of 350 mm×250 mm×3 mm. After surface treatment, the anti-corrosive primer, sealer coating, tie-coat and amphiphilic fouling release coating were coated on tested substrates in sequence. When the amphiphilic fouling release coating was hard-dry, test panels were immersed in shallow submergence [23,24] at a depth of one meter in Qingdao Port. Also the PDMS fouling release coating was employed as the blank. The gulf of Qingdao port has mature biomass and vigorous marine life, with the seawater flow less than 2 m/s. Observation by photographing was carried out to investigate fouling-resistance property of amphiphilic fouling release coating after two weeks.

3. Presentation and discussion of test results

3.1. Synthesis of Amphiphilic polymers

Amphiphilic polymers consisted of a block of PEG side chain and a block of fluoroalkyl side chain. In the structure of amphiphilic polymers [25,26], PEA6 acted as hydrophilic part while G04 gave hydrophobicity. This dual hydrophilic-hydrophobic nature provided an amphiphilic character to the material, and the amphiphilic fouling release coating derived therefrom were anticipated to respond to contact with water or seawater.

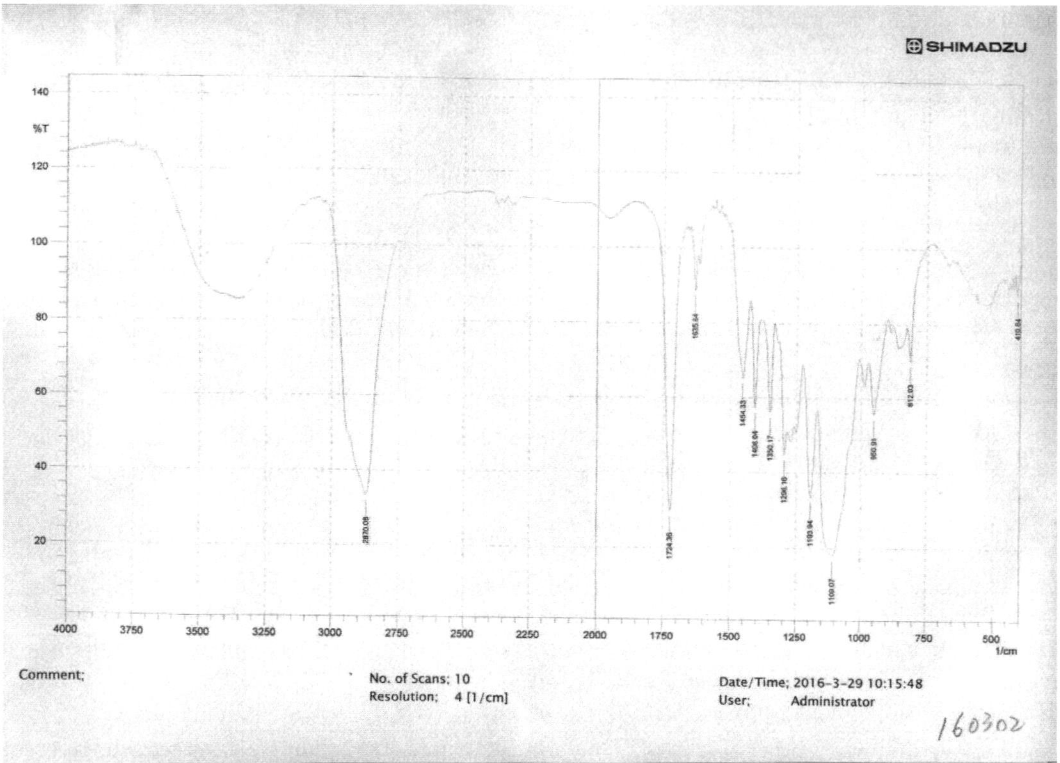

Figure 3. FTIR spectra of amphiphilic polymer.

The amphiphilic polymers were synthesized by a free radical solution polymerization method [27], whereby AIBN was used as initiator. In order to get homogeneous system, xylene and n-butanol were used as the mixed solvents. The FTIR spectra of amphiphilic polymers showed peaks corresponding to the functional groups in the polymeric chains in figure 3: the observed strong peaks at 2870 cm^{-1} and 1109 cm^{-1}, corresponding to Si-OMe stretching; 1724 cm^{-1}, corresponding to the C=O of A174; 1636 cm^{-1} corresponding to the C=O of PEA6 and G04; 1296cm^{-1} corresponding to the C-F stretching of G04; 1194 cm^{-1} corresponding to aliphatic ether C-O stretching of PEA6. Obvious C=C stretching was not found, which showed three monomers all contributed to the structure of amphiphilic polymer.

Accordingly, the number average molecular weight (Mn) and weight-average molecular weight (Mw) of amphiphilic polymer were 788 g/mol and 1597 g/mol, respectively. The polymer was monomodal and characterized by a polydispersity Mw/ Mn≈1.97.

3.2. Static contact angles and surface energy

Surface energy [28] is closely related to bacterial attachment behavior. Low surface energy could reduce attachment of bacterial and microbial and enhance fouling release properties. The static contact angles of water (θ_w), ethylene glycol (θ_e) and cyclohexane (θ_c) on amphiphilic fouling release coating were measured by the sessile drop technique. The measured values of θ were then used to obtain the surface energy of the polymer films referring to the so-called Owens-Wendt-Kaelble approach [29]. This regards the surface energy as being composed of two or more additive components, the dispersion (γ^d) and the hydrogen bonding and dipole-dipole(γ^p) components (equation (1)), and the greater difference in solvent polarity, the more accurate the results obtained. A geometric mean relationship is postulated both of the solid-liquid and liquid-liquid interfacial tensions (equation (2)).

$$\gamma=\gamma^d + \gamma^p \quad (1)$$

$$\gamma_{12}=\gamma_1+\gamma_2-2(\gamma_1^{\ d}\ \gamma_2^{\ d}\ )^{0.5} \quad (2)$$

Combination with the well-known Young's equation leads to equation (3).

$$\gamma_L(1+\cos\theta)=2[(\gamma_S^d\gamma_L^d)^{0.5}+(\gamma_S^p\gamma_L^p)^{0.5}] \qquad (3)$$

Where γ_S is the surface tension of the solid surface and γ_L is that of the wetting liquid. As there are two unknowns (γ_S^d and γ_S^p) of the solid surface, static contact angle of three different liquids with known γ_L^d and γ_L^p on the solid surface. After that, the surface energy of the amphiphilic fouling release coating was got through the three simultaneous equations.

Static contact angles of amphiphilic fouling release coating with three different solvents were tested; the average of three measurements for each solvent was showed in table 1, and also its surface energy. Blank referred to PDMS fouling release coating without graft modification using amphiphilic segment. By comparing the contact angle values of amphiphilic fouling release coating and the blank, all of them were smaller for amphiphilic fouling release coating than for the blank [30], and the difference was more pronounced for θ_w than for the other two angles.

Table 1. Contact angles and surface energy for amphiphilic fouling release coating.

Film type	θ_w/°	θe/°	θc/°	Surface energy/mJ·m^{-2}
amphiphilic fouling release coating	103.9	85.9	29.6	21.39
blank	111.0	88.9	31.5	21.24

This suggested that the inclusion of fluoroalkyl and PEG chains in amphiphilic fouling release coating caused the top layer turned to be both hydrophobic and hydrophilic at the same time, especially in the hydrophilic character. This was a further confirmation of the preferential and selective segregation of the fluoroalkyl and PEG segment at the top layer–air interface. And the chemical incompatibility of fluoroalkyl and PEG segment lowered hydrophobic property. And hydrophilic segment dominated in the interaction of polar material to the coating.

Measurements of liquid-solid static contact angles are commonly used to evaluate solid surface energy. The Owens-Wendt-Kaelble approach was used to extract the surface energy for amphiphilic fouling release coating and the blank from experimental θ values, and the results are shown in table 1. The surface energy of the blank (PDMS fouling release coating) was slightly lower than amphiphilic fouling release coating, which was due to the contribution of the PEG side chain permitted polar interaction with three solvents. And it also gave the occurrence of microphase separation of the incompatible blocks in the surface.

3.3. Surface morphology of amphiphilic fouling release coating

The use of the phase separation behavior of fluorinated polymer is well-suited to generate ordered, self-assembled, low surface energy materials. We used HIROX KH-7700 digital microscope to investigate the morphological features of amphiphilic fouling release coating in figure 4.

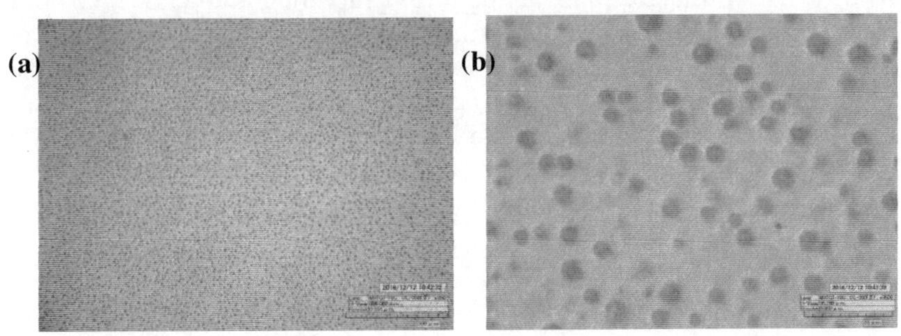

Figure 4. Morphology of coatings with different magnification.

The uniform and dense distributed pores are observed on the surface of amphiphilic fouling release coating (figure 4(a)), for the PEG side chain and fluoroalkyl segment could not merge in the silicone release coating system, tending to migrate out of the system. At figure 4(b), we could see the pore in the surface was regular circular at the magnification of 3500 times. Also we could get the actual size of the pore from figure 5.

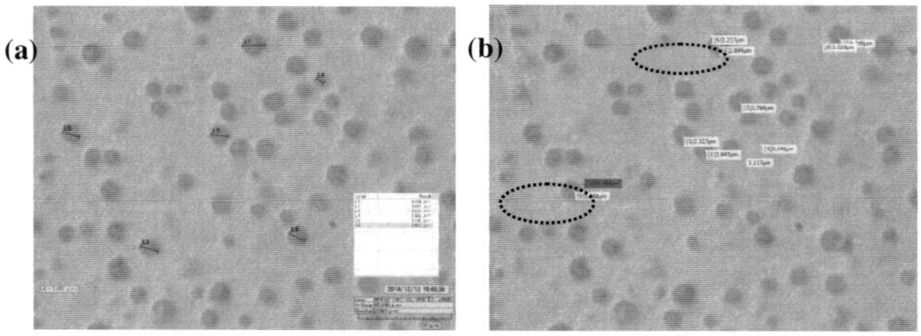

Figure 5. Size of pores for amphiphilic fouling release coating.

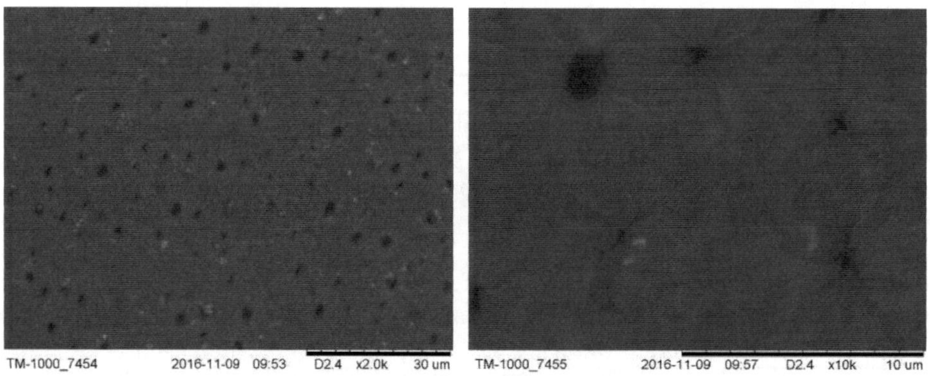

Figure 6. SEM images of amphiphilic fouling release coatings.

The diameter of the circular hole was measured; most of them were within a few microns. The measured values obtained in figure 5(a) were from 1.9 μm to 4.8 μm. The depth of the pores was got from figure 5(b), the values of them were within a few microns, too.

SEM images on amphiphilic fouling release coating were showed in figure 6. Many micro- and nanosized pores appeared on the surface in figure 6(a). And by magnification of ten thousand times in figure 6(b), several flower-like shapes present in front. Fluoroalkyl and PEG side chain like slender petals were grafted in the fouling release coating, while some of them were straight and others were bending, which gave the microstructure construction on amphiphilic fouling release coating.

Seawater exposure methods were carried out lasting two weeks during a period of mature biofilms. After that, test board with amphiphilic fouling release coating was taken out of the seawater. And photos were taken about marine biofilm on testing board as shown in figure 7. The biofilm was transparent just like jell. And then, the biofilm was sampled from the test board. Biological microscope KH7700 (HIROX) was used to observe marine organisms in the biofilm.

Figure 7. Biofilms on test boards.

Naviculaceae in elliptical solid line on figure 8(a) belongs to *Bacillariophyceae*, and the shape of *Naviculaceae* is just like a boat. Comparing two pictures of figure 8, it could be said that fewer diatoms *Navicula* were seen in biofilm of amphiphilic fouling release coating than in biofilm on blank sample (PDMS fouling release coating). The same tendency was observed for species of *Chlorella*. It is obvious that amphiphilic fouling release coating had certain inhibition during initial stage of biofilm formation.

(a)

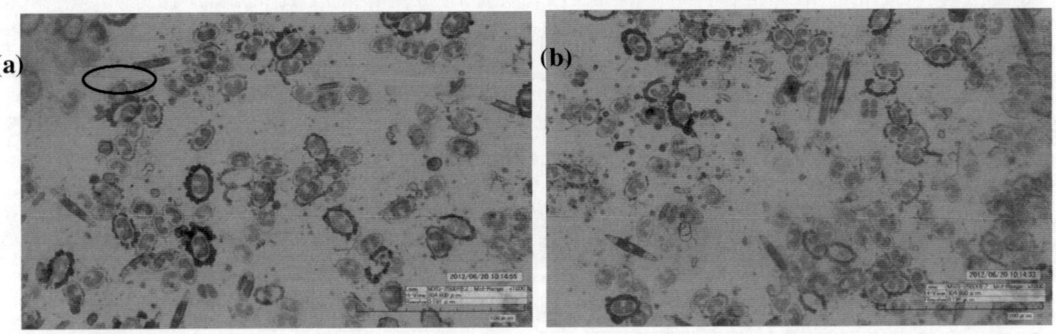

Figure 8. Biological community in biofilm of (a) amphiphilic fouling release coating and (b) PDMS fouling release coating.

4. Conclusions

Amphiphilic surface-active copolymers with fluroalkyl and PEG side chains were synthesized through the polymer modification reaction of a free radical solution polymerization method. The resultant polymer was characterized using FTIR spectra and gel permeation chromatography confirming the attachment of the hydrophobic fluoroalkyl segments and hydrophilic PEG side chains. The amphiphilic nature of the fouling release coating was characterized with static contact angle measurements and surface energy which suggested that the surface was highly populated with the functional side chains, that the PEG side chains migrated to the surface, and the fluoroalkyl chains became slightly buried. The seawater exposure method with immersed coating showed lower diatoms *Navicula* adsorption and confirmed that the hydrophilic PEG side chains and hydrophobic fluoroalkyl side chains migrated to the surface and substrate, respectively. The FR profile of amphiphilic fouling release coating was different from those of PDMS, which contained only PDMS side chains. The combination of fluoroalkyl and PEG chains at the coating surface offered potential advantages over conventional silicone-based FR coatings, which tend to perform poorly with respect to diatom fouling.

References

[1] Tenore K R and Coull B C 1980 *Marine Benthic Dynamics* (Columbia: University of South Carolina Press) pp 341-6
[2] Lee J W, Nam J H, Kim Y H, Lee K H and Lee D H 2008 *J. Microbiol.* **46** 174-82
[3] Parker J D, Duffy J E and Orth R J 2001 *Mar. Ecol. Prog. Ser.* **224** 55-67
[4] Rodriguez S R and Ojeda F P 1993 *Mar. Ecol. Prog. Ser.* **97** 193-207

[5] Zhao Q, Liu Y, Wang C, Wang S and Müller-Steinhagen H 2005 *Chem. Eng. Sci.* **60** 4858-65
[6] Schultz M P 2007*Biofouling* **23** 331-41
[7] Schultz M P, Bendick J A, Holm E R and Hertel W M 2011 *Biofouling* **27** 87-98
[8] Molnar J L, Gamboa R L, Revenga C and Spalding M D 2008 *Ecol. Environ.* **6** 485-92
[9] Konstantinou I K and Albanis T *A* 2004 *Environ. Int.* **30** 235-48
[10] Turner A 2010 *Mar. Pollut. Bull.* **60** 159-71
[11] Albanis T A, Lambropoulou D A, Sakkas V A and Konstantinou I K 2002 *Chemophere* **48** 475
[12] Majumdar P, Crowley E, Htet M, Stafslien S J, Daniels J, VanderWal L and Chisholm B J 2011 *Acs. Comb. Sci.***13** 298-309
[13] Molino P J and Wetherbee R 2008 *Biofouling* **24** 365-79
[14] Nurioglu A G, Catarina A, Esteves C and Gijsbertus D 2015 *J. Mater. Chem. B* **3** 6547-70
[15] Krishnan S, Wang N, Ober C K, Finlay J A, Callow M E, Callow J A, Hexemer A, Sohn K E, Kramer E J and Fischer D A 2006 *Biomacromolecules* **7** 1449-62
[16] Akuzov D, Vladkova T, Klöppel A, Brümmer F and Murthy S 2014 *Journal of Chemical Technology and Metallurgy* **49** 229-37
[17] Lejars M, Margaillan A, Bressy C 2012 *Chem. Rev.* **112** 4347-90
[18] Michel R, Pasche S, Textor M and Castner D G 2005 *Langmuir* **21** 12327-32
[19] Zhou Z L, Calabrese D R, Taylor W, Finlay J A, Callow M E, Callow J A, Fischer D, Kramer E J and Ober C K 2014 *Biofouling* **30** 589-604
[20] Good R J and Girifalco L A 1960 *J. Phys. Chem.* **64** 561
[21] Johnson K, Kendall L K and Roberts A D 1971 *Proc. R. Soc. Lond.: A Math. Phys. Sci.* **324** 301
[22] Berglin M and Gatenholm P 1999 *Journal of Adhesion Science and Technology* **13** 713-27
[23] Standard Test Method for Testing Antifouling Panels in Shallow Submergence *American Standard for Testing and Materials* ASTM D3623 1994
[24] Antifouling Panels in Shallow Submergence GB/T 5370-2007
[25] Christina F L, Tommy N, Patric J and Bengt W 1996 *Biomaterials* **17** 2199
[26] Krishnan S, Weinman C J and Ober C K 2008 *J. Mater. Chem.* **18** 3405-13
[27] Ahmad N M, Heatley F and Lovell P A 1998 *Macromolecules* **31** 2822-7
[28] Owens D K and Wendt R C 1969 *J. Appl. Polym, Sci.* **13** 1741-7
[29] Kaelble D H 1970 *J. Adhesion.* **2** 66-81
[30] Stafslien S J, Christianson D, Daniels J, VanderWal L, Chernykh A and Chisholm B J 2014 *Biofouling* **31** 135-9

IOP Conf. Series: Materials Science and Engineering **283** (2017) 012003 doi:10.1088/1757-899X/283/1/012003

Advances of the research evolution on aluminum electrochemical anodic oxidation technology

Z B Yang[1,2,*], J C Hu[2], K Q Li[2], S Y Zhang[2], Q H Fan[1,*] and S A Liu[2]

[1]Department of Electrical and Computer Engineering, Department of Chemical Engineering and Materials Science, Michigan State University, East Lansing, MI 48824, USA
[2]Henan Polytechnic University, Jiaozuo 454000, China
*E-mail: Yangzhibo2001@163.com/qfan@egr.msu.edu

Abstract. This article gives an overview on the development of aluminum anodization technique in terms of fundamental aspects and practical applications in the past decades. Besides, the formation mechanism and structural characteristics of anodic alumina films as well as the factors affected the formation of porous anodic alumina films are also discussed. Anodic aluminum oxide (AAO) prepared by the anodization method can be divided into two categories: dense anodic alumina (DAA) and porous anodic alumina (PAA). This article also summarizes the optical properties, magnetic properties, solar absorption properties, and catalytic properties of porous anodic alumina film and its applications in nanomaterials, optical materials, magnetic materials, biosensors, solar cells, and so on. In addition, future developmental trend of porous anodic alumina film is covered.

1. Introduction

Aluminum is an active element. When aluminum is exposed to air, it oxidizes rapidly and forms a thin surface layer of oxide film with a thickness of 0.01-0.10 μm. However, this oxide layer is amorphous, thin, porous, and shows low mechanical strength. Although this layer can prevent further oxidation of the aluminum, it fails to fulfill the demands of aluminum and its alloys in terms of decoration, protection, and function in various applications. The anodization of aluminum in an electrolyte, first developed in the 1920s, offers a solution to this problem. Development of this process has led to widespread applications of anodized aluminum. This article reviews the preparation and processing methods of porous anodic alumina films. The formation mechanism of anodization is also examined, as it is essential to this process. Finally, the practical applications and the future development of porous anodic alumina films are discussed.

2. Characteristics and preparation of porous anodic alumina

Anodic aluminum oxide (AAO) prepared by the anodization method can be divided into two categories: dense anodic alumina (DAA) and porous anodic alumina (PAA). When aluminum is anodized in neutral electrolytes (boric acid, decanedioic acid, hexanedioic acid, etc.), a dense oxide film with a compact structure is formed. DAA is a good insulator and can be used in capacitors. When aluminum is anodized in acids or weak bases (phosphoric acid, sulfuric acid, oxalic acid, etc.), porous oxide film is formed because aluminum oxides are easily etched by the electrolytes. This type of film has unique structural characteristics as it is a thin and dense barrier layer formed on the aluminum surface. PAA has a packed array of columnar hexagonal cells with uniformly sized holes that are

IOP Conf. Series: Materials Science and Engineering **283** (2017) 012003 doi:10.1088/1757-899X/283/1/012003

perpendicular to the alumina surface and parallel to each other. Its structure is illustrated in figure 1 [1-5].

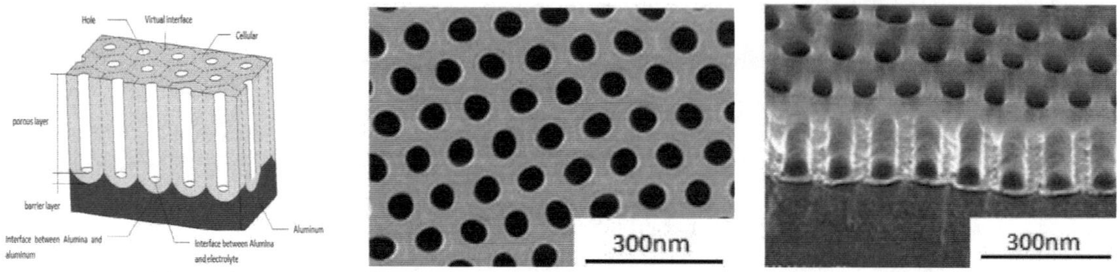

(a) (b) (c)

Figure 1. Porous structure of aluminum oxide film.

Studies of porous anodic alumina have identified several advantages: (1) barrier layer hardness that is higher than corundum; (2) good wear resistance, corrosion resistance, and chemical stability; (3) pore morphology and film thickness that are tunable by varying the electrolytic parameters; and (4) simple preparation without harsh reaction conditions or complicated experimental setups [3-8]. The following section will describe commonly used methods for preparing anodic oxide film in acidic electrolyte [2-19].

2.1. Sulfuric acid anodization

Films prepared with sulfuric acid anodization have the following characteristics: (1) thickness; (2) colorlessness; (3) high porosity (10-15% on average); (4) good absorption; (5) can be easily dyed; (6) high hardness; (7) high corrosion resistance; (8) high wear resistance; and (9) good colorability. Parameters affecting the film include the concentration of sulfuric acid, temperature, current density, oxidation duration, stirring, additives, and the composition of aluminum alloys. The advantages of this method are (1) simple and easy operation; (2) simple effluent handling; (3) low energy consumption; (4) low cost; (5) short oxidation duration; and (6) high production efficiency.

2.2. Chromic acid anodization

Chromic acid anodization produces films of inferior quality compared to films prepared with sulfuric acid. They are relatively thin (5-20 μm) and soft, with low porosity and high corrosion resistance. However, the films have low wear resistivity. They are less suitable for dyeing due to their gray color. These films offer good adhesion to organic coatings and can be used as the base for coatings, but the chromic acid anodization process requires high-energy consumption.

3. Oxalic acid anodization

Films prepared by oxalic acid anodization are relatively thick. They can be yellow or white, with color that darkens with increasing film thickness. They feature low porosity, high corrosion resistance, high wear resistance, and good insulation. However, the high-energy consumption of the anodizing process and the need for cooling lead to high cost. In addition, the electrolyte has low chemical stability.

Researchers have developed two-step anodization as an improvement on anodization using acid electrolytes to produce alumina with arrays of ordered pores over large areas.

In 1995, Masuda *et al* [14] demonstrated two-step anodization to prepare a PAA template. First, aluminum was sonicated in a mixture of acetone, ethanol, and distilled water to remove impurities from the surface. Then, a mixture of ethanol and perchloric acid in a volume ratio of 4:1 was used as the electrolyte for electropolishing before the first anodization. After anodization, the formed oxide layer was removed using a mixture of chromic acid and phosphoric acid under certain temperature conditions. Subsequently, a second anodization was carried out to obtain an ordered PAA template.

This discovery has had a significant impact on the preparation and applications of PAA. Fan *et al* [2,15] fabricated porous anodic aluminum oxide (AAO) on glass substrates using electrochemical oxidation and subsequent etching. The pore depth was controlled by electrochemical anodization and/or chemical etching time.

Two-step anodization has become the most commonly used method of preparing PAA. Li *et al* [16] successfully prepared highly ordered PAA film in oxalic acid electrolyte by modifying the two-step anodization method developed by Masuda. Later, Li *et al* [17] prepared a PAA template with cell diameters of 77 and 96 nm using oxidation voltages of 40 and 50 V in a H_2SO_4-$Al_2(SO_4)_3$-H_2O electrolyte system. In addition, Dou *et al* [18] successfully synthesized an ordered PAA template with nanopores using two-step anodization in oxalic acid electrolyte. Highly ordered PAA templates with larger pore diameter were demonstrated by Gao *et al* [19] using oxidation under constant current density. These recent improvements to the anodization process have further advanced its potential.

3.1. Hard stamping anodization

Masuda *et al* [20-22] developed hard stamping anodization based on the concept of acid dissolution of alumina to induce porous structure. In this method, a SiC mold with ordered arrays of convexes is pressed onto the aluminum surface using an oil press to generate the corresponding array of concaves on the surface of aluminum. This step is followed by a conventional anodization process to obtain an ordered PAA template, as shown in figure 2.

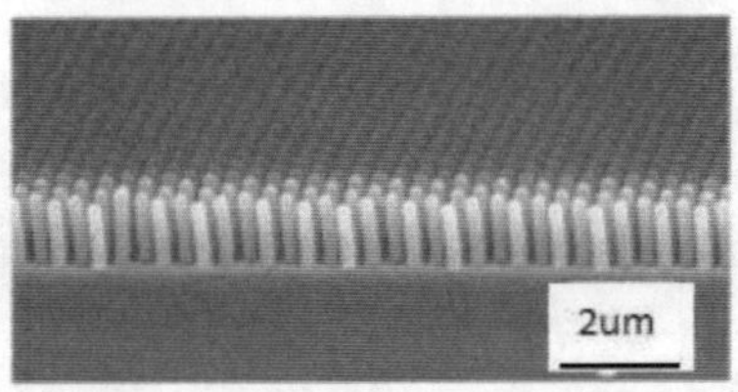

Figure 2. SEM images of polymer fiber arrays prepared by nanoimprinting using anodic porous alumina.

3.2. Y-shaped, tube-in-tube, bone-shaped, and parallel channels in PAA templates

Prior research has demonstrated that pore density is inversely proportional to the square of the anodizing voltage. Using this concept, Gao *et al* successfully obtained Y-shaped branched channels in PAA films by reducing the anodizing voltage in each step by $(V/2)^{1/2}$, as illustrated in figure 3. The discovery of one-dimensional Y-shaped nanostructures played a critical role in the development of electrical transport in nano-circuits [23,24].

Lee *et al* [24] prepared PAA film with a specific pore diameter and further anodized the film into a coaxial tube-in-tube structure with different inner and outer diameters under the same reaction conditions, as illustrated in figure 4.

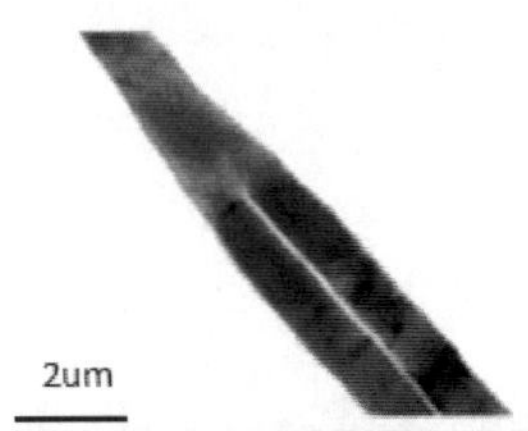

Figure 3. Y-junction copper nanowires.

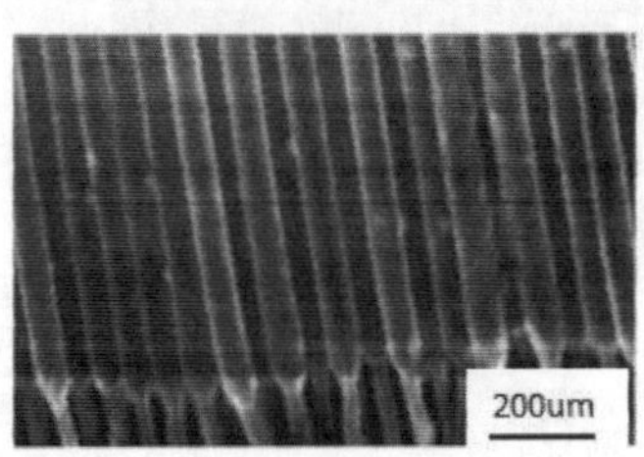

Figure 4. The linearly joined tubes with different diameters.

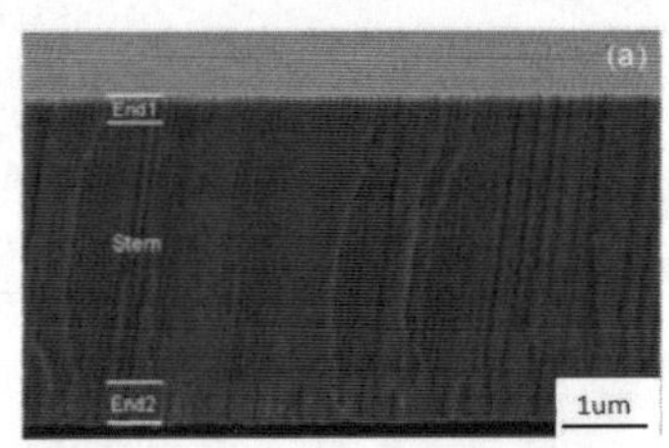

Figure 5. SEM micrographs of the bone-shaped PAA film cross-section.

IOP Conf. Series: Materials Science and Engineering **283** (2017) 012003 doi:10.1088/1757-899X/283/1/012003

Xu *et al* [25] obtained PAA film with bone-shaped pores using a four-step anodization. The first three anodization steps were conducted in 0.3 mol/L oxalic acid under a constant current of 40 V and for oxidation times of 5 h, 5 min, and 40 min, respectively. The last step was conducted in 0.04 mol/L oxalic acid at a temperature of 17°C, under constant current of 80 V, and for an oxidation time of 3 min. The bone-shaped pores are illustrated in figure 5.

PAA film with ordered nano-channels parallel to the base was synthesized by Costel *et al* [26] under low voltage (3 ~ 5 V). The length of the ordered channels was more than several hundred nanometers and the smallest pore diameter was about 3 ~ 4 nm.

3.3. Radial, triangle/diamond-shaped PAA template

Lee *et al* [26,27] used hard and mild anodization methods to prepare PAA templates with long channels and tunable pore diameters, as illustrated in figure 6. Ruy *et al* [28] employed a reoxidation method to anodize Al wires to prepare PAA templates with radial morphology, as illustrated in figure 7. The diameters of the Al wires were maintained within the range of millimeters or micrometers to ensure the intactness of the hexagonal columnar arrays of the PAA templates.

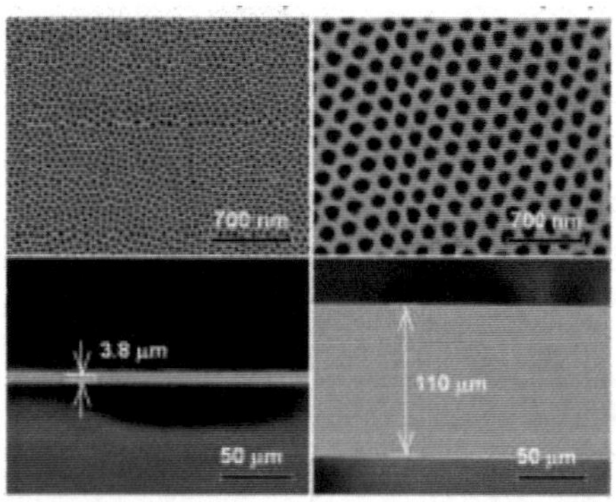

Figure 6. SEM micrographs of AAO membranes formed by MA for 2 h (left column) and by HA for 2 h (right column).

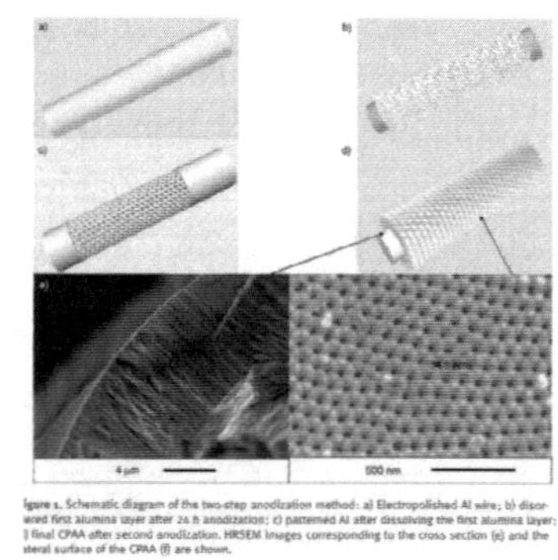

Figure 7. Radial morphology of the two-step anodization method.

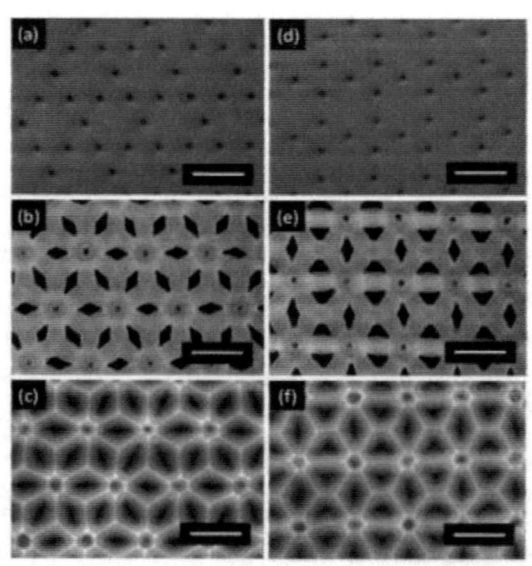

Figure 8. FESEM images of diamond and hybrid triangle-diamond patterns (All scale bars are 500 nm).

Smith *et al* [29] made use of the non-equilibrium dynamics present during the growth of PAA and the tessellation effects during cellular formation to prepare extended, uniform, and partially compressed hexagonal columnar structure, giving rise to the formation of diamond-shaped, spherical, and triangular pores, as illustrated in figure 8.

3.4. Zigzag-shaped PAA templates with conical channels

Lee *et al* [26,30] used pulsed anodization to prepare aluminum oxide nano-channels with tunable pore diameter, as illustrated in figure 9. The advantage of this method is that the electrolyte does not need to be changed because the nano-channels and the pore diameter are tuned by the pulse voltage and anodizing duration.

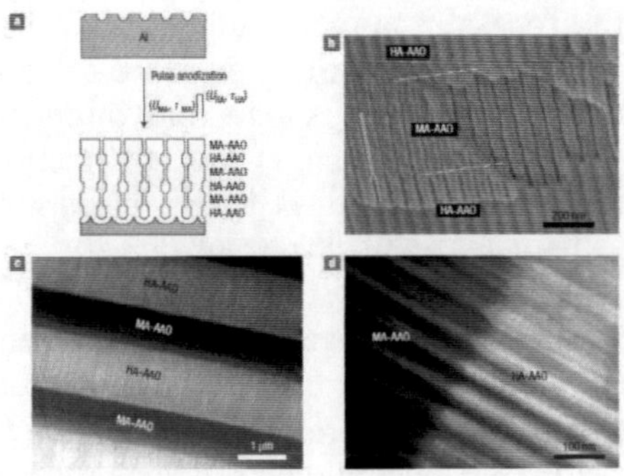

Figure 9. Modulation of pore diameter.

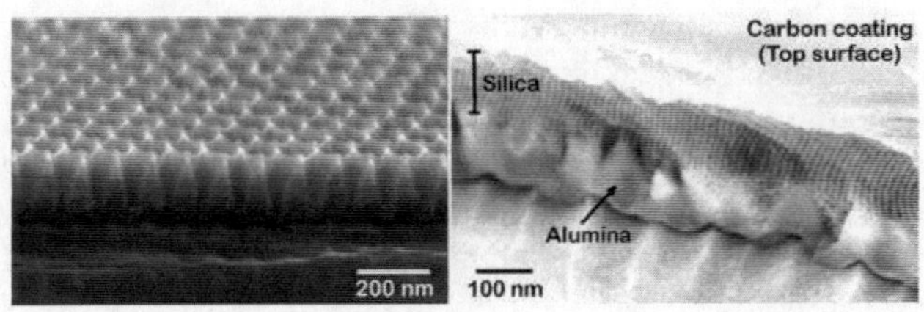

Figure 10. Cross-sectional SEM images of PAA substrates with conical holes with different aspect ratios.

Yusuke *et al* [31] prepared conical pores with inverted honeycomb structure and different aspect ratios, as illustrated in figure 10. The authors also proposed the concept of oriented growth to explain the important role of surface structure of PAA templates in the perpendicular growth of channels.

Losic *et al* [32] reported a new anodization process, termed cyclic anodization to achieve improved control over internal pore shape, as shown in figure 11. By changing the period of the current oscillation, control over the geometry of pore structures was achieved and AAO with asymmetrical (ratchet-type) and symmetrical (circular) periodic pore geometries, with different lengths, periodicity and gradients were fabricated [33]. Similarly, AAO membranes with conical holes were observed by Yamauchi *et al* [31]. The conical holes were achieved by repeating and alternating anodization at 40 V and chemical etching with 5% phosphoric acid solution at 30℃.

(a) (b)

Figure 11. (a)-structural modulation of AAO; (b)-Modulated AAO pores.

4. Mechanism of anodization

Anodic aluminum oxidation is an electrochemical process during which an external current passes through an electrolyte, in which aluminum serving as an anode is connected to the positive electrode of an external power supply and a cathode is connected to the negative electrode. The formation and dissolution of oxide film occur simultaneously during anodization but the dominating process is determined by their reaction rates. The formation mechanism of anodic alumina film is complicated and has not yet been fully understood. Based on the ion migration phenomenon observed during the oxide film formation in chromic acid, phosphoric acid, and oxalic acid, several models have been proposed to explain the reaction mechanisms of anodic alumina.

4.1. Critical current density model

In the 1970s, O'Sullivan and Wood [34] studied the morphology and formation mechanism of PAA. They proposed that the mechanism was associated with the formation and dissolution of anodic alumina under an electric field. In the 1980s, Xu *et al* proposed the critical current density model by investigating the ion migration phenomenon during the growth process of anodic alumina film. The authors also explained the formation mechanism of oxide films at the film/electrolyte interface using a dissolution-precipitation theory. The experiment showed that aluminum and oxygen ions migrated through the oxide film in opposite directions during the formation of anodic alumina film. The fraction of aluminum ions migrating in the same electrolyte remained constant. There is a critical current density in this model. A compact barrier layer was formed when the applied current density was greater than the critical current density, causing the aluminum ions to form new film material at the film/electrolyte interface. On the contrary, a loosely packed porous layer was formed when the current density was lower than the critical current density. In this case, aluminum ions migrated towards the electrolyte, leading to the formation of oxide film at the film/electrolyte interface. This theory, which describes the possible formation of porous anodic alumina in acidic and neutral electrolyte as well as the concept of barrier anodic film, challenges the conventional views on film morphology and electrolytes used for anodization.

4.2. Volume expansion stress model

In 1992, Shimizu *et al* [35] studied the structure of anodic alumina and proposed that the barrier layer was formed as the result of simultaneous inward migration of O^{2-} and outward migration of Al^{3+}. At this layer, O^{2-} filled up the vacant space after outward migration of Al^{3+} and formed Al_2O_3. The volume of newly formed Al_2O_3 was smaller than the volume of outward migrated Al^{3+}. As the oxidation reaction proceeded, the total volume of the aluminum foil decreased as the oxide layer continued to form, following the mechanism that involved the inward migration of O^{2-} and outward migration of Al^{3+}.

Tensile stress was formed when the volume of aluminum oxide film was reduced, leading to the formation of cracks in the oxide surface. In addition, temperature increased as the current density concentrated at the cracked areas, causing the existing cracks to close up. Repeated cracking and closing of the oxide surface eventually led to the formation of pores and the porous layer. These were the conditions of pore and layer formation in the volume expansion stress model.

In 1998, Jessensky *et al* used the anodization method to prepare porous oxide film with uniform hexagonal arrays and proposed the volume expansion stress model [36]. This model qualitatively analyzed the effects of self-assembly in the growth process of ordered pores and explained the formation mechanism of porous layers, as shown in figure 12. The results showed that highly ordered arrays of hexagonal pore structures were formed in oxalic acid or sulfuric acid electrolyte while the pore diameter was dependent on the anodizing voltage. The volumetric loss caused by the outward migration of Al^{3+} was offset by the inward migration of O^{2-}, leading to the formation of a barrier layer. The volume of the newly formed oxide layer was smaller than the volume of aluminum being consumed. Thus, the total volume became smaller and induced tensile stress in the surface. As the tensile stress continued to build, cracks formed in the barrier layer. As the current density was concentrated at the cracked areas, the temperature of those areas became high enough to cause volume expansion, closing up the existing cracks. Repeated cracking and crack closure led to the formation of pores and porous layers. Self-assembled pores in compact hexagonal arrays resulted in the lowest free energy. Conversely, because the spherical pore had the smallest surface area, it yielded the most stable structure.

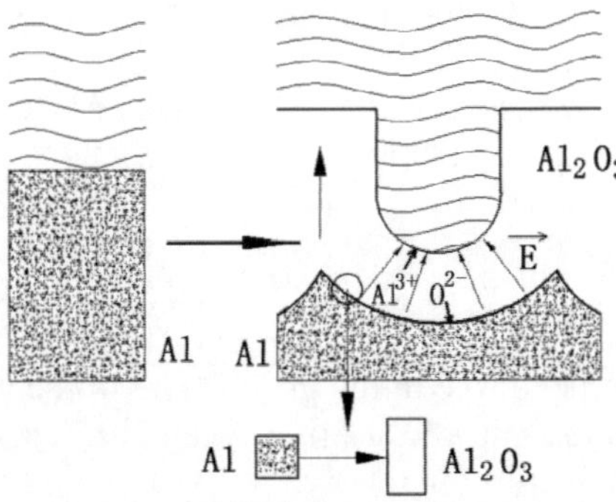

Figure 12. Schematic diagram of volume expansion.

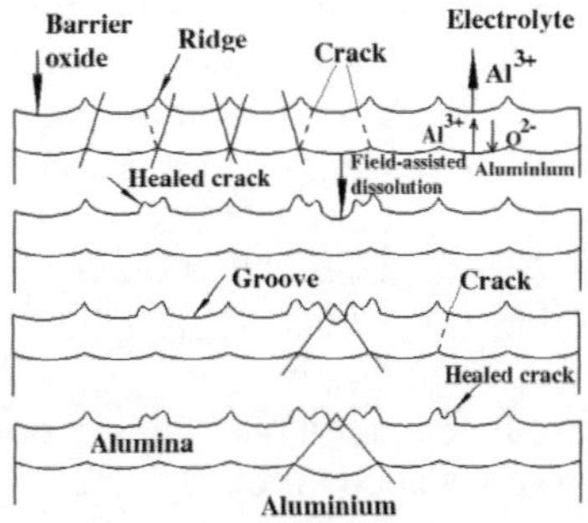

Figure 13. Schematic diagram of the development process in the early stage of pores.

4.3. Field-assisted dissolution model

In 1978, Thompson *et al* [37] proposed an acid-assisted dissolution model, which is also known as the field-assisted dissolution model. As illustrated in figure 13, the anodization process can be divided into three stages: the formation of the barrier layer, the formation of pores, and the dynamic equilibrium in dissolution. Experimental results showed that the anodization process required a dynamic equilibrium between the growth and dissolution of oxide film in an acidic electrolyte. A compact barrier layer was formed on the surface of aluminum when external voltage was applied, but this newly formed barrier layer was partially dissolved in acidic electrolyte. Pore nuclei started to form at the active sites on the film surface when the thickness of the barrier layer exceeded the critical value. As a result, the electric field was concentrated at the base of these areas. Rising temperature caused by increasing current at the base of the affected areas accelerated their dissolution rates. The final porous layer was formed when the pore nuclei continued to grow in a vertical direction. The barrier layer at the interface with the aluminum base was formed by the inward migration of O^{2-} and OH^{-} ions after passing through the film and by the outward migration of Al^{3+} ions under the influence of the electric field. The porous layer started to grow steadily when dynamic equilibrium was reached between the growth and dissolution of film, as illustrated in figure 13.

4.4. Quasi-static mathematical model of field distribution within the film

In a study reported by Xu *et al* [38,39], several assumptions related to the quasi-static field were made to explain the formation mechanism of porous layers. The author assumed that the geometrical and charge distributions at the film/electrolyte and film/aluminum interfaces varied over time during anodization. At the same time, charges at the interfaces could be rapidly re-distributed. Based on these assumptions, the term "quasi-static field" was used to refer to the two static fields in the film that possessed equipotential surfaces. Therefore, the strength of electric fields in the film and the potential of pore formation at different stages could be calculated using the finite element method. Experimental results showed that the field distribution in the film was non-uniform during the growth of the porous layer. In addition, the electric field played a critical role in the growth of pores: the pores grew along the field that possessed the highest intensity. The porous layer was formed when the anodizing current was lower than the critical value. The growth of pores went through a series of stages that included uniform dissolution, formation of pore nuclei, competition in pore development, expansion of pore diameter, and formation of a curved shape at the base of the pores, followed by steady growth.

4.5. Dynamics model of functional integrity

Patermarakis *et al* [40] studied the dynamics model of the functional integrity of PAA film in sulfuric acid electrolyte under a constant current condition. Experimental results showed that the transference numbers of Al^{3+} and O^{2-} ions at the metal/oxide interfaces were determined by the current density and temperature. The transference numbers of ions increased with the current density but dropped when the temperature was increased. These phenomena were associated with the density of the oxides at the metal/oxide interfaces and the oxidation conditions. Generally, aluminum was partly converted into unstable oxides, while its remaining part was converted into amorphous γ-Al_2O_3. This process that occurred at the oxide/electrolyte interfaces led to the porous layer formation.

4.6. Oxygen bubble mold effect model

In 1995, Palibroda *et al* [41,42] proposed that the formation of the porous layer could be divided into three stages by observing the anodization of aluminum under an electron microscope. The breakdown of the barrier layer under a high current condition played a crucial role in the growth rate of the porous layer. Electron avalanche breakdown resulted in the generation and release of oxygen gas, leading to pore formation. This concept differed from the acid-assisted dissolution theory.

Based on the Wood model and the electron avalanche breakdown theory, Zhu *et al* [43] proposed the oxygen bubble mold effect model after many years of effort, as illustrated in figure 13. The authors found that oxygen was released during the anodization of aluminum and the morphology of the oxide film was determined by the amount of oxygen. The Wood model proposed that the inner walls of pores were contaminated by electrolyte species, while Zhu *et al* proposed that the outermost layer of the oxide film was contaminated by electrolyte species. The contaminated layer in each model is illustrated in figure 14(a). The contaminated layer was caused by the growth and dissolution of the oxide layer. At the beginning of the reaction, cations passed through the oxide layer to enter the electrolyte while the anions passed through the electrolyte to enter the oxide layer. During these processes, the accumulation of anions and cations at the boundary of the oxide layer leads to the formation of a contaminated layer. Unlike the acid-assisted dissolution theory, the oxygen bubble mold effect model proposed that the pore formation was attributed to the evolution of oxygen gas, as illustrated in figure 14(b). Pores that connected to the electrolyte formed during the evolution of oxygen gas. As illustrated in figure 14(a), three interfaces were involved in the preparation of oxide film: the electrolyte/contaminated layer, the contaminated layer/barrier layer, and the barrier layer/aluminum interfaces.

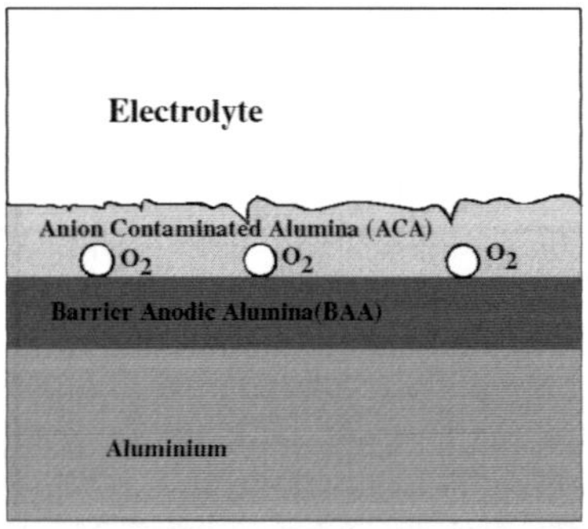

(a) Formation of contaminated layer

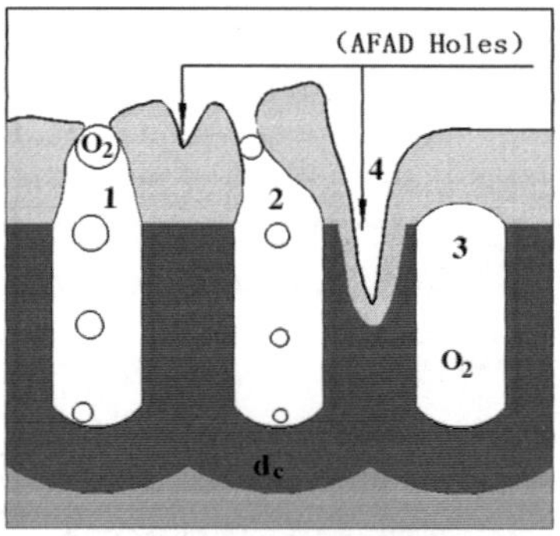

(b) Hole formation

Figure 14. Oxygen bubble mold effect growth model of porous template.

5. Factors affecting the formation of oxide film

Several reaction parameters in the oxidation process determine the properties and structure of oxide films. Previous experiments have established a close correlation between the structural properties of anodic alumina (e.g., pore diameter, distance between neighboring pores, pore density, thickness of oxide film, and crystallinity of oxide film) and the preparation conditions. Therefore, the structural properties of the AAO template can be tuned by the preparation conditions. Numerous factors affect the formation of oxide films, including anodizing voltage (or anodizing current), electrolyte type, electrolyte concentration, oxidation temperature, and oxidation duration.

5.1. *Effect of oxygen evolution on the formation of the porous layer*

Zhu *et al* [44, 46] studied the anodization process in three different electrolytes (ammonium dihydrogen phosphate solution, hexanediamine-boric acid-ethylene glycol solution, and mono-substituted phosphate-ethylene glycol solution). They proposed that oxygen evolution on the aluminum surface affected the formation of the porous layer. This was the reason that a barrier layer of oxide film can be eventually converted into a porous layer. Experimental results showed that the porous layer could still be formed without prior dissolution of the barrier layer in the same electrolyte. In addition, the formations of oxide film precursors and pores were attributed to oxygen evolution.

Wang *et al* [47] studied the anodization of hard aluminum alloys. Experimental results showed that a large amount of bubbles formed during oxide film formation; this phenomenon changed the current distribution on the surface of the oxide film, leading to a lower heat transfer rate between the electrolyte and the reaction surface. Furthermore, bubbles covering the surface of oxide film increased the local temperature, accelerating the electrochemical dissolution of oxide film. Hence, pores tended to be generated at those areas covered with bubbles.

5.2. *Effect of anodizing current (or anodizing voltage) on the formation of PAA*

There is a linear relationship connecting the distance between neighboring pores and the oxidizing voltage at which the oxide film reaches a stable stage, given that the electrolyte type and concentration are held constantly [45,46]. Several studies showed that an AAO template could only be successfully prepared when the oxidizing voltage (or current density) was controlled within a certain range [41-47]. The reaction was difficult to produce when the oxidizing voltage was too low. On the contrary, when the oxidizing voltage was too high, the dissolution rate of aluminum was higher than its formation rate. Therefore, a highly ordered template could only be achieved by controlling the oxidizing voltage within a certain range. Other studies showed that oxidizing voltage had a linear relationship with both the pore diameter and the distance between neighboring pores [45-48].

Vrublevskya *et al* [48] studied the effects of current density on the volume expansion of the PAA layer. Experimental results showed that the electrolyte concentration did not affect the volume expansion coefficient of the PAA layer under a constant current density condition. The volume expansion coefficient increased when the temperature of the electrolyte was decreased or when the current density was increased. The relationship between the volume expansion coefficient (K) of the PAA layer and the anodizing voltage (U) could be expressed as follows: $K = 1.092 + 0.007U$. Therefore, there was a linear relationship between the logarithm of current density and the volume expansion coefficient of the PAA layer. Montero-Moreno *et al* [49] studied self-assembled anodization under a constant current condition. The authors used two-step anodization to prepare high-quality oxide film and compared it with oxide film prepared under a constant current condition. Experimental results showed that the thickness of oxide film increased when the current density was increased. Additionally, when the current density was increased, the fraction of alumina with crystalline phase decreased while the fraction of alumina with amorphous phase increased.

Formation of oxide film is a competition between growth and dissolution processes. High current density can cause the temperature to rise and accelerate the dissolution of the oxide film. Under this circumstance, the dissolution rate is higher than the growth rate of oxide film, affecting the formation of film. Therefore, there is a threshold current density. When the current density is lower than the threshold value, the growth process of oxide film is slow, leading to a decrease in film thickness. In some cases, the film may not even be generated under such a condition.

5.3. *Effect of the composition of electrolyte*

The composition of electrolyte, including concentration and type, affects the properties of the oxide film. Generally, different types of electrolytes contain different types of anions when other reaction conditions remain constant. Experimental results showed that the properties of oxide film (color, thickness, and orderliness) were affected by the type of anion. For example, electrolytes such as sulfuric acid, chromic acid, and oxalic acid solutions showed good dissolving power. Therefore, the porous layer of anodic oxide film formed under this condition was relatively thick and the barrier layer was relatively thin. On the contrary, the oxide film was almost covered by the barrier layer when a boric acid electrolyte with less dissolving power was used.

Liu *et al* [50,51] studied the morphology and structure of oxide films prepared by anodizing aluminum in acidic electrolytes, such as sulfuric acid, oxalic acid, and chromic acid solutions. Experimental results showed that the electrolytes could be sorted by pore diameter in the following descending order: chromic acid > oxalic acid > sulfuric acid. However, the porosity was sorted in the opposite order. Additionally, oxide film prepared in different electrolytes varied widely in adhesive force and compactness. In 2003, Ono and Masuda [52] discovered the effect of electrolyte type on the distance between neighboring pores. Under the same reaction conditions, the electrolytes could be sorted by distance between neighboring pores in the following descending order: phosphoric acid > chromic acid > oxalic acid > sulfuric acid. Anodic alumina prepared in different electrolytes could be used for different applications best suited to their characteristics.

In 2008, Yang *et al* [52-55] found that the current density had a significant increment when a mixture of sulfuric acid and oxalic acid in a 1:1 ratio was used as the anodizing electrolyte. Under this condition, a thin barrier layer could be formed. If the anodizing voltage was further reduced, the pore diameter of the alumina film grew larger.

5.4. Effect of oxidation duration

The oxidation duration determines the thickness and compactness of the oxide film. Under longer oxidation duration, the film becomes thicker and more compact and shows higher corrosion resistance. However, heat released during the oxidation process accelerates the dissolution rate of the oxide film if the reaction duration is too long. As a result, the oxide film grows thicker in the initial stage, but eventually the growth process is retarded due to the dissolution process. In addition, long reaction duration affects the surface morphology of the film, reduces its smoothness, and hinders the binding behavior between the film and the base. More severely, a dusting phenomenon occurs, causing a significant drop in corrosive and wear resistances. Therefore, oxidation duration must be controlled in order to prepare an oxide film with the desired properties.

5.5. Effect of electrolyte concentration and temperature

Ion transfer rate is mainly affected by the electrolyte temperature. Consequently, the dissolution, formation, and heat transfer rates of oxide film are also affected by the electrolyte temperature. Therefore, electrolyte temperature is an important parameter in controlling the structure and distance between neighboring pores of oxide film. Under certain reaction conditions, the dissolution rate of oxide film increases when the electrolyte temperature is increased, causing the film to be thinner. Oxide film cannot be formed when the temperature is too high. On the contrary, the film thickness increases when the electrolyte temperature is decreased, producing oxide film with higher compactness and corrosive resistance. However, the duration required to prepare oxide film with the desired thickness becomes extremely long when the electrolyte temperature is too low. This is not suitable for mass production because the production efficiency is low and the energy consumption is high. Oxide film with different structures and properties can be obtained under different conditions and reaction parameters.

The intensity of the oxidation reaction is affected by the degree of erosion of the electrolyte, while the degree of erosion is affected by the concentration and temperature of the electrolyte. If the electrolyte concentration is too low, no oxidation reaction can be initiated; therefore, oxide film cannot be formed. An electrolyte concentration that is too high also inhibits the formation of oxide film because the dissolution rate is greater than the growth rate of the oxide film. Therefore, the electrolyte concentration must be controlled within an optimal range during the preparation of anodic alumina to ensure successful formation of pores. Zaraska *et al* [55] conducted anodization of aluminum using phosphoric acid solution and high anodizing voltage. Experimental results showed a regular pattern in current density under different concentrations of phosphoric acid: the current density decreased rapidly in the beginning of the reaction, followed by a subsequent increase before reaching a stable stage. This pattern was generally followed during the preparation of AAO film. The current density increased when the concentration of phosphoric acid was increased, requiring less time to reach the stable stage. The thickness of the barrier layer and the distance between neighboring pores increased as well. Xu *et al* [56-58] conducted the anodization of aluminum under different concentrations of phosphoric acid. They found that the distance between neighboring pores decreased when the concentration of phosphoric acid increased, but the increment was small. In summary, the distance between neighboring pores was affected by the electrolyte concentration, but different electrolytes could have different degrees of influence on the distance between neighboring pores.

Experimental results showed an initial increase followed by a decrease in pore diameter and the distance between neighboring pores when the temperature was increased. In an experiment conducted at high temperature, the distance between neighboring pores of the AAO film was larger than that obtained at low temperature. The distance between neighboring pores was even smaller when the oxidation temperature was set at zero. This was attributed to the greater influence of temperature on the dissolution of film than on the growth of film. The temperature used for anodization should be controlled because high temperature may reduce the orderliness of pores and the composition of the oxide film.

6. Application of alumina film

Based on previous investigations and understanding of the microstructures and properties of anodic alumina, future prospects of PAA film are focused on two areas: (1) utilization of its porous structure to develop new membranes for precise separation [54,55], and (2) development of new materials by depositing foreign substances such as metals, semiconductors, and polymeric materials into the nanopores of PAA film.

6.1. Nano-template application

Nanotechnology has emerged as a popular, advancing technology, with numerous studies investigating nanomaterials. Owing to the unique properties of nanomaterials, they have been widely used in various fields, including electronics components, medical equipment, and material synthesis. Highly ordered porous oxide film can be achieved through a simple anodization of aluminum foil. Moreover, the pore diameter of the oxide film can be tailored by changing the preparation conditions and parameters. Nanomaterials with various sizes and shapes can be prepared using this porous alumina film as the template. As a result, porous aluminum oxide film is used in the synthesis of nanostructured array systems with a variety of morphologies, such as nano-columns, nano-wires, and nano-dots.

6.2. Optical and optoelectronic applications

The current era is dominated by information technology (IT). The transmission and display of optoelectronic information is the central technology of the growing optoelectronic information science industry. The unique functions of PAA film make it a promising material for use in research and development of optoelectronic materials. In addition, nano-sized porous film prepared by the anodic oxidation method can be applied to the development and preparation of ultrafine optoelectronic material. Furthermore, luminescent films and photosensitive films used for imaging can be achieved by introducing fluorescent and photosensitive materials into PAA film. For example, fluorescent film that can emit green light can be achieved by introducing Tb3+ onto the oxide film. Through deposition of certain metals onto porous aluminum oxide film, new optical and optoelectronic properties can be achieved in the creation of optical phase plates, polarization plates, and other optoelectronic devices. Tang *et al* [58] prepared Cu/Al_2O_3 composites using the deposition method. Experimental results showed that the Cu composites induced a redshift in spectrum and exhibited better optical performance compared with single-phase porous aluminum oxide film.

In order to effectively utilize solar energy, materials should have high absorption coefficients in the visible range of the solar emission spectrum, but low emission rates in the range of the thermal emission spectrum. For example, functional film that can selectively absorb solar energy can be prepared by electrodepositing Ni on nanoporous aluminum oxide film in a phosphoric acid solution. The selective absorption characteristic of these films can be measured using reflectance.

6.3. Catalytic applications
Catalysis is useful for enhancing reaction efficiency, reducing energy consumption, and providing good selectivity and recyclability. Two commonly used catalysts are molecular sieve and ion exchange resins. However, these two types of catalysts show relatively low catalytic efficiency and are unstable at high temperatures, limiting further development of the technology. Aluminum substrate has good thermal conductivity. Owing to the unique structure of porous aluminum oxide film, it shows a good processability and its properties can be easily modified. In addition, the porous aluminum oxide film has a large inner surface area. Therefore, it can be used to develop functional film with good thermal conductivity and high catalytic efficiency. The Al_2O_3/TiO_2 functional thin films fabricated by Zhang *et al* [59] showed good photocatalytic properties and improved the catalytic efficiency of traditional porous aluminum oxide film.

6.4. Magnetic applications
Functional films with magnetic properties can be obtained by filling the pores of the ordered porous aluminum oxide film with magnetic metal or alloy materials. The obtained functional films can be further processed into a variety of magnetic products, such as tapes, disks, and magnetoresistance devices. Studies have shown that during the crystallization of magnetic metal located in the pores of PAA film, the orientation of the magnetic material was consistent with the direction of the magnetic axis, improving the retentivity and vertical magnetism of the functional film. Such films are commonly used as vertical magnetic recording media and magnetoresistance devices. Deng *et al* [60] fabricated highly ordered CoPt nanotube by introducing metallic magnetic material into the pores of the aluminum oxide film through an electrochemical method. Investigation with various characterization techniques (TEM, SEM, XRD, and VSM) showed that the CoPt nanotubes had good magnetic storage performance.

6.5. Separation membranes
Separation membranes have a wide range of applications, including wastewater treatment, ion separation, and biomedicine. By controlling the conditions of anodization in the production of porous aluminum oxide film, different thicknesses and sizes can be obtained, in addition to highly ordered pores. The film is a new material for separation membranes because it is selective and can be used to separate molecules with different sizes. Aluminum oxide film has several advantages over the conventional polymer membrane, such as good chemical stability, strong corrosion resistance, good thermal conductivity, and good recyclability. It can be used as a separation membrane for gas, liquid, and blood at room temperature. Furthermore, it can be used for gas separation at high temperatures in processes such as desulfurization and deoxygenation of flue gases.

El-Safty *et al* [61] used anodized aluminum oxide film as a separation membrane for sewage purification treatment. Experimental results showed that the removal rate of solid particles greater than 0.1 μm was close to 100%. The removal rates of BOD5 and CODCr reached 83% and 67%, respectively. After the anodized aluminum oxide film was used, the sewage turbidity dropped about 40° and the treated water became colorless. The used oxide film was easily washed without significant silting and was reusable. Additional uses of porous aluminum oxide film include wear- and corrosion-resistant materials, hard materials, and lubricants.

7. Future prospects
With its ordered porous structure, aluminum oxide film has potential as an effective material in a wide range of applications with high commercial value. Over the past several years, significant progress has

been made with regards to structural engineering and surface fication of nanoporous AAO material. Access to these structures is achieved by changing anodization conditions such as current, voltage and type of electrolytes during electrochemically self-ordering of AAO. However, the mechanism of pore formation is still not fully understood. In addition, there is still a need for further investigation of PAA and development of better PAA films. The preparation process should be optimized to reduce cost and increase efficiency. The physical and chemical properties of the PAA template should be investigated, and the parameters more finely adjusted.

Acknowledgments
The authors wish to thank the National Science Foundation Grants No. 1536209 and Chinese Visiting Scholarship Support Fund.

References

[1] Ren Y, Ma Z and Bruce P G 2012 *Chem. Soc. Rev.* **41(14)** 4909-27
[2] Fan Q H, Galipeau D, Ghimire B, Gupta R K, Dubey M and Mandal N 2014 *J. Coat. Sci. Technol.* **1** 69-77
[3] Martin C R 1996 *Chem. Mater.* **8(8)** 1739-46
[4] Yanagishita T, Nishio K and Masuda H 2005 *Adv. Mater.* **17(18)** 2241-3
[5] Colombo P 2008 *Science* **322(5900)** 381-3
[6] Thompson G E and Wood G C 1981 *Nature* **290** 230-2
[7] Bennett S P, Menon L and Heiman 2008 *J. Appl. Phys.* **104** 024309
[8] Poodt P, Lankhorst A, Roozeboom F, Spee K, Maas D and Vermeer A 2010 *Adv Mater.* **22(32)** 3564-7
[9] Li A P, Muller F, Birner A, Nielsch K and Gosele U 1999 *J. Vac. Sci. Technol.* **17(4)** 1428-31
[10] Nielsch K, Choi J, Schwirn K, Wehrspohn R B and Gosele U 2002 *Nano Lett.* **2(7)** 677-80
[11] Jani A M M, Losic D and Voelcker N H 2013 *Prog. Mater. Sci.* **58(5)** 636-704
[12] Keller F, Hunter M S and Robinson D L 1953 *J. Electrochem. Soc.* **100(9)** 411-9
[13] Thompson G E, Furneaux R C, Wood G C, Richardson J A and Goode J S 1978 *Nature* **272** 433-5
[14] Masuda H and Fukuda F 1995 *Science* **268(5216)** 1466-8
[15] Hu Z, Shrestha M and Fan Q H 2016 *Thin Solid Films* **598** 131-40
[16] Li F Y, Zhang L and Metzger R M 1998 *Chem. Mater.* **10** 2470-80
[17] Li Y, Ling Z Y, Chen S S and Wang J C 2008 *Nanotechnology* **19(22)** 225604-9
[18] Dou W H, Xu Q M and Zhao H L 2007 *Dvpt. Appl.Mater.* **22(5)** 49-53
[19] Gao Y B, Peng Z, Song G J, Yong-Wei M A and She X L 2007 *Mater. Rev.* **21(7)** 136-8
[20] Yanagishita T and Masuda H 2015 *Mater. Lett.* **160** 235-7
[21] Li J, Papadopoulos C and Xu J 1999 *Nature* **402** 253-5
[22] Gao T, Meng G, Zhang J, Sun S and Zhang L 2002 *Appl. Phys. A* **74(3)** 403-6
[23] Tian Y H, Meng G W, Gao T, Sun S H, Xie T, Peng X S, Ye C H and Zhang L D 2003 *Nanotechnology* **15(1)** 189-91
[24] Lee J S, Gu G H, Kim H, Jeong K S, Bae J and Suh J S 2001 *Chem. Mater.* **13(7)** 2387-91
[25] Xu T T, Fisher F T, Brinson L C and Ruoff R S 2003 *Nano. Lett.* **3(8)** 1135-9
[26] Cojocaru C S, Padovani J M, Wade T, Mandoli C, Jaskierowicz G, Wegrowe J E, Fontcuberta Morral A and Pribat D 2005 *Nano .Lett.* **5(4)** 675-80
[27] Lee W, Ji R, Gösele U and Nielsch K 2006 *Nat. Mater.* **5** 741-7
[28] Sanz R, Hernández-Vélez M, Pirota K R, Baldonedo J L and Vázquez M 2007 *Small* **3(3)** 434-7
[29] Smith J T, Hang Q L, Franklin A D, Janes D B and Sands T D 2008 *Appl. Phys. Lett.* **93** 043108-12
[30] Lee W, Schwirn K, Steinhart M, Pippel E, Schol R and Gösele U 2008 *Nat. Nanotechnol.* **3** 234-9
[31] Yamauchi Y, Nagaura T, Ishikawa A, Chikyow T and Inoue S 2008 *J. Am. Chem. Soc.* **130**

10165-71
[32] Losic D, Lillo M and Jr L D 2009 *Small* **5(12)** 1392-7
[33] Jani A M M, Losic D and Voelcker N H 2013 *Prog. Mater. Sci.* **58(5)** 636-704
[34] O'Sullivan J P and Wood G C 1970 *Proc. R. Soc. Lond.* **317(1531)** 511-43
[35] Shimizu K, Kobayashi K, Thompson G E and Wood G C 1992 *Philos. Mag. A* **66(4)** 643-7
[36] Jessensky O, Muller F and Gösele U 1998 *Appl. Phys. Lett.* **72(10)** 1173-5
[37] Thompson G E, Furneaux R C and Wood G C 1978 *Nature* **272(565)** 433-5
[38] Xu Y, Thompson G E and Wood G C 1989 *J. Chin. Soc. Corr. Pro.* **9(1)** 1-10
[39] Xu D S, Chen D P, Xu Y J, Shi X S, Guo G L, Gui L L and Tang Y Q 2000 *Pure Appl. Chem.* **72** 127-35
[40] Patermarakis G and Moussoutzanis K 2009 *Electrochim. Acta* **54(9)** 2434-43
[41] Palibroda E 1995 *Electrochim. Acta* **40(8)** 1051-5
[42] Palibroda E, Farcas T and Lupsan A 1995 *Mat. Sci. Eng. B* **32(1-2)** 1-5
[43] Zhu X F, Liu L, Song Y, Jia H B, Yu H D, Xiao X M and Yang X L 2008 *Monatsh. Chem.-Chem. Month.* **139(9)** 999-1003
[44] Zhu X F, Han H, Song Y, Ma H T, Qi W X, Chao L and Chen X 2012 *Acta Phys. Sin.* **61(22)** 228202-7
[45] Cao G B, Zhu W, Li J R and Li Z X 2014 *Mater. Rev.* **28** 153-9
[46] Zhu X F, Li D D and Meng D W 2006 *J. Nanjing Univ. Sci. Tech.* **30(5)** 644-8
[47] Wang P and Wei X W 2005 *Sur. Tech.* **34(6)** 28-32
[48] Vrublevsky I, Parkoun V, Schreckenbach J and Mark G 2003 *Appl. Sur. Sci.* **220** 51-9
[49] Montero-Moreno J M, Sarret M and Müller C 2010 *Micropor. Mesopor. Mat.* **136** 68-74
[50] Liu F X and Xia Z C 1994 *Mater. Prot.* **27(1)** 19-22
[51] Yang P X, Zhang X M and An M Z 2008 *Electroplat. Pollut. Control* **28(4)** 28-30
[52] Ono S, Saito M and Asoh H 2005 *Electrochim. Acta* **51(5)** 827-33
[53] Michalska-Domańska M, Norek M, Stępniowski W J and Budner B 2013 *Electrochim. Acta* **105** 424-32
[54] Hussain T, Shah A T, Shehzad K, Mujahid A, Farooqi Z H, Raza M H, Ahmed M N and Nisa Z U 2015 *Int. Nano Lett.* **5** 37-41
[55] Zaraska L, Jaskula M and Sulka G D 2016 *Mater. Lett.* **171** 315-8
[56] Zaraska L, Brudzisz A, Wierzbicka E and Sulka G D 2016 *Electrochimi. Acta* **198** 259-67
[57] Xu Y F, Liu H, Li X J, Kang W M, Cheng B W and Li X J 2015 *Mater. Lett.* **151** 79-81
[58] Tang H J, Wu F Q, Wang H L, Wei Y H and Li Q S 2006 *J. Appl. Phys.* **100(6)** 064316-22
[59] Zhang X Y, Chen T Q and Wang C Y 2005 *J. South China Univ. of Techno.(Nat. Sci. Ed.)* **33(9)** 67-72
[60] Deng C H, Qiao X Y, Yan Y N, Wang F, Fn J P, Zeng H and Xu X H 2016 *IEEE T. Magn.* **52(4)** 1-5
[61] El-Safty S A, Sakai M, Selim M M and Alhamid A A 2015 *Chem.Asian J.* **10(9)** 1909-18

IOP Conf. Series: Materials Science and Engineering **283** (2017) 012004 doi:10.1088/1757-899X/283/1/012004

Mechanical activation effect on structure, physicochemical, and biological properties of potassium/magnesium orotates

O M Kanunnikova[1*], V V Aksenova[1], O V Karban[2], V V Muhgalin[1], B V Senkovski[3] and V I Ladjanov[1]

[1]Physicotechnical Institute, the Ural Branch of the Russian Academy of Sciences, 426000 Izhevsk, Russian Federation
[2]Izhevsk State Agricultural Academy, 426069, Izhevsk, Russian Federation
[3]Russian-German Laboratory at BESSY II (Helmholtz Zentrum), 12489 Berlin, Germany
*E-mail: olam313597@gmail.com

Abstract. In the present paper, the possibility of the formation of the tautomeric structure of potassium orotate and magnesium orotate in the solid state is discussed. Using ball milling, we have received the orotate samples with the predominant content of hydroxy and dihydroxy forms. The crystalline and tautomeric structures have been analyzed by XRD, IR, NEXAFS and XPS spectroscopy. There is a slight difference in the water solubility of the orotate tautomers, but the physicochemical properties (viscosity and volumetric thermal expansion coefficients) of tautomers' water solutions significantly differ. The above differences are related to variation of hydration area of orotate anions due to the composition of the functional groups being changed together with the structure of tautomers. The residual differences in the tautomeric structure of solid orotates formed during the milling process are observed in the orotate water solutions for 1.5-2.5 hours. The biological activity of the water solution of a hydroxy form relative to erythrocytes and epithelial cells is higher than that of oxo- and dihydroxy-form water solutions.

1. Introduction

Due to the well-known notion that the basis of pharmacy is the synthesis of new biologically active substances, leading pharmaceutical companies actively improve the existing pharmaceutical forms. The advantages of known and tested drugs are that their side effects are already identified, while their therapeutic effect is proved by long-term practice. The improvement of drug properties is mainly attained by the formation of crystalline, amorphous, or isomeric structures with a high biological activity without changing the drug composition.

In recent years, a considerable part of the pharmaceutical research has benefited from the advance made in the physics of polymorph transformations [1, 2]. There are many examples of radical differences in physicochemical and biological properties of various crystalline polymorphs sharing the same chemical composition. Amorphous solids are characterized by higher solubility and chemical reactivity [3-5]. Multiple biologically and pharmaceutically important chemical substances are known to exist in different isomeric forms [6]. Given this, many scholars focused their attention on understanding the formation of optical isomers in saccharides [7-9].

Tautomerism is a form of isomerism, in which the intramolecular migration of a hydrogen atom or proton takes place accompanied by the transformation of a single bond into a double one or vice versa

[10]. The influence of tautomerism on a particular structure of drugs and their potential to interact with biological systems is discussed in [11, 12].

The tautomeric forms of the DNA and RNA have been extensively studied [8]. It has been found that the misuse of the purine and the pyrimidine bases can lead to errors in replication and to point mutations [13-15]. A number of papers are dedicated to the keto-enol tautomerism of p-hydroxyphenylpyruvic acid, which is used to diagnose a congenital metabolic defect [16].

Tautomerism takes place mainly in a solution, gas phase, or amorphous state because the proton transfer occurs easily in such states. In the crystalline state, tautomerism is also possible but only under particular conditions, e.g. at high temperatures [17-20]. Moreover, in this case, the tautomerism rate is much smaller than in a supercooled liquid, solution, or gas phase. Unlike other classes of isomers, tautomers exist in the dynamic equilibrium between one another; therefore, the attempts to prepare separate substances usually result in the formation of a mixture that has all chemical and physical properties of the base components [21].

Recently, "green chemistry" has become popular, which methods involve the synthesis or modification of chemical compounds without the use of solvents or gaseous media, which can harm the environment or contaminate the reaction final product. In this connection, the use of solid-phase methods for modifying properties of medicinal substances is crucial. As a solid-phase method, the method of mechanical activation in high-energy dispersing devices, including ball milling, is very practical [22-28].

In this paper, we investigate the possibility of preparing various tautomeric forms of the solid-state orotates of magnesium and potassium by milling using a planetary ball mill, and compare the biological properties of the prepared tautomers. The lifetime of different solid state tautomers is of special interest from the standpoint of practice. Since dissolution is a limiting stage of an effect of drugs on an organism, such factors as the solubility of preparations and the preservation of differences between tautomers in solution are also very important.

2. Experimental

2.1. Materials
To carry out the work, we used potassium orotate and magnesium orotate obtained from Sigma-Aldrich Co. (US) without any additional purification.

2.2. Sample preparation
For milling substances, a planetary ball mill AGO-2 (Russia) was used. The rotation speed of the drums was 600 rpm. The power density was 2кJ/g. The dose of mechanical energy supplied to a sample after 1 hour of milling was 7 kJ/g, after 3 hours - 21 kJ/g, after 6 hours - 42 kJ/g. The temperature of the vial walls did not exceed 60°C during the milling procedure due to the forced water-cooling.

2.3. Investigation methods
The elemental composition and solubility of the samples were analyzed by inductively coupled plasma atomic emission spectroscopy (ICP-AES) using a Spectroflame Modula spectrometer (Germany). According to the data of the analysis there were no impurities from the materials of the balls and vials.

The morphology of the prepared powders was examined by atomic force microscopy (AFM) with an Integra Prima (NT-MDT) scanning probe microscope in the semi-contact mode in the air. The powder has been preliminarily attached to a polystyrene film produced by evaporating ethyl-acetate in which polystyrene was dissolved. The film was applied on glass-ceramics with the subsequent fixation of the powder by ultraviolet emission.

The X-ray diffraction patterns (XRD) were obtained using a D8 Advance diffractometer (Bruker AXS) in the parallel-beam geometry (MoKα and CuKα). Calculations of the structure characteristics were made using Topas 4.2. Structure refinement was performed by the Rietveld method [29] using

MgO as a standard sample [30]. The calculation of the elementary unit cell parameters was based on 205 reflexes. For the Rietveld refinement, the structural data (the initial coordinates of atoms) were obtained from elsewhere [31].

The spectra obtained by Near-edge X-ray absorption fine structure (NEXAFS) and X-ray photoelectron spectroscopy (XPS) were measured at the Russian-German (RG-PGM) dipole beamline at BESSY II, Helmholtz-Zentrum, Berlin. The beamline uses the plane-grating monochromator with the 400 and 1200 lines/mm gratings to select the photon energies in the range of 40 - 1500 eV, with the resolving power of up to $E/\Delta E = 100\,000$. A detailed description of the beamline can be found elsewhere [32]. All spectra were taken under ultrahigh vacuum conditions (5×10^{-10} Torr).

The NEXAFS data were acquired in a total electron yield (TEY) mode by measuring the sample drain current with varying photon energy. The spectra were normalized to the incident photon flux which was controlled by registering a complete electron exit from the clean surface of the gold crystal mounted on the manipulator holder.

The *C* K-edge absorption spectral features were calibrated with respect to the energy position of the $C1s \rightarrow \pi^*$ resonance of the measured absorption spectrum of the highly oriented pyrolitic graphite (HOPG) at 285.35 eV [33].

The XPS spectra were measured at the photon energy varied in the range of 200 - 1000 eV with the monochromator energy resolution of 125–500 meV. The energy scales of the spectra were calibrated to eliminate the charging effect by referencing the $C1s$ peak of hydrocarbons to 285.0 eV [34]. The high-resolution spectra were analyzed in order to determine the presence of the various chemical substances. Each spectrum was deconvoluted using the methods based on Fourier transform with improved convergence procedure [35].

The shapes of both the NEXAFS and the XPS spectra strongly indicated that no damage of samples under the synchrotron radiation exposure occurred, irrelevant of the irradiation time.

The diffuse reflectance infrared Fourier transform (DRIFT) spectroscopy spectra and the transmission spectra were recorded with the use of an FSM-1202 spectrometer (Russia) with the resolution of 1 cm^{-1} in the 4000-450 cm^{-1} region. The transmission measurements were performed on pellets made of the orotate samples and pure KBr powder. The DRIFT spectra were obtained using a diffuse reflectance accessory. The samples were analyzed at room temperature.

The biological properties of the source and milled orotates were analyzed by the microelectrophoresis method, where the amplitude of the cell movement in the field of the microscope was measured. In the electrophoresis chamber, the cells exhibited a constrained reciprocating motion during the stress reversal on the electrodes at 10 V with frequency of 0.1 Hz. The frequency of such motion was equal to the latter one, but its amplitude was different, being controlled by the cell surface charge, which was the marker of physiological state of a cell [36]. The study was conducted using a Cytoexpert complex (Russia).

3. Results and discussion

3.1. Potassium orotate

When the milled powder imparts the mechanical energy dose of 7 kJ/g, it represents separate spherical particles mainly of size 60 nm. The maximum size of some large particles does not exceed 250 nm. When the mechanical energy dose is 21 kJ/g, the particles have the form of ellipsoid of size 80-150 nm. Unstable aggregates up to 1.5 μm are formed (figure 1). The latter can be destroyed under scanning by the silicon probe. After the mechanical energy dose of 42 kJ/g has been exerted to the powder, laminated aggregates of particles of the size over 5 μm are formed. Some layers are formed of disc-shaped particles with the average thickness of 100 nm and diameter of 400-900 nm.

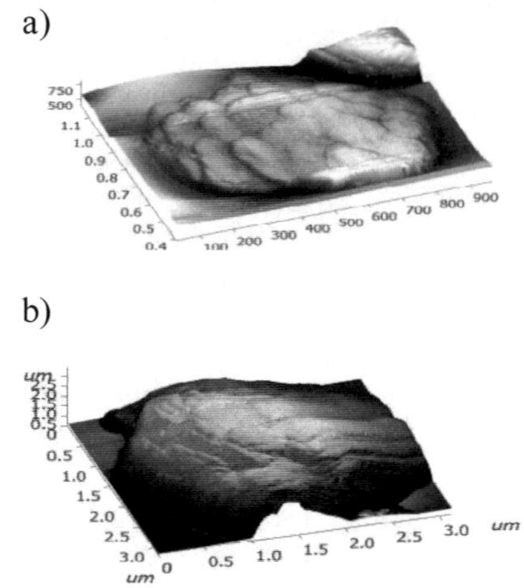

Figure 1. AFM-image of potassium orotate powder after milling with the mechanical energy dose of 21 kJ/g (a) and 42 kJ/g (b)

In table 1, the crystallographic parameters of the source and milled potassium orotate powder are shown. Indexing the diffraction pattern (LP-search [37]) shows that the unit cell space-group symmetry of the source powder corresponds to orthorhombic *Cmma* with the unit cell dimensions a = 6.842(1) Å, b = 12.535(2) Å and c = 7.685(1) Å. The type of the crystal structure remains unchanged (figure 2); however, the main parameters of an elementary unit cell change: a and b increase, and c decreases. The broadening of the diffraction lines indicates the decrease of the crystallites size and the growth of the lattice microstrain (*e0*). When the mechanical energy dose is 7 kJ/g, the development of the preferred orientation is observed: the relative intensity of the (*200*) and (*201*) reflections increases. The molecules (packed in the molecular crystal as lamels) of potassium orotate are oriented parallel to the (*100*) plane. The increased intensity of the (*h00*) reflections may arise because of the potassium orotate dispersion with the formation of anisotropically shaped crystallites. It can be assumed that milling at the initial stages takes place due to the shift and detachment of lamels, and the following deforming impactions result in the gradual resizing of the unit cell.

Table 1. Crystallographic parameters of source and milled potassium orotate

Parameter	Source	Mechanical energy dose, kJ/g		
		7	21	42
Space group		Cmma		
Crystal system		Ortorombic		
a, Å	6.842(1)	6.844(2)	6.848(3)	6.856(5)
b, Å	12.535(1)	12.538(2)	12.533(3)	12.549(5)
c, Å	7.685(1)	7.687(2)	7.679(2)	7.673(4)
Z		4		
Crystal size, nm	140	28	24	19
Lvol-IB, nm	47	18	15	12
StrainL, %	0.25	0.52	0.89	0.99
e0, %	0.06	0.13	0.22	0.24
R_{wp}	7.22	7.7	6.2	6.1
GOF	1.93	2.04	1.65	1.63
Vol, Å^3	659.1(2)	659.6(3)	659.4(3)	660.0(5)

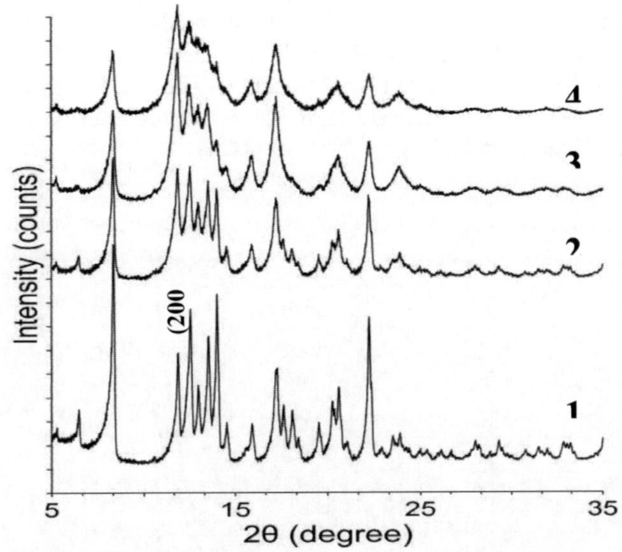

Figure 2. X-ray diffraction patterns of potassium orotate: source (1) and after milling with the mechanical energy dose of 7 kJ/g (2), 21 kJ/g (3) and 42kJ/g (4).

The tautomeric transformations affect the structure of the hydrogen bonds and the interaction of the double bonds of the pyrimidine base. The tautomerization results in the changes in the IR bands associated with the heterocycle vibrations in the region of "fingerprints". Figure 3 demonstrates the IR spectra of source potassium orotate and potassium orotate milled with the supplied mechanical energy dose of 42 kJ/g. The absorption in the regions 750 to 800 cm^{-1}, 850 to 950 cm^{-1}, and 990 to 1020 cm^{-1} is characteristic of the deformational vibrations of the pyrimidine ring. The IR spectrum of the milled sample shows the variation of the intensity ratio of the bands and the change in the width of the bands in comparison with the source sample.

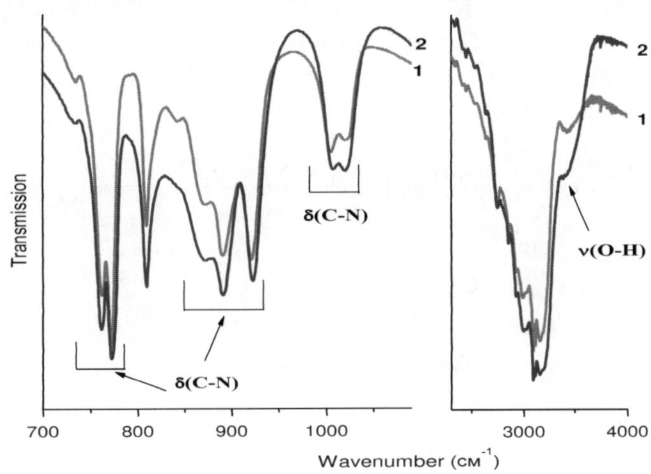

Figure 3. Transmission IR spectra of potassium orotate source (1) and after milling with the mechanical energy dose of 42 kJ/g (2).

The nature of the skeleton oscillations of the heterocycle changes due to the formation of the hydroxyl groups. However, in the sample, all tautometric forms of potassium orotate are present simultaneously, which complicates the detailed interpretation of the spectra.

IOP Conf. Series: Materials Science and Engineering **283** (2017) 012004 doi:10.1088/1757-899X/283/1/012004

In addition, the band of stretch vibrations of the associated hydroxyl groups in the region 2500-3600 $см^{-1}$ appear and correspond to the hydroxyl and C-N group stretch vibrations.

Figure 4 shows the DRIFT spectra of potassium orotate. The band of the vibrations of the NH, CN, and C=O (1650 to1750 cm^{-1}) loses its fine structure due to the tautomeric transformations after milling with the mechanical energy dose of 42 kJ/g.

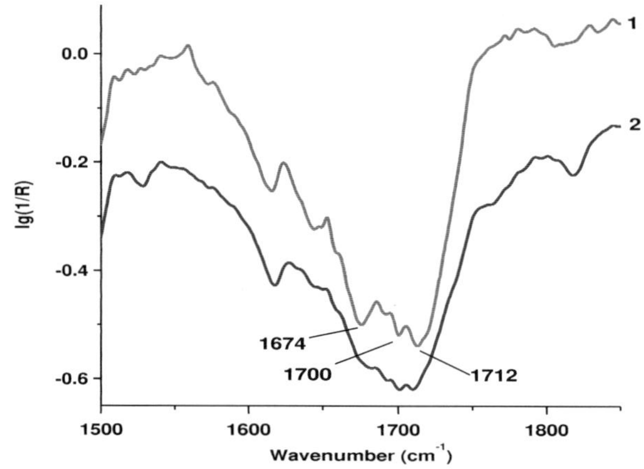

Figure 4. DRIFT spectra of potassium orotate source (1) and after milling with the mechanical energy dose of 42 kJ/g (2).

In table 2 the comparative data of NEXAFS and XPS are presented. The changes in the NEXAFS spectra structure (figure 5) correlate with those in the XPS spectra structure (figure 6). Milling with the mechanical energy dose of 7 kJ/g results in the decrease of the intensity of the peaks for C=O and –NH- groups, as well as the formation of the lines of C-OH and –N= groups. After milling with the mechanical energy dose of 42 kJ/g, the peaks of C-OH and –N- dominate, this is the evidence of the domination of the dihydroxy- form in the potassium orotate sample. The sharp resonance (doublet) near 300 eV is assigned to the potassium $L_{2,3}$ –edge absorption. These resonances are particularly high in the milled samples because of the preferential orientation of the potassium orotate molecules in the flat scaly crystalline samples.

Table 2. Interpretation of NEXAFS and XPS spectra of orotates

XPS				NEXAFS		
E_b, ±0.1, eV			Group	E_{ph}, eV		Group
C1s	O1s	N1s		C	N	
285.0			$(CH_2)_n$	284.2		C1s→π (C=C)
286.2	533.4		C-OH	285.0		C1s→π (C=C)
289.0	531.8	400.5	O=C-N	288.5		C1s→π (C=O)
290.0			shake-up	290.0		C1s→π (C-OH)
		398.4 - - 399.0	heterocycle		398.0	N1s→π (-N=; C-N)
					399.7- -400.0	N1s→π (-NHC=O)

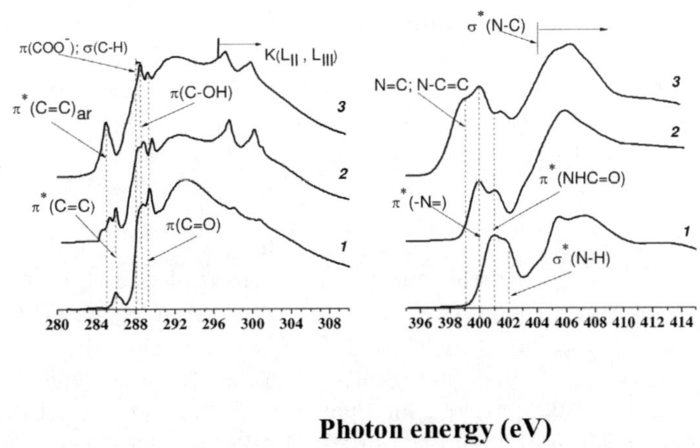

Figure 5. NEXAFS C1s K-edge and N1s K-egde spectra of potassium orotate source (1); after milling with the mechanical energy dose of 7 kJ/g (2) and 42 kJ/g (3).

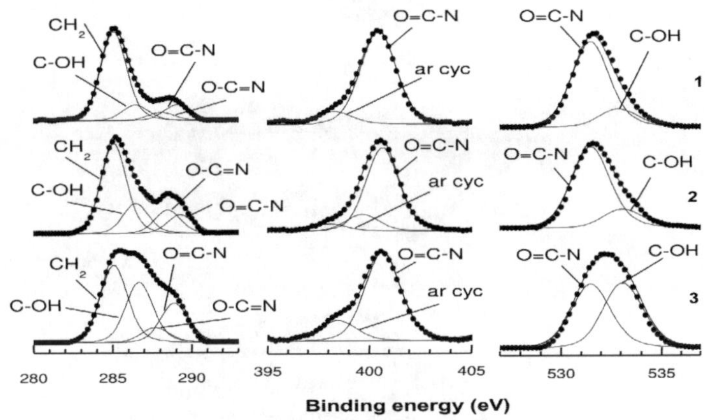

Figure 6. XPS spectra of potassium orotate source (1); after milling with the mechanical energy dose of 7 kJ/g (2), and 42 kJ/g (3)

In the XPS C1s-spectra, the component of the $(CH)_2$ groups of the adsorbed layer on the surface of the particles is dominant (E_b= 285.0±0.2 eV). The (C-OH) groups are at E_b (C1s) = 286.2±0.2 eV. The component at 289.0±0.2 eV refers to the carbon atoms bonded with the oxygen and nitrogen atoms (O=C-N) in the oxo- and hydroxy-forms of the orotate anions. The fraction of the component of the C-OH groups increases from 12% in the source sample to 20% and 36% in the samples after milling with the mechanical energy dose of 7 kJ/g and 42 kJ/g, respectively. In the O1s-spectra, there are components at 531.8±0.2 eV and 533.4±0,2 eV corresponding to the oxygen atoms in the carbonyl (C=O) and hydroxyl (C-OH) groups. The fraction of the components at E_b = 533.4 eV increases from 18% to 50% (figure 6). In the N1s-spectra, the intensive component at 400.5±0.2 eV is attributed to the NHC=O group. The components at E_b=398.4 ±0.2 eV and E_b=399.6 ±0.2 eV refer to the nitrogen atoms in the aromatic and nonaromatic heterocycles. The fraction of the components of the nitrogen atoms in the aromatic heterocycle increases from 8% to 19.5% (figure 6).

Thus, milling with the mechanical energy dose up to 42 kJ/g does not change the potassium-orotate crystalline structure. The decrease in the size of particles and crystallites is observed. However, the

IOP Conf. Series: Materials Science and Engineering **283** (2017) 012004 doi:10.1088/1757-899X/283/1/012004

milling invokes a change in the tautomeric structure of the orotate anion, i.e., the subsequent transformation of the oxo-form into hydroxy- and dihydroxy- forms takes place.

The hydroxy- and dihydroxy- forms of potassium orotate are unstable. Both forms turn into the oxo-form after 2-month's storage in the exsiccator (apparatus for drying substances or preserving them from moisture) at room temperature.

3.2. Magnesium orotate

Figure 7(a) shows the AFM-image of the magnesium orotate powder after milling with the mechanical energy dose of 7 kJ/g. The powder particles are spheroid-shaped of size 100-200 nm. Single particles of size 40 nm are also observed. The crystalline grains in the particles are 40 nm. The particles are integrated in loosely bound aggregates of up to 5 nm.

When the mechanical energy dose is increased to 21 kJ/g, the aggregate dimensions are almost similar to the above (figure 7(b)); however, the density and shape change. The aggregates become laminated with layers of about 70-180 nm (figure 7(c)). The average particle size is 162 nm. The particles do not demonstrate grain structure.

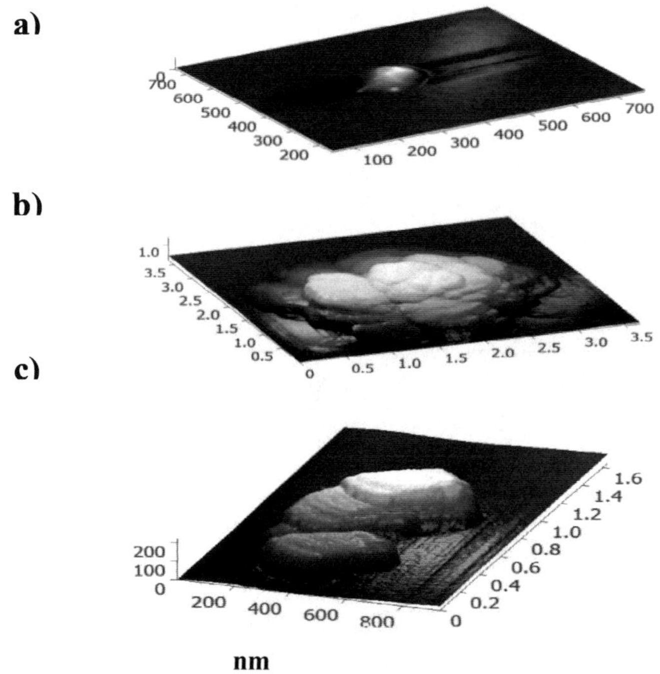

Figure 7. AFM images of magnesium orotate after milling with the mechanical energy dose of 7kJ/g (a); 21 kJ/g (b) and 42 kJ/g (c)

Upon further milling (the mechanical energy dose is more than 42 kJ/g), the aggregation process prevails over dispersion. The particles are of 300-1000 nm. Particles and dense small aggregates of about 1 nm cannot be distinguished. The particles represent laminated aggregates with particles of the size about 50 nm. Being scanned with an AFM cantilever, such aggregates do not decompose. The milling of magnesium orotate with the mechanical energy dose of 7 kJ/g and 21 kJ/g results in the reflection diffusion (with the preservation of the main set of lines), which is associated with an increase of the structure imperfection and the crush of crystallites (figure 8). These changes are accompanied by an increase in the intensity of the amorphous halo at 19 and 27 deg. Strong reflections of magnesium orotate at 43 and 50 deg can be due to a change in the phase composition of the sample. These lines persist after milling with the mechanical energy dose of 42 kJ/g, when the crystalline

IOP Conf. Series: Materials Science and Engineering **283** (2017) 012004 doi:10.1088/1757-899X/283/1/012004

structure of the source magnesium orotate is destroyed (a new set of reflexes appears instead of the initial ones). The complete rearrangement of the initial crystal structure and the formation of a multiphase amorphous-crystalline mixture are observed after milling with the mechanical energy dose of 42 kJ/g.

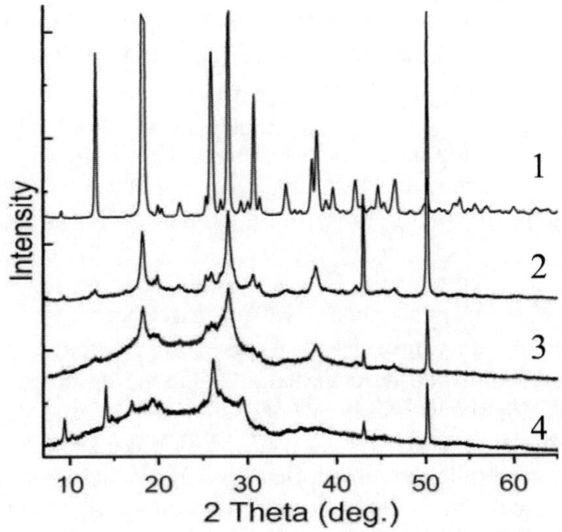

Figure 8. X-ray diffraction pattern of magnesium orotate source (1) and after milling with the mechanical energy dose of 7 kJ/g (2), 21 kJ/g (3) and 42kJ/g (4)

Figure 9 shows the IR spectra of the source magnesium orotate and the milled magnesium orotate. The increase in the milling time decreases the band intensity at 533 cm^{-1} due to the vibrations of δ(CH)+δ(NH). This can be a proof of the deprotonation of NH groups. The absorption between 600 and 800 cm^{-1} is caused by the deformational vibrations of the N-H pyramidine ring (there are out-of-plane wagging NH vibrations near ~700 cm^{-1}). Therefore, during milling, the appearance of various (hydroxy- and dihydroxy-) tautomeric forms of magnesium orotate influence the absorption in this region of the IR spectrum. The "breathing" vibrations of the pyrimidine ring in the spectrum region between 1000 and 1050 cm^{-1} are also changed with the milling time.

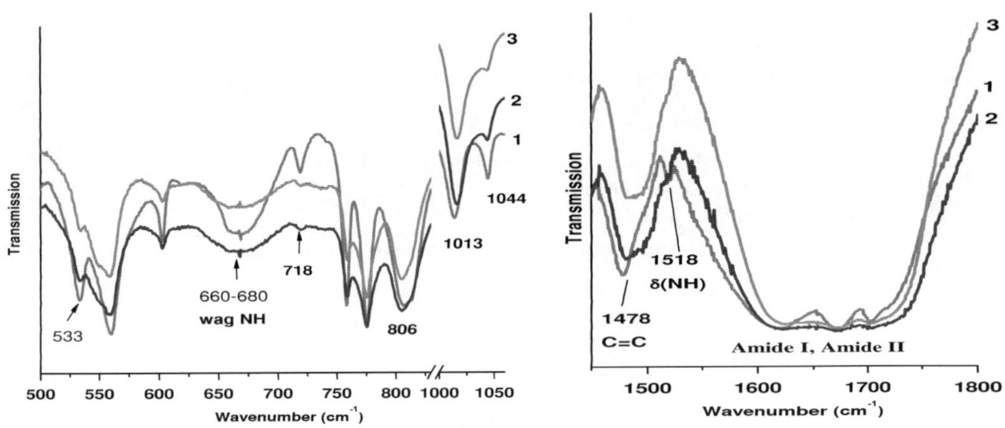

Figure 9. Transmission IR spectra of magnesium orotate source (1) and after milling with the mechanical energy dose of 21 kJ/g (2) and 42kJ/g (3)

The band at 1478 cm^{-1}, which is ascribed to the ν(C=C) vibrations, expands due to a shoulder at ~1500 cm^{-1}. This indicates the lactam-lactimic tautomeric transformations, which lead to the formation

IOP Conf. Series: Materials Science and Engineering **283** (2017) 012004 doi:10.1088/1757-899X/283/1/012004

of the aromatic structures. Milling decreases the intensity of the band at 1518 cm^{-1} associated with the deformational vibration of δ(NH) the pyrimidine ring. After milling with the mechanical energy dose of 42 kJ/g, the band is not observed. This indicates the deprotonation of NH groups due to milling.

The gradual transition of the pyrimidine ring to the completely aromatic form changes the shape of the IR-bands between 1500 and 1600 cm^{-1} which have appeared due to the NH deformational vibrations. The overlapping of the absorption of the carbonyl groups complicates finding point correlations. For the source sample, in the region 1600-1700 cm^{-1}, the high-resolution bands the amides I (ν(C=O)) and the amides II (δ(NH)+ν(CN)) are observed. With an increase in the milling time (and the mechanical energy doze), the bands merge. This indicates the tautomeric transitions which result in protonization of carbonyls. And, since the region 1600-1700 cm^{-1} is characterized by skeleton vibrations of a completely aromatic pyrimidine ring, it can also evidence that an aromatic structure appears.

The NEXAFS spectra of the K-edge of the C1s of pyrimidine bases have almost the same structure [15]. There are four well-resolved π*-resonances, two σ*-resonances near 296 and 304 eV. The peaks of π*-resonances correspond to the electron transitions from unequal core levels C1s of different carbon electrons in the molecules to the delocalized π*-nonbonding orbitals. In the uracil spectrum, there are peaks at 284.6 eV, 286.0 eV, 288.0 eV, and 289.5 eV. In the thymine spectrum, the low-energy peak shifts to the higher energy side by 0.4 eV (285 eV). It takes place due to the appearance of the methyl substitute. Still, it should be noted that the figure demonstrating the uracil chemical composition shows the structures corresponding to the oxo-forms and provides no data on the isomeric state of uracil and thymine. The heterocycle containing nitrogen has bonds similar to aromatic heterocycles, but the explicit aromatic nature of the bonds is common only for the dihydroxy-forms.

Figure 10 presents the NEXAFS spectra of the C1s-edge of absorption of magnesium orotate. In all the spectra, there is an intensive peak at 285.0 eV corresponding to the electron transition C1s→π* (C=C) and indicating the presence of an aromatic cycle. The electron transitions C1s→π* in C=O and C-OH contribute to the peaks at 288.5 eV and 290.5 eV. This strongly suggests that magnesium orotate has several tautomeric forms.

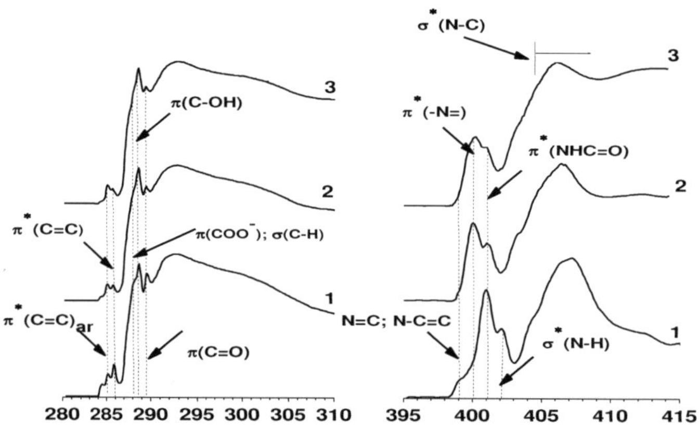

Figure 10. NEXAFS spectra of Cs1-edge of absorption of magnesium orotate: source (1) and after milling with the mechanical energy dose of 7kJ/g (2) and 42kJ/g (3).

The shape of the N1s spectra of magnesium orotate and the change of these spectra during milling are very similar to the above-described potassium orotate spectra.

The XPS C1s-, N1s-, O1s- spectra (figure 11) of magnesium orotate are similar to those of potassium orotate. In the XPS C1s-spectra, the component of the $(CH)_2$ groups dominates in the adsorbed layer on the surface of the particles (E_b= 285.0±0.2 eV). The (C-OH) groups have E_b (C1s) = 286.2±0.2 eV. The component at 289.0±0.2 eV refers to the carbon atoms bonded with the oxygen and the nitrogen atoms in the (O=C-N) group in the oxo- and hydroxy-forms of the orotate anions. The fraction of the component of C-OH groups increases from 25% in the source sample to 33% and 36% in the samples after milling with the mechanical energy dose of 7 kJ/g and 42 kJ/g, respectively. In the O1s-spectra, there are components at 531.8±0.2 eV and 533.4±0,2 eV corresponding to the oxygen atoms in the carbonyl (C=O) and the hydroxyl (C-OH) groups. The fraction of the components at E_b = 533.4 eV increases from 26% to 47% (figure 11). In the N1s-spectra, the intensive component at 400.5±0.2 eV is attributed to the nitrogen atoms in the NHC=O group. The components at E_b=398.4 ±0.2 eV and E_b=399.6 ±0.2 eV refer to the nitrogen atoms in the aromatic and nonaromatic heterocycles. The fraction of the nitrogen components in the aromatic heterocycle increases from 15% to 23.5% (figure 11).

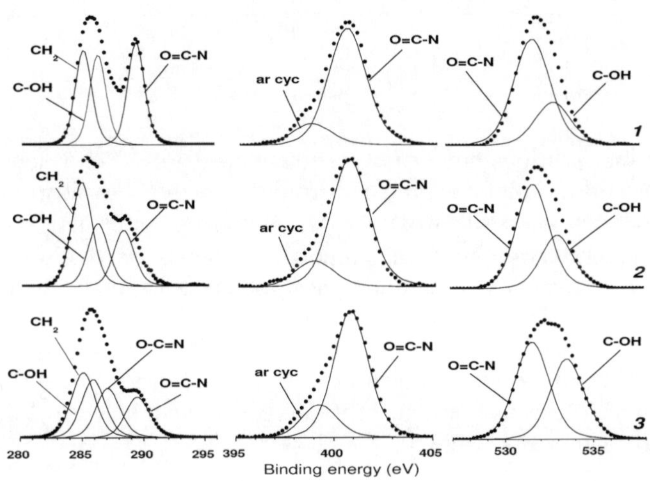

Figure 11. XPS spectra of magnesium orotate source (1); after milling with the mechanical energy dose of 7 kJ/g (2), and 42 kJ/g (3)

3.3. Orotates water solutions

The pH value of the water solutions of magnesium and potassium orotates decreases in the following sequence: oxo-form > hydroxy-form > mixture of hydroxy-form + dihydroxy- form > dihydroxy-form (table 3). The observed effect is caused by the strong acid properties of the orotate dihydroxy-form (which is OH-acid) rather than by the acid properties of the oxo-form (which is NH-acid). The hydroxy-form is in the intermediate position as it has one OH- and one NH- group. During the storage for a few hours at room temperature, the pH value of the water solutions becomes equal and corresponds to the pH of the oxo-form solution.

Table 3. Viscosity (Pa·s) of magnesium orotate and potassium orotate water solutions (0.1 wt.%).

Mechanical energy dose, kJ/g	Magnesium orotate			Potassium orotate		
	0	7	42	0	7	42
t, °C						
10	1.3298	1.3242	1.3165	1.2807	1.2689	1.1732
20	1.0265	1.0100	0.9989	0.9977	0.9482	0.8323
30	0.5302	0.8102	0.8002	0.8221	0.7999	0.6929
40	0.6777	0.6897	0.6777	0.6102	0.6809	0.5821

The capillary viscosity and the expansion coefficients (at 15°-25°C) of the solutions of the tautomers prepared by milling are presented in tables 3 and 4. The differences in the capillary viscosity and expansion coefficients are due to the change in the area of the hydration of orotate anion because of the changes in the composition of the functional groups (=O and –OH) caused by the change of the tautomeric structures of the orotate anions.

Table 4. Solubility of orotates, volumetric thermal expansion coefficient (Pa ·s) of water solutions of orotates

Mechanical energy doze, kJ/g	Solubility (25°C), $g/100gH_2O$		Volumetric thermal expansion coefficient (15-25°C, 0.1wt.%), K^{-1}	
	Magnesium orotate	Potassium orotate	Magnesium orotate	Potassium orotate
0	0.12	0.25	0.165	0.155
7	0.15	0.26	0.157	0.140
42	0.19	0.29	0.154	0.135

The differences in the solutions properties are maintained for 1.5-2.5 hours. Then, the properties become similar to those of the solutions prepared from the orotate oxo-form. It can be concluded that in 1.5-2.5 hours, the hydroxy - and dihydroxy – form solutions transform into the oxo-form solution.

The duration of the preservation of the properties depends on the solution concentration and temperature. The systematic studies of the duration of the preservation of the properties of the orotate solutions are still under way.

3.4. Biological properties of potassium orotate and magnesium orotate tautomers

The difference in the microelectrophoretic mobility of live cells in the water solutions reflects different biological activity of the tautomers (table 5).

Table 5. Epiteliocytus microelectrophoresis mobility in water solutions of orotates

Mechanical energy doze, kJ/g	Magnesium orotate		Potassium orotate	
	Fraction of active cells, %	Amplitude of cell vibrations, μm	Fraction of active cells , %	Amplitude of cell vibrations, μm
0	52±3.2	2.1±0.3	36±4.1	1.±3.7
7	84±2.8	3.7±0.5	100±12.0	3.7±0.7
42	36±3.6	0	45±4.7	1.5±0.3

The studies on laboratory rats have shown the following. After the injection of the hydroxy-form, the amount of magnesium in the blood plasma of rats increased by 64%, but remained unchanged when the oxo- and dihydroxy- forms were injected.

For the laboratory rats with experimental steroid osteoporosis, the content of the overall collagen was lower by 42.8% ($p<0.01$) than that in the control group of healthy rats. The magnesium orotate injection balances these changes. The usage of oxo-form and dihydroxy-forms of the magnesium orotate makes it possible to reach the level of collagen characteristic of the reference group. The usage of hydroxy-form of magnesium orotate produced the collagen level exceeding that of healthy rats in the control group by 65.3 ($p<0.001$).

4. Conclusions

In this paper, the effect of milling on the tautomeric structure of magnesium orotate and potassium orotate was investigated. In the source samples, the oxo-forms of orotates were dominant. The hydroxy- and dihydroxy- tautomeric forms were obtained depending on a dose of mechanical energy.

It is shown that the increasing dose of mechanical energy during milling leads to the successive transformations of the chemical structure of orotates: oxo→hydroxy→dihydroxy forms. The solid tautomers preserve their structure from three to four months. After that, the sample mainly transforms into the oxo-forms of orotates. The tautomers solubility varies slightly. The physicochemical properties (viscosity, volumetric thermal expansion coefficients) of the water solutions of the tautomers differ. The differences are due to the change in the area of the hydration of the orotate anion because of the changes in the composition of the functional groups caused by the change of the tautomeric structures. The isomeric differences formed during milling remain in the orotate water solutions for 1.5 -2.5 hours.

Acknowledgments
The authors are grateful to Prof. N. Chuchkova and Dr. N. Savinova for their participation in the experiments and discussion. The work is supported by the program of the Russian–German laboratory (RGL) BESSY II (Helmholtz Zentrum, Berlin).

References
[1] Angell C A 1995 Formation of glasses from liquids and biopolymers *Science, New Series* **267** 1924-35
[2] Angell C A 1991 Relaxation in liquids, polymers and plastic crystals-strong/fragile patterns and problems *J Non-Cryst Solids* **131-133** 13-31
[3] Raza K, Kumar P, Ratan S, Malik R and Arora S 2014 Polymorphism: The phenomenon affecting the performance of drugs *SOJ Pharm Pharm Sci* **1(2)** 10-20
[4] Saifee M, Inamdar N, Dhamecha D and Rath A 2009 Drug polymorphism: A review *Int J Health Res.* **2(4)** 291-306
[5] Omar M, Makary P and Wlodarski M 2015 A review of polymorphism and the amorphous state in the formulation strategy of medicines and marketed drugs UK *J Pharm and Biosci* **3(6)** 60-6
[6] Naveen C and Madan L A 2013 Deepak padmanabhan. A review of drug isomerism and its significance *Int J Appl Basic Med Res* **3(1)** 16-8
[7] Wlodarczyk P, Kaminski K, Paluch M and Ziolo J 2009 Mutarotation in d-fructose melt monitored by dielectric spectroscopy *J. Phys. Chem. B* **113(13)** 4379-83
[8] Morpurgo S, Brahimi M, Bossa M and Morpurgo G O 2000 A theoretical study on proton transfer in the mutarotation of sugars *Phys. Chem.Chem. Phys.* **2** 2707-13
[9] Nguyen L A, Hua H and Chuong P-H 2006 Chiral drugs: An overview *Int. J. Biomed Sci.* **2(2)** 85-100
[10] Martin V C 2009 Let's not forget tautomers *J Comput-Aided Mol Des.* **23** 693-704
[11] Katritzky A R, Hall C D, El-Dien B, *et al* 2010 Tautomerism in drugs discovery *J Comput-Aided Mol Des.* 24(6-7) 475-84
[12] Negi A and Gill B S 2013 Success stories of enolate form of drugs *PharmaTutor* **1(2)** 45-53
[13] Podolyan Y, Gorb L and Leszczynski J 2003 Ab initio study of the prototropic tautomerism of cytosine and guanine and their contribution to spontaneous point mutations *Int. J. Mol. Sci.* **4(7)** 410-21
[14] Daniels M 1972 Tautomerism of uracil and thymine in aqueous solution: spectroscopic evidence *Proc. Natl. Acad. Sci. U.S.A.* **69(9)** 2488 -91
[15] Feyer V, Plekan O, Richter R, Coreno M, *et al* 2009 Tautomerism in cytosine and uracil: An experimental and theoretical core level spectroscopic study *J Phys Chem A.* **113(19)** 5736-41
[16] Huang L, Huang Y, Chi Y, *et al* 2007 On the kinetics of tautomerism in drugs: New application of broadband dielectric spectroscopy *J Chromatogr A.* **1175(2)** 283-8
[17] Godsi O, Turner B, Suwinska K, *et al* 2004 Enol–enamine tautomerism in crystals of 1,3-Bis(pyridin-2-yl) propan-2-one: A combined crystallographic and quantum-chemical investigation of the effect of packing on tautomerization processes *J Am Chem Soc.* **126(41)** 13519-25
[18] Tombari E, Ferrari C, Salvetti G and Johari G P 2007 Configurational specific heat of molecular liquids by modulated calorimetry *J Chem Phys* **126** 021107-16

[19] Wojnarowska Z, Adrjanowicz K, Kaminski K, *et al* 2014 Effect of pressure on tautomers equilibrium in supercold glibenclamide drug: Analysis of fragility behavior *J Phys Chem B.* **114** 14815-20
[20] Wojnarowska Z, Wlodarczyk P, Kaminski K, *et al* 2010 On the kinetic of tautomerism in drugs: New application of dielectric spectroscopy *J Chem Phys.* **133(9)** 094507-1-9
[21] Smith M B and March J 2001 *Advanced Organic Chemistry* 5th ed (New York: Wiley)
[22] Drebushchak V A, Boldyreva E V, Drebushchak T N and Shutova E S 2002 Synthesis and calorimetric investigation of unstable β –glycine *J Cryst Growth.* **241** 266-8
[23] Boldyreva E V and Boldyrev V V 2010 Mechanochemistry and mechanical activation of solids, Part I. *Experimental and Theoretical Studies in Modern Mechanochemistry* (Kerela (India): Transworld Research Network)
[24] Tumanov I A, Achkasov A F, Boldyreva E V and Boldyrev V V 2011 Following the products of mechanochemical synthesis step after step *Cryst Eng Comm.* **13** 2213-6
[25] Rybin D S, Konygin G N, Porsev V E, *et al* 2014 Deformation-induced structural transformations in molecular crystals *Acta Phys Pol A* **126(4)** 1014-8
[26] Shakhtshneider T P 1998 The use of mechanical activation for modification of the properties of pharmaceuticals and immobilization of enzymes *Preprints of 1st Europ. Symp. Process Technology in Pharmaceutical and Nutritional Sciences* (Nrnberg (Germany): NrnbergMesse Gmbh)
[27] Shakhtshneider T P and Boldyrev V V 1999 *Mechanochemical Synthesis and Mechanical Activation of Drugs / Reactivity of Molecular Solids* Ed. Boldyreva E, Boldyrev V. (Chichester: Wiley)
[28] Shakhtshneider T P, Dande F, Capet F, *et al* 2007 Grinding of drugs with pharmaceutical excipients at cryogenic temperatures *J Therm Anal Calorim.* **89(3)** 699-715
[29] Rietveld H M 1969 A profile refinement method for nuclear and magnetic structures *J Appl Crystallog* **2(2)** 65-71
[30] Pratapa S and O'Connor B 2002 Development of MgO ceramic standards for x-ray and neutron line broadening assessments *Adv X-ray Anal* **45** 41-7
[31] Bekiroglu S and Kristiansson O 2002 J Chem Soc Dalton Trans 1330-5 http://crystallography-online.com.
[32] Gorovikov S A, Follath R, Molodtsov S L and Kaindl G 2001 Optimization of the optical design of the Russian-German soft-X-ray beamline at BESSY II. *Nucl Instr and Meth A.* **467-468** 565-8
[33] Batson P E 1993 Carbon 1*s* near-edge-absorption fine structure in graphite *Phys Rev B.* **48** 2608-10
[34] Beamson G and Briggs D 1992 High resolution XPS of organic polymers *The Scienta ESCA300 Database* (New York: Wiley)
[35] Povstugar V I, Shakov A A, Mikhailova S S, *et al* 1998 Chemical derivatization techniques in the determination of functional groups by X-ray photoelectron spectroscopy *J Anal Chem.* **53** 697
[36] Haramonenko S S and Rakitjanskaja A A 1974 Jelektroforez kletok krovi v norme i patologii. Minsk.
[37] Zubavichus J V 2012 Mjagkaja rentgenovskaja sinhrotronnaja spektroskopija bioorganicheskih materialov, vody i vodnyh rastovorov. Avtoreferat diss. d. f.-m.n. Moscow.

Analysis of BJ493 diesel engine lubrication system properties

F Liu[1*]

[1]Technical Department, Yinjian Automobile Repair Co., Ltd., BJ 100070, Beijing, China
*E-mail: 773539103 at qq.com

Abstract. The BJ493ZLQ4A diesel engine design is based on the primary model of BJ493ZLQ3, of which exhaust level is upgraded to the National GB5 standard due to the improved design of combustion and injection systems. Given the above changes in the diesel lubrication system, its improved properties are analyzed in this paper. According to the structures, technical parameters and indices of the lubrication system, the lubrication system model of BJ493ZLQ4A diesel engine was constructed using the Flowmaster flow simulation software. The properties of the diesel engine lubrication system, such as the oil flow rate and pressure at different rotational speeds were analyzed for the schemes involving large- and small-scale oil filters. The calculated values of the main oil channel pressure are in good agreement with the experimental results, which verifies the proposed model feasibility. The calculation results show that the main oil channel pressure and maximum oil flow rate values for the large-scale oil filter scheme satisfy the design requirements, while the small-scale scheme yields too low main oil channel's pressure and too high. Therefore, application of small-scale oil filters is hazardous, and the large-scale scheme is recommended.

1. Introduction

Lubrication system is the key system of diesel [1,2]. BJ493ZLQ4A diesel engine is based on the primary BJ493ZLQ3, through the improvements and redesigns of combustion and injection systems, and upgraded to exhaust level regulated by the National GB5 standard [2]. Diesel timing transmission adopts gear transmission scheme; according to the requirements for the entire vehicle's brake, air compressor for braking has been added to the diesel, the injection oil pump adopts Bosch CB18 pump with the reinforced model EGR cooler. So the lubrication system of diesel has been much improved, of which changes require in-depth analysis to guarantee the engine operation safety and reliability.

According to the lubrication system structure, technical parameters, and indices, its model was constructed using the Flowmaster flow simulation software. Such properties as the oil flow rate and pressure at different rotation speeds were analyzed for the schemes involving large- and small-scale oil filters. The calculated values of the main oil channel pressure are in good agreement with the experimental results, which verifies the proposed model feasibility. The calculation results show that the main oil channel pressure and maximum oil flow rate values for the large-scale oil filter scheme satisfy the design requirements.

The National GB5 exhaust standard of China has adopted the national emission standard level V, which requires that the automobiles must satisfy the pollution exhaust rules. As compared with the former standard, the new one is more rigorous: the CO_2 emission must be reduced by 25% for automobiles and by 43% for buses and trucks.

2. Diesel lubrication system and arrangement

As shown in figure 1, the BJ493ZLQ4A diesel lubrication system includes oil pump, oil filter, oil cooler, main shaft, connecting rod shaft, cam shaft bearing, turbo lubrication, piston cooling jet (PCJ), air compressor lubrication, timing gear shaft lubrication, and oil pump lubrication. Among these, the lubrication of the main shaft, connecting rod shaft, cam shaft bearing, gears, cylinder head, and PCJ: these parts are provided through the main and cylinder body inner oil channels [3]; while the turbo and air compressor are lubricated through the external oil pipes of the cylinder body.

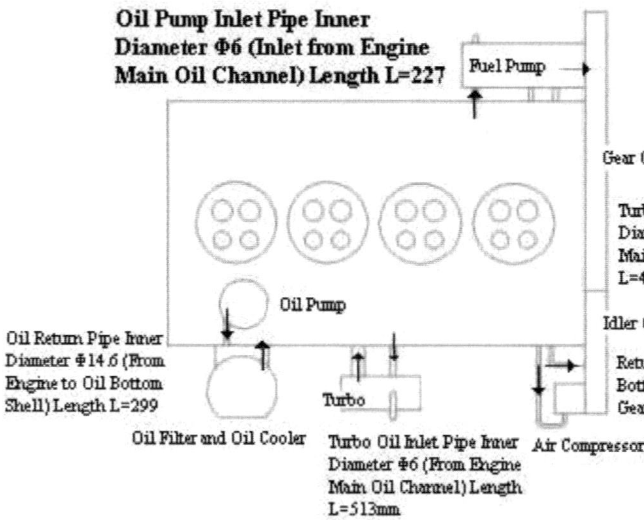

Figure 1. The BJ493ZLQ4A diesel engine lubrication system.

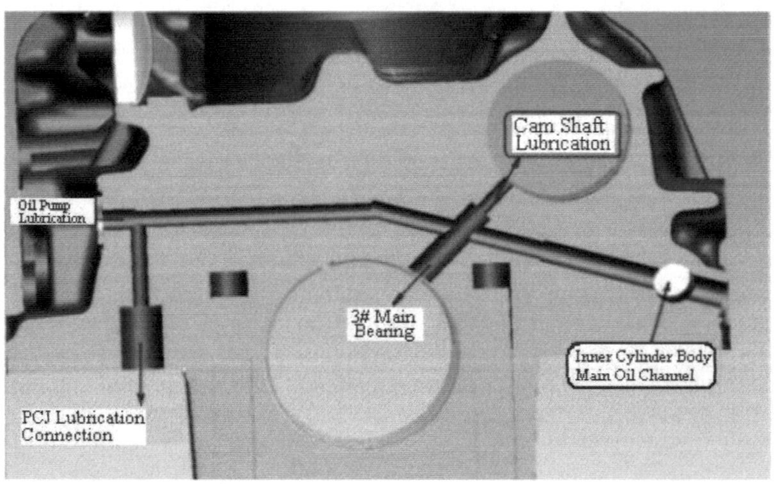

Figure 2. Lubrication oil channel and distribution of No. 3 main bearing and No. 2 cam shaft.

2.1.Lubrication of parts via main and inner cylinder body oil channels

Figure 2 depicts the conditions of the 3# main bearing and main oil channel connections. The main oil channel in the cylinder body of the diesel BJ493ZLQ4A is located at the right bottom part of the cylinder body (viewed from the back of the engine), while the main shaft neck is lubricated by a changeable oil channel (of 8.4 mm or 7 mm in diameter) with the inclination angle of 15° to the horizontal plane, and another downward oil channel (of 10 mm in diameter) with the inclination angle of 45° to the horizontal plane. The intrinsic feature of 3# main bearing oil channel arrangement is that the main oil channel supplies oil to 3# main bearing and 2# cam bearing through the inner cylinder body oil channel, while providing oil directly to the PCJ and diesel oil pump. Figures 3-6 depict the oil-connecting conditions for 1#, 2#, 4#, 5# main bearings and main oil channels.

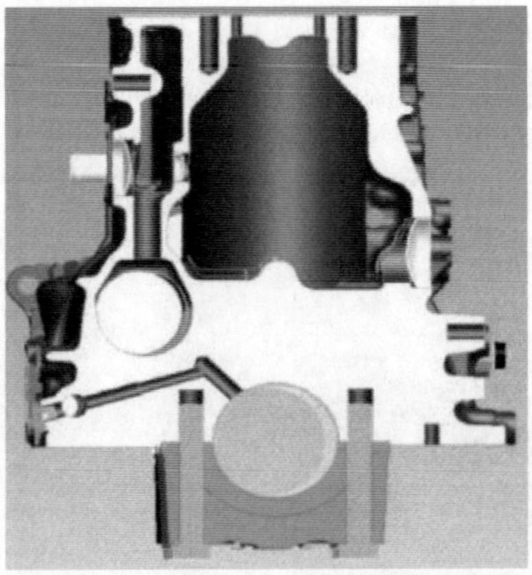

Figure 3. No. 1 main bearing and main oil channel.

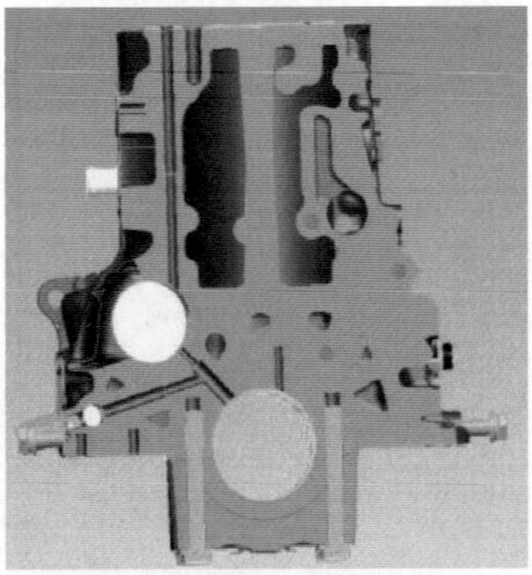

Figure 4. No. 2 main bearing and main oil channel.

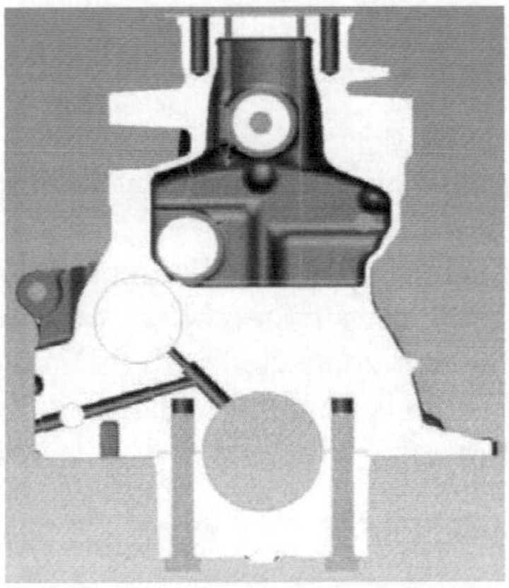

Figure 5. No. 3 main bearing and main oil channel.

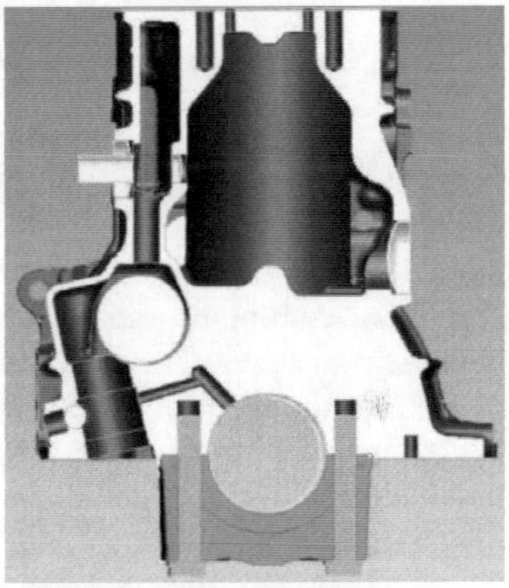

Figure 6. No. 4 main bearing and main oil channel.

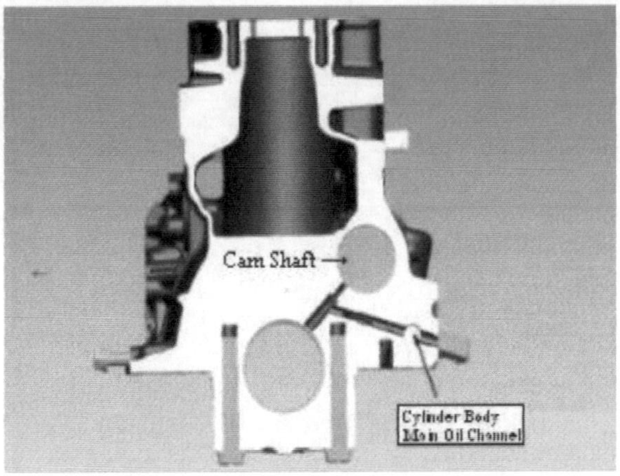

Figure 7. Cam shaft lubrication channel.

IOP Conf. Series: Materials Science and Engineering **283** (2017) 012005 doi:10.1088/1757-899X/283/1/012005

2.2.Cam shaft lubrication

As seen from figure 7, the diesel engine cam shaft is arranged at the right side of its middle line (viewed from the diesel's back), and provides oil for the main oil channel via the inner oil channel of cylinder body at 5#, 3#, 1# main shaft bearings, then returns to the oil bottom shell by the natural oil re- circulation method. The diameter of oil channel connected to the bearing bush is 6 mm, and its length is 15.5 mm.

2.3.PCJ bearing lubrication

As seen from figure 8, oil flows into the connecting pipe of piston cooling jet (PCJ) bearing, which protrudes from the cylinder body [4]. The area of injection outlet nozzle is 1.77 mm^2, and the diameter of PCJ pipe is 3 mm.

Figure 8. PCJ bearing lubrication.

2.4.Gear lubrication

As seen from figure 9, gears lubrication mainly relies on the oil provided via the main oil channel, while air compressor, oil pump and idler gear of the air compressor are supplied by oil via the external oil pipe, which is connected with the main oil channel. Idler gear A is connected with the main bearing bush oil nozzle. The diameter of the external oil pipe of the oil pump is 6 mm, the diameter of the idler gear near air compressor is 4 mm, while the diameter of spatter oil supplying pipe is 1 mm.

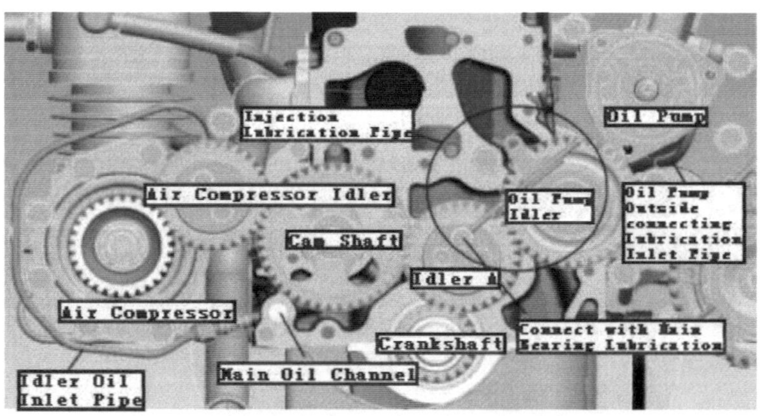

Figure 9. Gear lubrication.

2.5.Turbo

As shown in figure 10, the external inlet oil pipe of the turbo is connected with the main oil channel directly [5]. The pipe inner diameter is 6 mm, the length is about 510 mm.

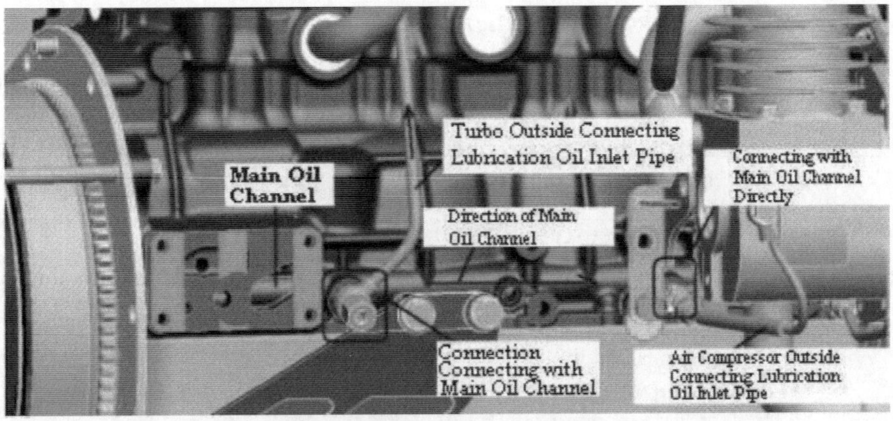

Figure 10. Turbo lubrication.

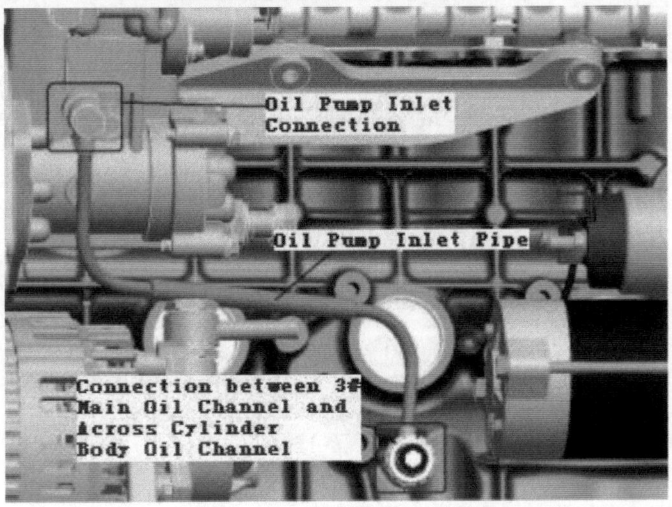

Figure 11. Oil pump lubrication pipe.

2.6.Oil pump

The diameter of oil pump external connection oil pipe is 6 mm, as is shown in figure 11.

2.7.Shaft main bearing bush

Shaft main bearing has 5 channels, its diameter is 70 mm and width is 21.9 mm, while the width of oil tank is 180°×3 mm, the depth is 0.85 mm; standard gap is 0.033-0.079 mm, under the limited size conditions, the relative gap of bearing is 0.113%.

2.8.Shaft connecting rod neck bearing bush

Connecting rod neck bearing has 4 channels, its diameter is 53 mm and width is 24.7 mm; there is no oil tank; standard gap is 0.029-0.075 mm, under the limited size conditions, the relative gap of bearing is 0.140%; the axial gap between connecting rod and shaft is 0.175-0.29 mm.

2.9.Cam shaft bearing

Cam shaft bearing of 50 mm in diameter has 3 channels of width equal to 26, 234, and 23 mm for the first, second, and third channels, respectively. There is no oil tank; the standard gap between cam bearing bushing nozzle and cam shaft bearing neck is 0.015-0.085 mm, the limit of gap abrasion is 0.12 mm, and under the limited size conditions, the relative gap of bearing is 0.17%.

3. Elaboration of models for software simulation

According to the structural features and the arrangement conditions of lubricated parts of BJ493ZLQ4A diesel, the simulation models [6] were elaborated, as shown in figures 12 and 13. The structures and principles of the lubrication models correspond to the actual engine set-up, such as the main shaft bearing, cam shaft bearing, connecting bearing, the main oil channel, and the engine's components' lubrication: turbo, air compressor, cylinders, oil pump, PCJ, positive crankcase ventilation valve (PCV), etc. The data on their sizes and properties, especially for lubrication materials and systems, can be input into the model, and they fully correspond to the actual engine operating conditions. Insofar as the simulation results are also close to the actual engine test data, the elaborated model feasibility is confirmed.

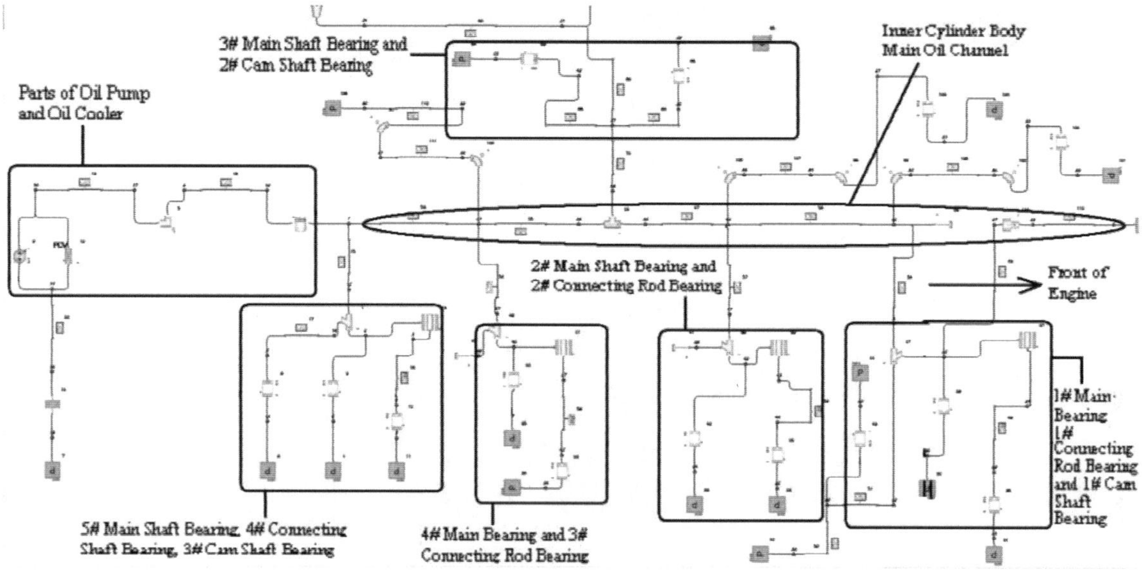

Figure 12. Model #1 of the 493 diesel lubrication system.

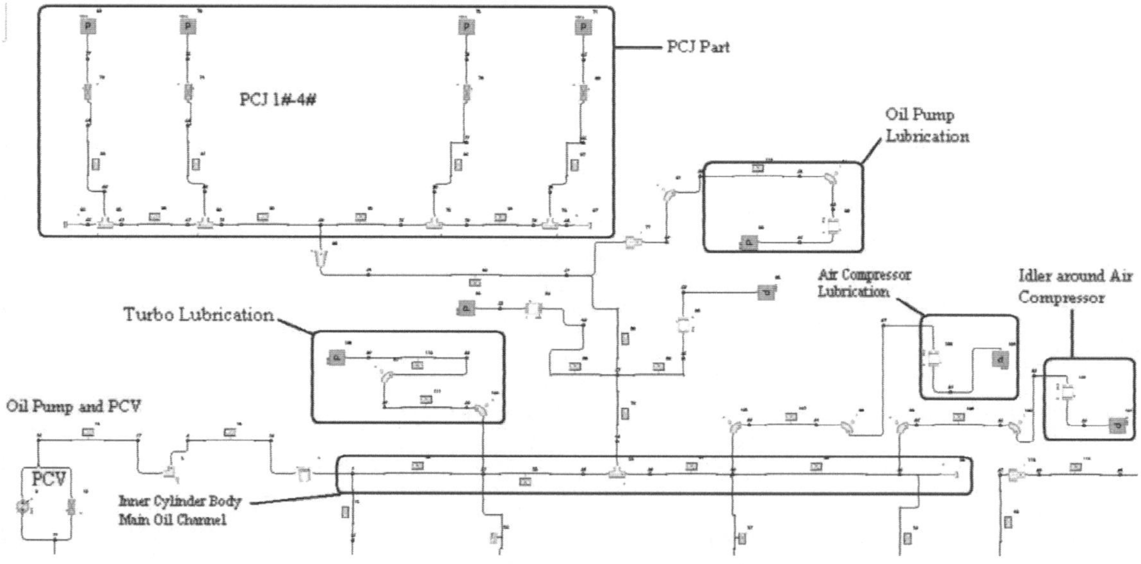

Figure 13. Model #2 of the 493 diesel lubrication system.

4. Loads and boundary conditions

4.1. Boundary conditions

The boundary conditions used in the model are listed in table 1.

Table 1. Boundary conditions.

Subject	Value	Property
Oil type	CF5W40	No changes
Oil temperature	125℃	No changes
Environmental pressure	1bar=0.1MPa	No changes
Bearing external pressure	1bar=0.1MPa	Same as the environmental one
Engine speed (rpm)	800,1200, 1500, 2000, 2800, 3200, 3600	
Oil pump speed (rpm)	400, 600, 750, 1000, 1400, 1600, 1800	Half of the engine speed
Oil pump	Nominal volume 22.70 mL/rev, Flow Rate: 37.12 L/min at 3200 rpm/6 bar=0.6 MPa	Design parameters and experimental data
Oil filter and oil cooler	0.22 m^2 0.38 m^2	Small-scale oil filter scheme Large-scale oil filter scheme
Bearing speed	Main shaft bearing: same as the engine speed	Changes according to operating mode
	Large head connecting rod: same as the engine speed	Changes according to operating mode
	Cam shaft speed: same as oil pump speed	Changes according to operating mode
Bearing relative gap	Main shaft bearing: φ70x21.9 mm, relative gap: 0.00113	No changes
	Connecting rod bearing: φ53x24.7mm, relative gap: 0.0014	No changes
	Cam shaft bearing: φ50x23mm, φ50x23mm, φ50x26mm, relative gap: 0.0017	No changes
PCJ	PCJ opening pressure: 2.0+/-0.25bar at 2000rpm	

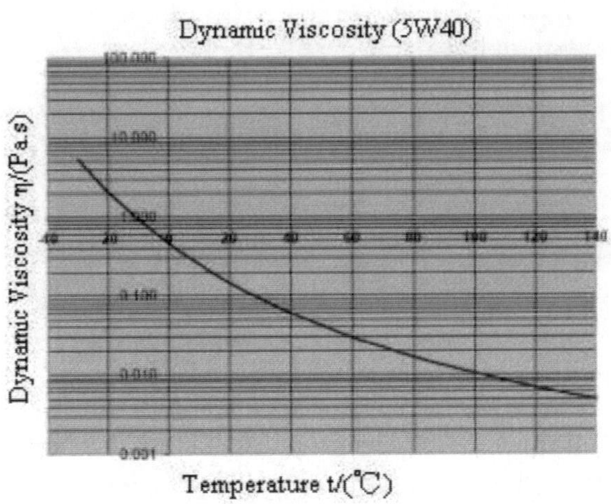

Figure 14. Dynamic viscosity vs. temperature curve of 5W40 oil.

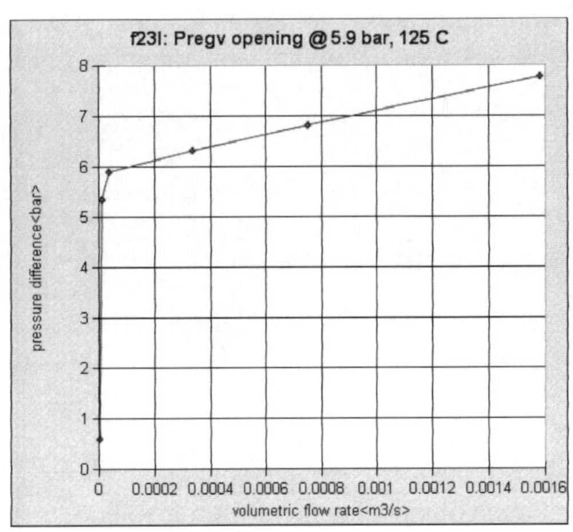

Figure 15. Pressure difference vs. flow rate curve of PCV at 0.59MPa and 125℃.

4.2. Features of medium and parts

IOP Conf. Series: Materials Science and Engineering **283** (2017) 012005 doi:10.1088/1757-899X/283/1/012005

- Properties of 5W40 oil

The 5W40 oil viscosity-temperature curve is shown in figure 14. The curve and its relevant data refer to the actual conditions obtained from the suppliers [7] and were exported from the Flowmaster database, of which data exhibit a good fit with the experimental results. Positive Crankcase Ventilation Valve (PCV) PCV feature's curve is plotted in figure 15.

- Features of Oil Pump

Oil pump's curves are plotted using the available experimental data. Figures 16(a) and 16(b) depict the volume efficiency-pressure and flow rate-pressure curves, respectively, of the oil pump at different rotation speeds.

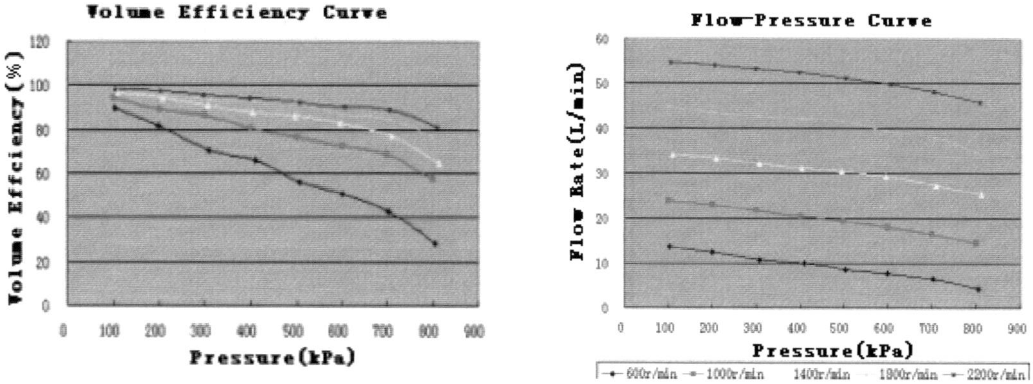

Figure 16. Volume efficiency vs. pressure (a)-L and flow rate vs. pressure (b)-R curves.

- Properties of Oil Filter and Oil Cooler

Large- and small-scale oil filter schemes [8,9] have been adopted in the oil filter design. For both schemes, the simulation of the diesel's lubrication system operation was performed separately. According to the data provided by suppliers, the effective area of small-scale oil filter is 0.22 m^2, standard flow rate is 33 L/min, the recommended flow rate is below 30 L/min; while the effective area of large-scale oil filter is 0.38 m^2, standard flow rate is 45 L/min, the recommended flow rate is below 40 L/min.

- Piston Cooling Jet (PCJ) features of the injection nozzle.

The pressure difference-flow rate curve of PCJ is shown in figure 17.

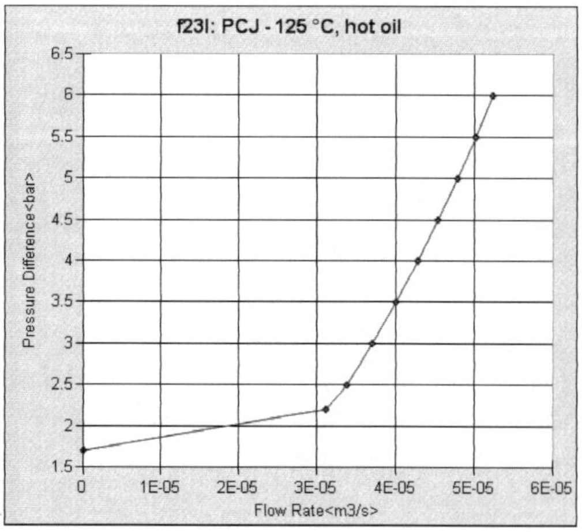

Figure 17. PCJ pressure difference vs. flow rate curve (input) at 25℃.

5. Selection of benchmarks for calculation

For analyzing the effects of two oil filter and oil cooler parts on the diesel's lubrication system, separate calculations were performed for the oil cooler with the large- and small-scale oil filters [10].

The boundary conditions for the both schemes are basically the same, except for the oil filter and oil cooler curves. Seven benchmark rotation speeds were selected, namely 800, 1200, 1500, 2000, 2800, 3200, 3600 rpm, whereas 800 rpm corresponds to the idle speed point, 1200, 1500, 2800 rpm are transition speed points, 2000 rpm is a predicted PCJ full point; 3200rpm is the rated point; 3600 rpm is the primary optimization point. For these benchmarks, which characterize diesel's properties and features, the calculations of the lubrication system's total oil flow rate, branch flow rate of the parts and each point's pressure values were conducted.

6. Calculation results and discussion

6.1.Calculation results of small-scale oil filter scheme

6.1.1. Pressure distribution. As shown in figure 18, with an increase in the diesel engine rotation speed, the parts' oil pressure increased constantly. When the engine speed was below 2800 rpm, the pressure of the parts increased rapidly. When it exceeded 2800 rpm, the oil re-circulation flow rate of PCV increased (between 3.2-6.3 L/min) with speed, the pressure increase of parts slowed down. For rotation speed between 3200-3600 rpm, the pressure of the inner main oil channel remained unchanged at about 0.376 MPa, while the oil pressure of parts exhibited a steady state [11,12].

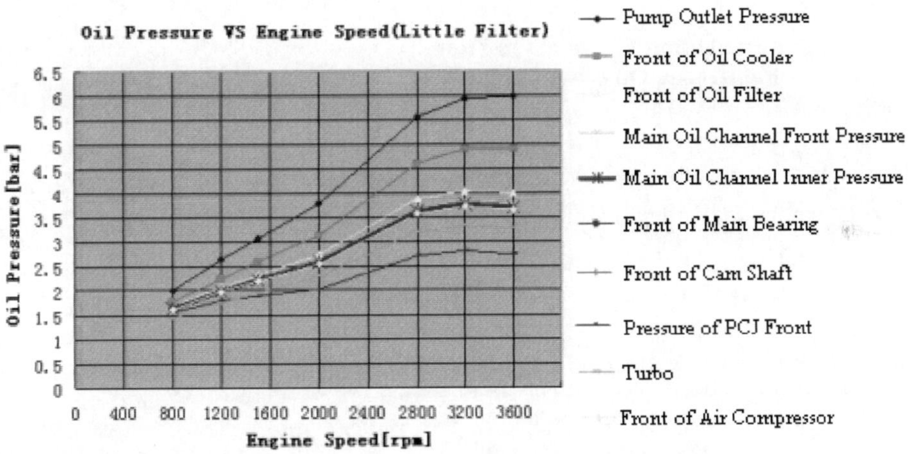

Figure 18. Oil pressure- engine speed curves for various parts of lubrication system with a small-scale oil filter.

Table 2 lists the main parts' oil pressure design requirements and simulated contrasting conditions. The simulated results indicate that under small-scale oil filter conditions, the pressure of the main oil channel cannot satisfy the design requirements for the diesel optimized condition and requires the experimental verification.

6.1.2. Flow rate distribution conditions. Figure 19 depicts the oil flow rate distribution conditions of the lubrication system parts for the small-scale oil filter scheme. With an increase in the engine rotation speed, the total oil flow rate is increased from 7.18 L/min to 38.8 L/min. While the engine speed is between 2800 and 3600 rpm, the main shaft bearing oil flow rate is in the range of 12.5 to 14.95 L/min; connection rod large head shaft bearing oil flow rate is from 5.7 to 6.5 L/min; the PCJ effective flow rate is between 8.2 and 8.52 L/min, which exceeds the requirement of 8 L/min for the piston cooling.

The maximum values of oil flow rate are: 1.16 L/min for turbo (at 3200 rpm) and 0.15 L/min for air compressor (at 3600 rpm), while the maximum value of the oil pump quantity rate is 0.35 L/min (at 3600 rpm). Under the increased rotation speed condition of 3600 rpm, the oil flow rate of oil filter is 32.5 L/min, which exceeds the maximal value of 30 L/min recommended by the producer, but this value is slightly less than the small-scale oil filter's standard flow rate of 33 L/min. Therefore, application of the small-scale oil filter scheme under increased rotation speed conditions may be hazardous due to inadequate oil supply.

Table 2. Design requirements and calculation results for small-scale oil filter conditions.

Requirements	Calculated value (bar)	Conclusions and proposals
Pressure of main oil channel exceeds 1.5 bar at 125°C	1.64	Approved
Pressure of main oil channel exceeds 3.85 bar at 125°C for the engine speed of 3200 rpm	3.76	Experimental verification is proposed
Turbo: the oil pressure is larger than 1.5 bar at 125°C for the engine speed above 1000 rpm	1.8	Approved
Air compressor: the oil pressure is larger than 3.0 bar at 125°C for the engine speed of 3600 rpm	3.6	Approved
Pressure from main shaft bearing to connecting rod shaft neck exceeds 2.2-2.8 bar at 125°C for the engine speed of 3200 rpm	3.7	Approved

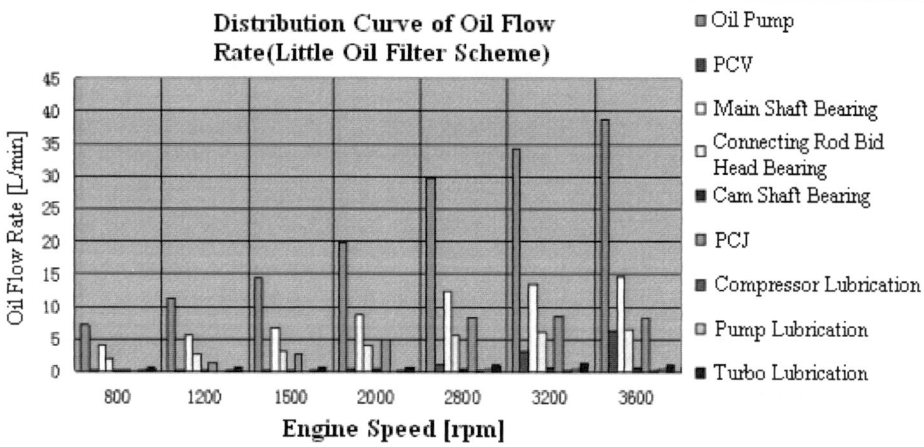

Figure 19. Oil flow rate distribution of various parts of the lubrication system (small-scale oil filter).

Flow rate design requirements and calculation results for the small-scale oil filter scheme are listed in table 3.

Table 3. Flow rate design requirements and calculation results for mall-scale oil filter scheme.

Requirements	Calculated value	Conclusions and proposals
Oil pump: effective exhaust rate 37.12 L/min at 3200rpm for the oil pressure of 6 bar		Needs to be approved
PCJ open pressure position: 2+/-0.25 bar; engine speed of 2800 rpm; recommended flow rate of 8.0 L/min or 2.0 L/min for each PCJ	8.2-8.5 L/min	Needs to be approved
Main oil channel's inner diameter is 15 mm, the inner diameter from oil pump to main oil channel needs to ensure that the oil flow	About 3 m/s	Approved

velocity is below 5-6 m/s, to ensure that the pressure loss is minimal.		
About 3 m/s. Under condition of increased rotation speed at 3600 rpm, the oil filter flow rate is less than 30 L/min	32.5 L/min	Hazardous, requires test verification

6.2.Calculated results for the large-scale filter scheme

6.2.1. Pressure distribution conditions. As seen from figure 20, in case of the large-scale oil filter and oil cooler scheme application, the oil pressure in lubrication system parts increased constantly with the diesel engine speed. Thus, at rotation speed below 2800 rpm, pressure in all parts increased rapidly, and the main oil channel inner pressure rose from 0.164 MPa to 0.259 MPa; at the speed above 2800 rpm, pressure rise slowed down due to the increased PCV leak rate; while at the speed in the range of 3200-3600 rpm, the main oil channel inner pressure remained in the range of 0.38-0.39 MPa, and steady pressure values in all parts were observed.

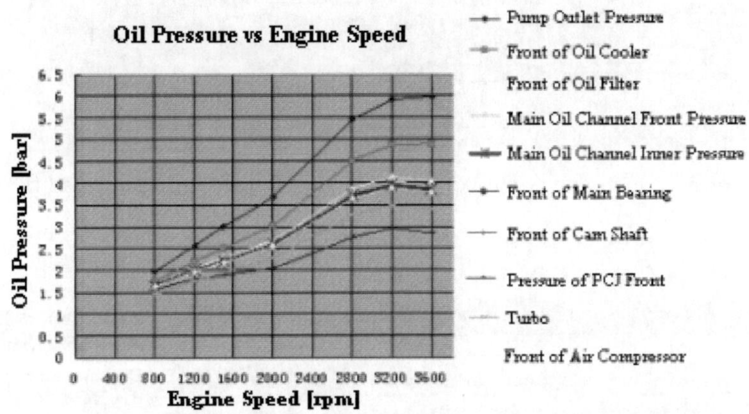

Figure 20. Oil pressure-engine speed curves of various parts for the large-scale oil filter scheme.

The oil pressure requirements and calculated values for the large-scale oil filter conditions are listed in table 4.

Table 4. Design requirements on the oil pressure and calculation results for the large-scale oil filter.

Requirements	Calculated value (bar)	Conclusions and proposals
Pressure of main oil channel exceeds 1.5 bar at 125°C in idle speed	1.64	Approved
Pressure of main oil channel exceeds 3.85 bar at 125°C for the engine speed of 3200 rpm	3.9	Approved
Turbo: the oil pressure exceeds 1.5 bar at 125°C for the engine speed above 1000 rpm	1.8	Approved
Air compressor: the oil pressure exceeds 3.0bar at 125°C for the engine speed of 3600 rpm	3.9	Approved
Pressure from main shaft bearing to connecting rod shaft neck exceeds 2.2-2.8 bar at 125°C for the engine speed of 3200 rpm	3.95	Approved

As seen from figure 21, prior to reaching the oil cooler, oil flows into through a pipe of 10 mm in diameter and makes a turn at the right angle, which results in the pressure drop of about 0.045 MPa (at 3600 rpm), according to the calculations, In addition, a pressure drop in the oil pump transfer pipe is estimated as 0.05 MPa (at 3600 rpm), which is due to the fact that the oil transfer pipe diameter of 10 mm is too small, and the oil flow velocity is too high. Since the combination of these two geometric features leads to the total pressure drop of 0.095 MPa (at 3600 rpm), it is proposed to increase the oil channel diameter in these two points under the condition that the cylinder body structural integrity is not violated.

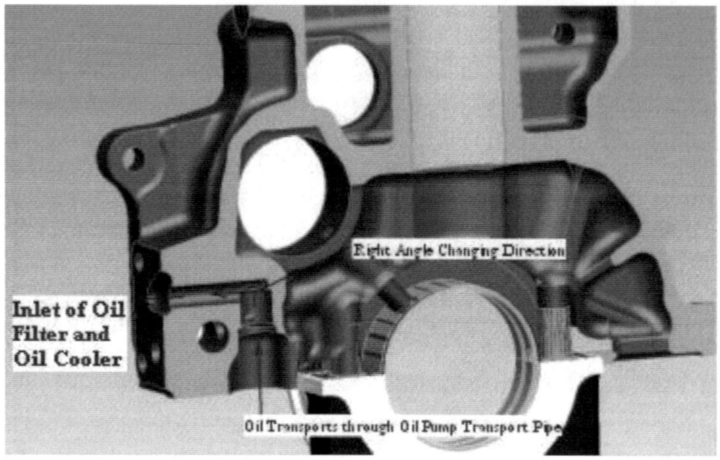

Figure 21. Oil channel geometry prior to oil flow to oil cooler and filter.

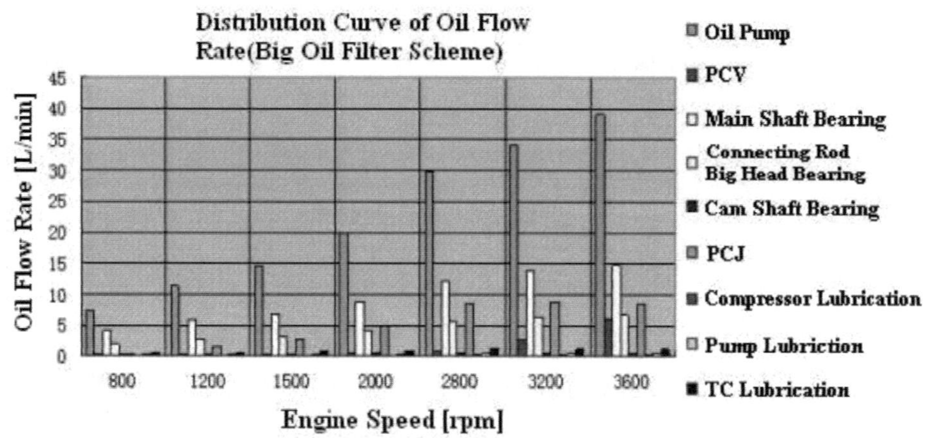

Figure 22. Oil flow rate distribution of various parts of the lubrication system (large-scale oil filter).

6.2.2. Flow rate distribution conditions. As shown in figure 22, the flow rate in lubrication system parts increased in the range from 7.19 to 39.2 L/min with the diesel engine speed: at 2800 rpm, the oil pump outlet pressure is 0.547 MPa, which exceeds the value of 0.54 MPa, which triggers the PCV opening, so the PCV begins to release at 2800 rpm, and the oil flow rate of PCV is 0.855 L/min, the main oil channel inner pressure being 0.37 MPa after releasing.

While the engine speed is below 2800 rpm, the main bearing oil flow rate is 4.07-8.75 L/min, connecting rod big head bearing oil flow rate is 1.89-4.02 L/min, cam shaft bearing flow rate is below 0.31 L/min, PCJ effective flow rate is 2.78-4.92 L/min, turbo compressor (TC) oil rate is below 0.8 L/min, oil pump flow rate is below 0.21 L/min, and air compressor flow rate is less than 0.084 L/min.

After the engine speed reached 2800 rpm, the PCV recirculation oil flow rate increases to the maximum of 6.17 L/min (at 3600 rpm), the total flow rate of the main oil channel inlet is 12.3-14.8 L/min, the oil flow rate of connecting rod big head bearing is 5.63-6.65 L/min, the oil flow rate of cam shaft is about 0.5 L/min, the maximum value of PCJ effective flow rate is 8.75 L/min (at 3200 rpm), the maximum oil flow rate of the turbo is 1.22 L/min (at 3200 rpm), the maximum oil flow rate of the pump is 0.355 L/min (at 3600 rpm), and the maximum oil flow rate of the air compressor is 0.15 L/min (at 3600 rpm).

The large-scale oil filter's design requirements and calculation results are listed in table 5.

Table 5. Design requirements and calculation results for the large-scale oil filter scheme.

Requirements	Calculated value	Conclusions and proposals
Oil pump: effective exhaust rate of 37.12 L/min at 3200 rpm for the oil pressure of 6 bar		Needs to be approved
PCJ open pressure position: 2+/-0.25 bar; engine speed of 2800 rpm; recommended flow rate of 8.0 L/min or 2.0 L/min for each PCJ	8.5-8.75 L/min	Needs to be approved
Main oil channel's inner diameter is 15 mm, the inner diameter from oil pump to main oil channel needs to ensure that the oil flow velocity is below 6 m/s, to make sure that the pressure loss is minimal.	About 3m/s	Approved
About 3m/s At rotation speed of 3600 rpm, the oil filter flow rate is below 30 L/min	33.1L/min	Approved

6.3.Comparison of two oil filter schemes and the experimental results

6.3.1. Comparison of two oil filter schemes. Comparison of the simulation results and recommendations for the large- and small-scale oil filter design schemes is shown in table 6.

Table 6. Comparison of the large- and small-scale oil filter design schemes.

Subject	Small-scale oil filter	Large-scale oil filter
Maximum pressure of the main oil channel	3.77 bar at 3200 rpm	3.94 bar at 3200 rpm
Pressure drop of oil cooler	0.86 bar at 3200 rpm	0.74 bar at 3200 rpm
Pressure drop of oil filter	0.21 bar at 200 rpm	0.11 bar at 200 rpm
PCJ front pressure	1.89413 bar at 1500 rpm	1.89454 bar at 1500 rpm
Maximum oil provided by oil pump	38.8 L/min at 3600 rpm	39.17 L/min at 3600
Flow rate of parts at 3200 rpm	Oil flow rate Oil pump: 34.2 L/min PCV: 3.21 L/min Main bearing: 13.58 L/min Large head connecting rod bearing: 6.15 L/min Cam shaft: 0.46 L/min PCJ: 8.52 L/min TC: 1.16 L/min Oil pump oil quantity: 0.326 L/min Air compressor oil quantity: 0.134 L/min	Oil flow rate oil pump: 34.2 L/min PCV: 2.6 L/min Main bearing: 13.8 L/min Large head connecting rod bearing: 6.25 L/min Cam shaft: 0.46 L/min PCJ: 8.75 L/min TC: 1.28 L/min Oil pump oil quantity: 0.332 L/min air compressor oil quantity: 0.134 L/min
Oil filter recommended maximum flow rate	30 L/min	40 L/min
Comparison of calculated flow rates and recommended limits	32.5>30 L/min (exceeds the recommended flow rate)	33.1<40 L/min (approved)

As follows from table 6, for the small-scale oil filter scheme, the maximum oil flow rate is 32.5 L/min (at 3600 rpm), which exceeds the flow rate margin of 30 L/min recommended by the producer.

This can lead to a dangerous situation with engine overheating, so the cooling ability needs to be experimentally verified. In case of the large-scale oil filter scheme, the maximum oil flow rate is 33.1 L/min (at 3600 rpm), which is less than the flow rate margin of 40 L/min recommended by the producer, and thus satisfies the design requirements. Therefore, the large-scale oil filter scheme should be finally adopted.

6.3.2. Comparison of calculated and experimental results. The comparison of calculated results with the available experimental data on the actual engine performance makes it possible to verify the applicability and accuracy of the simulations related to the oil pressure behavior in the main oil channel of BJ493ZLQ4A diesel engine. As seen from figure 23, the calculated and experimental results exhibit a very close correlation, which proves the applicability and efficiency of the elaborated lubrication system model.

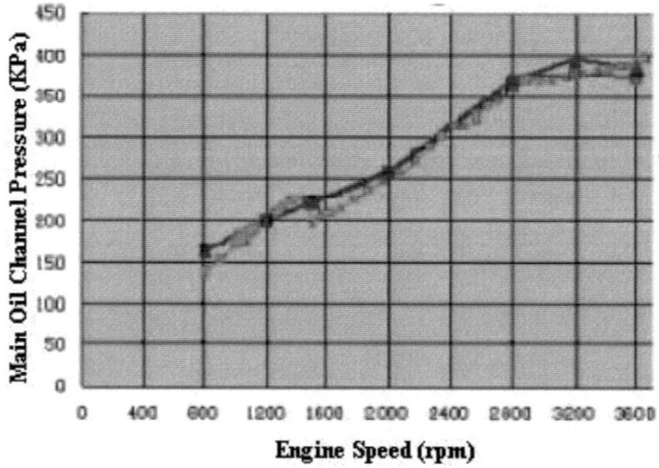

Figure 23. Comparison of the simulated and experimental results of the oil pressure in main oil channel: small-scale oil filter simulation (■), large-scale oil filter simulation (▲), experimental data (●).

Noteworthy is that the simulations involve input data obtained from different engines, so there might be some difference between the diesel engine actual condition and calculated one. During the computation process, the model can be refined and adjusted according to the accurate data on properties, e.g., of PCV or PCJ, which are provided by suppliers of diesel engine parts.

7. Conclusions

- Two simulation models of the 493ZLQ4A diesel lubrication system have been elaborated for the small- and large-scale oil filter and oil cooler schemes. The comparison of simulated and experimental results on oil pressure behavior of the main oil channel has revealed their good fit and confirmed the feasibility of the elaborated models.
- In the small-scale oil filter scheme, the main oil channel's pressure is somewhat lower than that of the design requirements, while the oil filter's maximum oil flow rate exceeds that recommended by the producer, which implies some risks in the application of this scheme. Therefore, the small-scale scheme cooling ability should be experimentally verified. In the large-scale oil filter scheme, the main oil channel inner oil pressure and the maximum oil flow rate fully satisfy the design requirements. In both schemes, the air compressor's front oil pressure exceeds the margin.
- Two geometric features were found to provide the excessive pressure drop between the oil pump outlet and the oil cooler inlet. It is proposed to increase the oil channel diameter in these

two points under the condition that the cylinder body structural integrity is not violated.

References

[1] Guo L S and Hu P 2014 Experimental study of lubrication system and oil pressure in main lubricating gallery of YZ485ZLQ diesel engine *J. Internal Combustion Engine.* **43** 50-3

[2] Liu F 2016 Properties analysis of BJ493 diesel engine lubrication system *J. Lubrication Engineering* **41** 132-9

[3] Zhang M 2016 Design SF33900 truck centralized lubrication system *J. Opencast Mining Technology* **46** 16-9

[4] Fang C C 2017 Design and test of lubrication system for turbocharged gasoline engine *J. Internal Combustion Engine and Powerplant* **36** 21-3

[5] Qian Y, Wang X Y and Hu Y C 2016 Simulation analysis of the lubrication system of a diesel engine *J. Internal Combustion Engine and Powerplant* **35** 31-3

[6] Solghar A A and Gandjalikhan Nassab S A 2013 Numerical analysis of turbulent lubrication in plain full journal bearings *J. Industrial Lubrication and Tribology* **65** 40-3

[7] Sautermeister F A, Priest M, Lee P M and Fox M F 2013 Impact of sulphuric acid on cylinder lubrication for large 2-stroke marine diesel engines: Contact angle, interfacial tension and chemical interaction *J. Tribology International* **59** 45-9

[8] Minet C, Brunetiere N and Tournerie B 2012 On the lubrication of mechanical seals with rough surfaces: A parametric study Proceedings of the Institution of Mechanical Engineers, Part J *J. Journal of Engineering Tribology* **226** (12) 32-6

[9] Javorova J, Alexandrov A, Andonov I, *et al* 2015 Stability characteristics of rigid rotor on elastohydrodynamic turbulent journal bearings *J. Journal of the Balkan Tribological Association* **21** 23-5

[10] Bloch H P 2015 Lubrication update for rotating equipment *J. Hydrocarbon Processing* **94** 15-8

[11] Chervyakov K 2015 Compressor Lubrication - the challenges and solutions *J. Maintenance & Engineering* **15** 13-6

[12] Tornehed P and Olofsson U 2013 Modeling lubrication oil particle emissions from heavy-duty diesel engines *Int J Engine Research* **14** 12-4

IOP Conf. Series: Materials Science and Engineering **283** (2017) 012006 doi:10.1088/1757-899X/283/1/012006

Study on the application of permanent magnet synchronous motors in underground belt conveyors

S H Ma[1*]

[1]Mechanical and Electronic Engineering Department, Shanxi Institute of Technology, Yangquan, Shanxi Province 045000, China
*E-mail: msh8535@163.com

Abstract. This paper analyzes and compares the advantages and disadvantages of several kinds of drive devices of belt conveyors from the angle of energy saving, and summarizes the application advantages and using problems of permanent magnet motor variable frequency drive system in belt conveyors. An example is given to demonstrate the energy saving effect of this system compared with other driving methods. This paper points out the application prospect of permanent magnet motor variable frequency drive system on belt conveyors and other large mining machines in coal mine. This paper is aimed to provide the design direction for the designer and the choice basis for the user on belt conveyor.

1. Introduction

After the integration of coal resources, China's coal mine production capacity has been greatly improved, and small and scattered production capacity has been phased out. For example, in Shanxi Province known for the coal production, the lowest production capacity of a single mine cannot be less than 300,000 tons, and the minimum production capacity standards have been gradually enhanced. Less than 600,000 tons of the production capacity of the mine coal will soon be eliminated this year, which has greatly promoted the development of the coal mine to the direction of large-scale and modernization. It improves the comprehensive benefits of mine production that the large, efficient, highly automated equipment has been input and used. Belt conveyor has gradually become the preferred equipment for coal transportation in coal mines, because of its simple structure, convenient installation, high efficient transportation, and other advantages. It has become the starting point and the foothold of designers and users to improve the stability and efficiency of equipment operation. The research on the belt conveyor mainly focuses on the dynamic characteristics, monitoring technology, starting characteristics, power balance, component reliability and service life, etc. [1]. All these make belt conveyor continuously turn to the direction of far distance, large volume, high speed, large dip angle, and high load development [2]. The whole system is toward the direction of high performance, high life, and the best operation of low energy consumption technology has become a hot topic of domestic and foreign scholars.

2. Energy-saving operation of belt conveyor

Belt conveyor working principle is shown in figure 1: the conveyor belt 3 is rotated around the transmission drum 1 and the turnabout drum 2 to form a closed ring, and the drive unit 4 drives the

IOP Conf. Series: Materials Science and Engineering **283** (2017) 012006 doi:10.1088/1757-899X/283/1/012006

transmission drum 1 to rotate, and the conveyor belt is continuously running in one direction depending on the friction between the drum surface and the conveyor belt, and the material on the conveyor belt is continuously transported from the loading end to the unloading end. The efficient and energy-saving operation of belt conveyor is mainly determined by its drive mode. The first energy-saving operation of belt conveyor is realized by reasonable optimization of the starting and running control mode of the conveyor motor [3], and 10% -50% of the energy savings can be achieved [4]; another way to save energy is to optimize or design the structure of the motor, which essentially improves the motor performance [3], 2% -8% of the energy savings can be achieved [4].

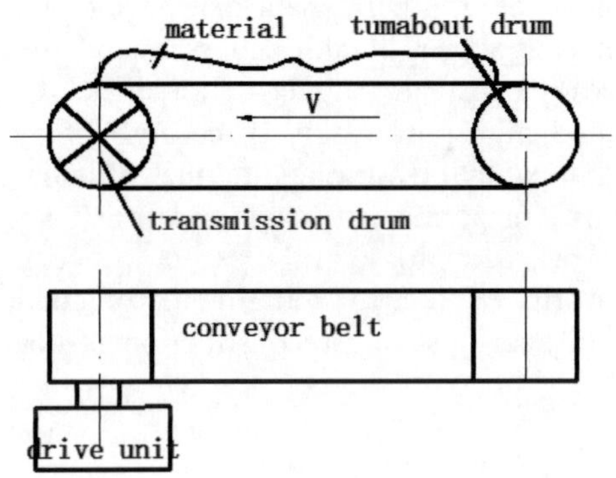

Figure 1. Working principle diagram of belt conveyor.

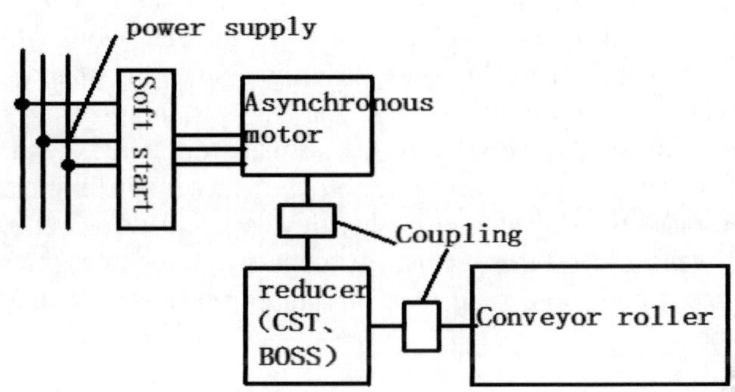

Figure 2. Schematic diagram of asynchronous motor drive system of belt conveyor.

2.1. Optimize start-up and control methods

The driving system of belt conveyor is generally composed of AC motor + connecting device + reducer + drive roller, as shown in figure 2. There are two categories of the optimization methods in start and control performance of the driving system, one is the use of electrical control circuit between electric power supply and the motor, such as adding step-down starting circuit, initiating device, soft starting switches, and inverters, etc.; the other is the use of different application of mechanical or hydraulic speed regulating device between motor and drive roller, such as the commonly used speed-type hydraulic coupler, slip clutch, CST, BOSS and other devices [5]. The better performance of starting and controlling device is the inverter and CST, and the inverter is more prominent in control performance [3]. Because of its complex structure, high maintenance costs, relatively high price and occupying larger space and so on, the application scope of CST is mainly concentrated on the places of the mining area of the groove and inclined shaft hoisting where has the advantage of the transportation convenience and the large installation space, while the frequency of converter is distributed in the mining area, roadway and inclined shaft hoisting and other kinds of the environment. Take the example of Erkuang Coal Mine in Yangquan Coal Industry (Group) CO., LTD, the whole mine has two levels, and the two main incline wells that are all promoted by the powerful belt conveyor. In the coal transportation system from coal mining working face to ground, the delivery is all using belt conveyor. Among 23 belt conveyors of 8 mining areas (not including the mining area of the groove), there are 10 using inverters with large power and long distance transportation. The others use soft starting switch and hydraulic coupler to start and control. In the lane transport and inclined well, there are 14 belt conveyors, among which one drive device is CST; one conveyor that can operate for a longer time and a shorter distance uses hydraulic coupler; four that run the shorter distance using the way of direct starting, and the remaining eight conveyors all use converters. Therefore, it can be seen that the application of converter as drive device in belt conveyor is approved. The inverter not only has a good speed performance and rich communication functions, but also can make the motor in the best power output through the speed matching, and achieve the energy saving in the operation.

2.2. Use high performance motors

Optimizing the system operation through the control strategy can improve the system efficiency to a certain extent, but it cannot compensate for the energy loss caused by the low efficiency of the design of the motor itself. Therefore, the use of high-performance motors is another way to achieve energy-saving operation of belt conveyors. There are many efficient energy-saving motors such as rare earth permanent magnet motors, high efficiency three-phase cage induction motors, switched reluctance motors, etc. [6]. Although they are a little expensive, for example, the efficiency of three-phase cage induction motors is 3% higher than that of ordinary motors, and the cost is higher than 25%-30%, and the efficiency of super-efficient cage induction motors is 5% higher than that of ordinary motors, and the cost is higher than 40%-60% [4]; the price of a permanent magnet motor is about 3-10 times as much as that of the asynchronous AC motor, as shown in table 1, it is much better than the AC motor in energy saving. Especially the rare earth permanent magnet motor, it has the advantages of simple structure, small loss, high efficiency and high power density because of the magnetic field generated by the permanent magnet, without the excitation winding and the excitation power supply, and with the improvement of the thermal stability of permanent magnet materials and corrosion resistance, and further development of power electronic technology, In the aspect of energy efficiency, rare earth permanent magnet motor shows the superiority of electric excitation motor, and it is more and more used in coal transportation belt conveyor in energy saving reconstruction of coal mine.

Table 1. Energy loss and various cost calculation for two configurations.

	Traditional drive system	Permanent Magnet Synchronous Variable Frequency Drive System
Drive motor parameters	2×250 kW Asynchronous motor	2×250 kW Permanent magnet motor
Motor efficiency,%	80	93
Reducer efficiency,%	85	None
Hydraulic coupler efficiency,%	90	None
Coupling efficiency,%	96	96
Total efficiency of transmission,/%	58.8	89.3
Drum power, kW	209.1	209.1
Total input power, kW	340.9	269
5 years' electricity consumption, kWh	1963584	1549440
5 years' electricity bill, yuan	1963584	1549440
5 years' maintenance fee, yuan	5000000	25000
Equipment purchase cost, yuan	150000	1600000
Total cost of running time, yuan	7113584	3174440
Saving power(kWh) /cost, yuan)	414144/3939144	

2.3. Permanent magnet motor variable frequency drive system

Whether the optimization of the starting control system or the use of efficient motor will greatly improve the efficiency of belt conveyor system, and achieve better energy efficiency. Combining two energy saving modes can form the permanent magnet variable frequency drive system, as shown in figure 3, that is, the use of permanent magnet motor + inverter drive system to replace the previous drive control system, the maximize energy saving effect can be achieved in the whole system by adding two energy saving benefits.

The permanent magnet motor used in the belt conveyor is a built-in rotor magnetic circuit structure. The rotor surface places the rat cage winding, and the motor's starting relies on the asynchronous interaction starting torque between rotor conducting bar and the stator rotating magnetic field, which is known as asynchronous start of permanent magnet synchronous motor. Permanent magnet motor and frequency converter are combined to form permanent magnetic variable frequency drive system. Through sensor-less vector control method, the starting control of belt conveyor and frequency conversion speed regulation are realized. The rotor of the motor is directly connected with the roller shaft of the belt conveyor, and the deceleration part in the prior drive mode is omitted. The schematic diagram of the drive system is shown in figure 2, and this drive is also known as direct drive. It not only has the advantage of transducer control, but also has the advantage of permanent magnet motor. It is the best drive system of the belt conveyor with optimal driving performance at present.

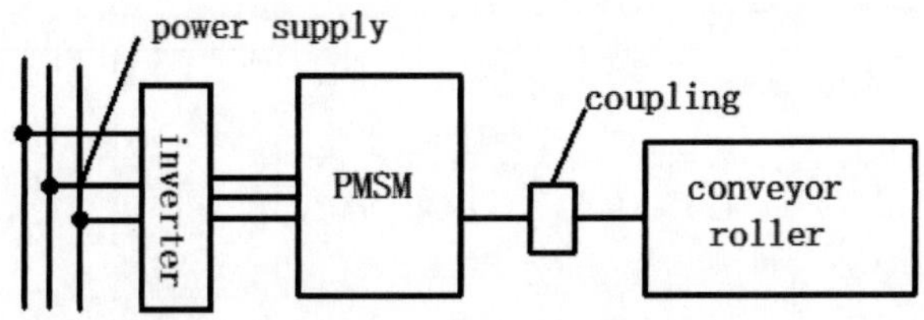

Figure 3. Schematic diagram of permanent magnet motor variable frequency drive system.

2.3.1. Advantages of permanent magnet motor drive system. Taking the application of SSJ-120 belt conveyor as an example, the advantages of using permanent magnet motor drive system are demonstrated.

- Simplified driving structure. The basic structure of permanent magnet motor variable frequency drive system is shown in figure 3. The motor is driven by the power converter through the inverter, and the permanent magnet motor is directly connected with the roller of the belt conveyor through the coupling. Compared with the traditional driving method, as shown in figure 2, the mechanical transmission link with soft start and speed regulation is reduced.
- Higher transmission efficiency and obvious energy saving benefits. SSJ-120 type belt conveyor is driven by two 250 KW three-phase asynchronous motors with 80% efficiency, and replaced by 2 sets of 250 kW permanent magnet motor with 93% efficiency in 2012.These two kinds of configuration of energy loss and various cost calculation is shown in table 1.

It can be seen from table 1 that the total efficiency of the permanent magnet drive system is about 30% higher than that of the original drive system. This is not only because of the permanent magnet motor using permanent magnet to generate the main magnetic field, and there is no exciting current, and power factor approximation is 1, so the rotor loss and the stator loss are small, which makes the motor itself have high efficiency, but also because of its simple drive system and transmission link is less; In the running time of the motor for five years, compared with the traditional driving system, the total cost saving of the permanent magnet synchronous variable frequency drive system is 3939144 yuan or 588,000 USD, which has obvious energy-saving benefit; {(1600000-150000)/[(1963548+5000000) -(1549440+25000)]/5}*12=16.14, that is, the installation cost of more than the traditional driving system can be recovered after 16.14 months of operation, and the efficiency can be highlighted by continuing operation.

Tables 2 and 3 of 160 kW and 315 kW permanent magnet motor drive system energy saving analysis is shown that under the same load and running time, the permanent magnet motor capacity is bigger, and the energy saving benefits are more obvious. Compared with the traditional driving system, the permanent magnet drive system can choose the motor with smaller capacity when driving the same load [7]. Combined with tables 1-3, it also can be seen that the permanent magnet drive systems generally need to experience about 8-17 months running time, energy-saving benefit and installation cost can balance, the greater the power, the shorter the time to recover the cost of installation, the more obvious energy conservation benefit. As compared to the traditional asynchronous motor + CST driven system, its energy saving rate reaches 22.3% [8].

Table 2. 160 KW energy-saving analysis of permanent magnet drive system (1000 KAV, primary voltage of 10000V and secondary voltage of 690 V).

Test motor type	Motor input technical data					Analog load (generator output technical data)					Same load output, input power difference
	Supply voltage	Operating voltage	Operating current	Input power	Power factor	Supply voltage	Operating voltage	Operating current	Input power	Analog load	
Permanent magnet drive system	710	599	13.5	13.03	0.93	605	0	0	0	No load	6.67
Asynchronous motor + reducer + hydraulic coupler	710	705	33.6	19.69	0.48	605	0	0	0	No load	
Permanent magnet drive system	710	599	81.2	78.35	0.93	592	85.4	78.8	1	half load	14.83
Asynchronous motor + reducer + hydraulic coupler	710	704	89.9	93.17	0.85	589	85.8	78.8	1	half load	
Permanent magnet drive system	710	598	163.9	156.18	0.92	604	160.4	150.9	1	full load	36.21
Asynchronous motor + reducer + hydraulic coupler	710	704	179.3	192.39	0.88	604	160.1	150.8	1	full load	
Saving electricity costs (10,000 yuan/year)	26.0712	Same power output, the permanent magnet motor and the ordinary three-phase asynchronous motor power difference 36.21 kW, calculate according to 360 days per year, 20h a day, 1 yuan per kWh: 36.21*360*20=26.0712									
Overhaul costs (10,000 yuan/year)	7.26	Overhaul maintenance cost: (2 sets traditional motor+2 sets reducer+2 sets+hydraulic coupler) cost price*30%=(12+13+3.2)*30%=8.46; (Permanent magnet motor drive system)cost price*1%=(2*60)*1%=1.2									
Replacement costs(10,000 yuan/year))	21.7	Replacement costs: 2 sets motor per year+1 set reducer per year+2 sets+hydraulic coupler 2 per year=12+6.5+3.2=21.7									
Labour charges (10,000 yuan/year)	10	(Car labour charges+Maintenance labour charges)*2 persons=10									
Total: (yuan/year)	65.0312	Time of cost recovery: permanent magnet motor drive system 60*2-(traditional motor 6*2+reducer 6.5*2+hydraulic coupler 1.6*2)=91.8; (91.8/65.0312)*12=16.9, that is, the cost can be recovered in 16.9 months									

Table 3. 315 KW Energy saving analysis of permanent magnet drive system (1000 KAV, primary voltage of 10000 V, secondary voltage of 690 V).

Test motor type	Motor input technical data					Analog load (generator output technical data)					Same load output, input power difference
	Supply voltage	Operating voltage	Operating current	input power	power factor	Supply voltage	Operating voltage	Operating current	input power	Analog load	
Permanent magnet drive system	710	595	35.6	34.12	0.93	590	0	0	0	No load	10.93
Asynchronous motor + reducer + hydraulic coupler	710	709	56.2	45.05	0.65	590	0	0	0	No load	
Permanent magnet drive system	710	594	170	162.65	0.93	595	171.7	159.2	1	half load	37.86
Asynchronous motor + reducer + hydraulic coupler	710	709	188.8	200.51	0.86	587	174.7	159.9	1	half load	
Permanent magnet drive system	710	591	336	319.86	0.93	590	341	313.4	1	full load	84.63
Asynchronous motor + reducer + hydraulic coupler	710	709	396.3	404.49	0.83	589	341	313	1	full load	
Saving electricity costs (Ten thousand yuan/year)	60.9336	Same power output, the permanent magnet motor and the ordinary three-phase asynchronous motor power difference 36.21 KW, calculate according to 360 days a year, 20 hours a day, 1 yuan per kilowatt hour: 84.63*360*20=60.9336									
Overhaul costs(Ten thousand yuan/year)	28.1	Overhaul maintenance cost: (2 sets traditional motor+2 sets reducer+2 sets+hydraulic coupler) cost price*30%=(32+64+5)*30%=30.3; permanent magnet motor drive system) cost price*1%=(2*110)*1%=2.2									
Replacement costs(Ten thousand yuan/year)	69	Replacement costs: 2 sets motor per year+1 set reducer per year+2 sets+hydraulic coupler 2 per year=32+32+5=69									
labour charges(Ten thousand yuan/year)	10	(Car labour charges+Maintenance labour charges)*2 persons=10									
Total: (Ten thousand yuan/year)	168.0336	Time of cost recovery: permanent magnet motor drive system 110*2-(traditional motor 16*2+reducer 32*2+hydraulic coupler 2.5*2)=119; (119/168.0336)*12=8.5, that is, the cost can be recovered in 8.5 months									

- Realize soft start and big starting torque. Through the inverter control starting, the inverter output frequency of permanent magnet motor changes from 0 to the operating frequency when starting, and the motor rotates synchronously following the frequency of the inverter output, therefore, without instantaneous high current, there will not be the bad impact on the grid; slow uniform starting lowers the maximum tension of belt and roller, reducing the impact on the idler, frame and other mechanical structures. Permanent magnet synchronous motor drive system can constantly output the starting torque, which is 2.8 times of rated load torque, but under the same power conditions, the starting torque is 55% of the rated load torque by using asynchronous motor drive system. In order to meet the starting torque, it is necessary to increase the motor capacity, and in the normal operation, it will waste more electricity, just as “a big horse pulls a small carriage”.

- Multimachine power balance. If the system uses the form of multiple permanent magnet motor drive, it can achieve multimachine power balance, to avoid multi-motor output imbalance caused by reduced motor life, burned and waste of electricity through the converter master-slave control mode.
- Reducing the system transformer capacity and volume. The structural characteristics of the permanent magnet motor determine its own high power factor that can be approximated as 1. In the case of the same power demand, compared with the asynchronous motor system, it can be driven with a smaller capacity motor, so the required capacity of the transformer will be reduced correspondingly.

Due to the use of inverter power supply, the use of high-frequency power supply can greatly reduce the size of the transformer.

2.3.2. Main problems in the operation of variable frequency drive system of the permanent magnet motor. Electromagnetic interference. As the frequency conversion to start the permanent magnet motor, the inverter control mode is the use of sine pulse width modulation (SPWM), but the output voltage is non-sinus, hence, it will generate a lot of time harmonics that the motor current contains a large number of harmonic components, which not only causes the increase of the motor's current and loss, the reduction of efficiency and power factor, but also with the increase of the motor power, this effect is more serious, and interferes with the adjacent electrical equipment [10]. It is possible to reduce the influence of harmonics by adding various harmonic suppression devices, such as AC reactors, DC reactors, filters, and other reactors. The effective control of interference can also be achieved by rational wiring and effective isolation.

- Irreversible demagnetization. Permanent magnet motor’s temperature stability is poor. When designing or using it improperly, under too high (NdFeB permanent magnet) or too low (ferrite permanent magnet), the permanent magnet motor is possible to produce irreversible demagnetization or misconduct with the impact of the current generated by the armature reaction or in the chemical corrosion, severe mechanical vibration and the use of time and other comprehensive factors. Consequently, the motor performance is reduced, or it is even impossible to use. All these reduce the reliability of motor operation [9]. To prevent the demagnetization, the first one is to take good cooling measures, because for the motor using on the underground belt conveyor, it has explosion-proof demand, the best way is the water-cooling; the second is to use the optimal motor structure for preventing demagnetization, as the results show when the maximum demagnetization point of the permanent magnet of the motor that has rotor with pole shoe structure is the highest, the anti-demagnetization ability is the strongest [10].

3. Conclusions

From the perspective of the application place, permanent magnet motor variable frequency drive system which has been put into use is mainly used for the modification and the new gate road, tunneling faces, inclined shaft hoisting, etc. And in the mining area and poor conditions of the alley, the use of permanent magnet motor variable frequency drive system has a certain limit. On the one hand, it is because of its stand-alone volume and weight, so underground transportation is not convenient, and the inverter also needs to occupy the corresponding chamber as the installation space; on the other hand, the coupler, reducer, asynchronous motor and other equipment that have been put into use are still in service, and are playing a corresponding role. From the application of power, because of its high price, the advantages in the use of low-power belt conveyor are not more obvious than in the large conveyor, so the use rate in permanent magnet drive system in small and medium-sized belt conveyor is very low. From the use effect of permanent magnet drive system, it can be trouble-free within a year, therefore, it greatly reduces the failure rate of belt conveyor drive system, ensures the belt conveyor running time and efficiency, and lightens the labor intensity of the maintenance and repair workers, so the permanent magnet drive system is very popular among the workers.

Through the comparison with the traditional driving system and the analysis of the advantages and disadvantages of practical application, it shows that the permanent magnet motor variable frequency drive system is much better than other drive control systems in the market, whether it is startup control, maintenance, failure rate and energy saving benefit. And through hardware and software settings, permanent magnet motor variable frequency drive system can not only realize the single or multimachine control, but also achieve the energy saving of the whole transportation system according to the optimal power matching speed. Through controlling the communication port of the system, combined with the monitoring system, it can achieve remote control of equipment, and further facilitate the realization of mine transportation system automation. Especially from the long-term operation effect of view, it is the best cost-effective driving system. This driving system is the design direction of the designers and the preferred choice of the users.

In short, the permanent magnet drive systems with the improving of the single coal mine production, the needs of national energy-saving and effect-raising driven, the advancement of mine automation level, the system performance further improving, will gradually replace other drive way and become a main drive system of the belt conveyor, will also be applied to the other large coal mining equipment such as scraper conveyor, fan, water pump, etc.

References

[1] Wang Z H 2014 Key Technology research of optimization control on mine coal flow conveyor system (Xuzhou, P. R. China: China University of Mining and Technology) pp 4-5
[2] Sun W P 2015 Studies of long distance belt conveyor and energy saving optimization strategy (Hefei, P. R. China: Hefei University of Technology) p 1
[3] Feng Y J 2012 Research on the key techniques and method of design for permanent magnet synchronous motors (Wuhan, P. R. China: Huazhong University of Science & Technology) p1
[4] Chen W H, Li X Y and Yao P 2008 Electrical technology development of energy saving technology for motor and its system *Summary* **9** 13-22
[5] Ma S H 2015 Coal mine machinery analysis on the selection of soft starting device for mine belt conveyor **36** 140-1
[6] Zhang Y J 2014 Belt conveyor energy-saving control system technology research (Xi'an, Shanxi, P. R. China: Xi'an University of Science and Technology) p 7
[7] Yuan C Y 2015 Coal mine machinery application of 400 kW permanent magnet synchronous motor in belt conveyor and economic analysis **36** 258-60
[8] Tang H H 2015 Colliery mechanical & electrical technology research on belt conveyor control system based on variable frequency driving of permanent magnet synchronous motor **4** 77-9
[9] Lu W P 2013 The demagnetization mechanism and anti-demagnetization method for line-start permanent magnet synchronous motors (Beijing, China: School of Electrical and Electronic Engineering) p 50
[10] Zhao G X 2011 Study on several key techniques of permanent magnet synchronous motor supplied by inventer (Shenyang, P. R. China: Shenyang University of Technology) pp 3-7 68

IOP Conf. Series: Materials Science and Engineering **283** (2017) 012007 doi:10.1088/1757-899X/283/1/012007

Effect of plasma arc welding variables on fusion zone grain size and hardness of AISI 321 austenitic stainless steel

S P Kondapalli[1*]

[1]Department of Mechanical Engineering, Anil Neeukonda Institute of Technology & Sciences, India
*E-mail: kspanits@gmail.com

Abstract. In the present work, pulsed current microplasma arc welding is carried out on AISI 321 austenitic stainless steel of 0.3 mm thickness. Peak current, Base current, Pulse rate and Pulse width are chosen as the input variables, whereas grain size and hardness are considered as output responses. Response surface method is adopted by using Box-Behnken Design, and in total 27 experiments are performed. Empirical relation between input and output response is developed using statistical software and analysis of variance (ANOVA) at 95% confidence level to check the adequacy. The main effect and interaction effect of input variables on output response are also studied.

1. Introduction

Austenitic stainless steel (ASS) is 300 series steels having high toughness and strength at both high and low temperatures. AISI321 is an austenitic stainless steel with excellent strength and good ductility at high temperature. It is widely used in bleed air components of aero-engine and cryogenic components. It can be joined using variety of welding process, both arc and non-arc welding process. Plasma arc welding (PAW) is one of the most important processing which is convenient to use in thin sheet joining. It is available in two forms: continuous current (CC) mode and pulsed current (PC) mode. Pulsed current MPAW is recommended for joining thin sheets because of stable nature of arc [1, 2]. Based on the work published by earlier researchers [3-8], four independent variables that influence the process are peak current, base current, pulse rate and pulse width, which are dominating variables influencing the weld quality.

In the present work, experiments conducted using design of experiments, developed mathematical models to predict the output responses. Most of the works reported by researchers studied the bead geometry, heat-affected zone, bead volume, etc., using mathematical models for various welding processes [9-11]. However, the desired welding input process variables are found based on the experience of the welder or from the data available in the handbook and one cannot relay the results obtained as the results are close to optimal [12]. Many researchers adopted statistical tools and other optimization like genetic algorithm, neural networks, etc. [13,14].

From the literature, it is understood that Kim *et al.* revealed that regression modeling, neural network, and Taguchi methods are helpful in attaining optimal solution only when the welding process was carried out at near the optimal conditions or at a stable operating range [15], however, near-optimal conditions cannot be easily found through full-factorial experiments when the number of experiments and levels of variables are more. Because of nonlinear characteristics of steepest ascent method, the results obtained can lead to an incorrect direction of the welding process. Modern optimization techniques like genetic algorithm (GA) and simulated annealing algorithm can overcome the above problems incurred with full-factorial experiments [16].

The work presented in this paper discusses the effect of welding variables such as peak current, base current, pulse rate and pulse width on grain size and hardness at fusion zone (FZ) of PAW AISI 321 sheets.

2. Welding procedure

Weld specimens of 100 x 150 x 0.3 mm size are prepared from AISI 321 sheets and joined using square butt joint. Tables 1 and 2 indicate the chemical composition and tensile properties of AISI 321 stainless steel sheets Argon is used as a shielding gas and a trailing gas to avoid contamination from outside atmosphere. The welding conditions adopted during welding are presented in table 3. From the works carried out by researchers [3-8], it is revealed that the peak current, back current, pulse rate and pulse width are the important variables which affect the weld quality characteristics in PAW process. The values of process variables considered in the present work are based on trail experiments. Peak current, base current, pulse rate and pulse width are chosen as variables and their levels are presented in table 4. Details about experimental setup are shown in figure 1. Four factors and three levels are considered and 27 experiments are performed as per Box-Benhken Design matrix shown in table 5.

Table 1. Chemical composition of AISI 321 (weight %).

C	Si	Mn	P	S	Cr	Ni	N
0.05	0.52	1.30	0.028	0.021	17.48	9.510	0.04

Table 2. Mechanical properties of AISI 321.

Elongation (%)	Yield strength (MPa)	Ultimate tensile strength (MPa)
53.20	272.15	656.30

Table 3. Welding conditions.

Power source	Secheron microplasma arc machine (Model: PLASMAFIX 50E)
Polarity	DCEN
Mode of operation	Pulse mode
Electrode	2% thoriated tungsten electrode
Electrode Diameter	1 mm
Plasma gas	Argon & hydrogen
Plasma gas flow rate	6 L per minute
Shielding gas	Argon
Shielding gas flow rate	0.4 L per minute
Trailing gas	Argon
Trailing gas flow rate	0.4 L per minute
Copper nozzle diameter	1 mm
Nozzle to plate distance	1 mm
Welding speed	260 mm/minute
Torch position	Vertical
Operation type	Automatic

IOP Conf. Series: Materials Science and Engineering **283** (2017) 012007 doi:10.1088/1757-899X/283/1/012007

Table 4. Process variables and their limits.

Input Factor	Units	Levels -1	0	+1
Peak Current	Amps	6	7	8
Base Current	Amps	3	4	5
Pulse rate	Pulses /Second	20	40	60
Pulse width	%	30	50	70

Figure 1. Microplasma arc welding setup.

Table 5. Design matrix with experimental results.

Experiment No.	Peak Current (Amps)	Base current (Amps)	Pulse Rate (Pulses/second)	Pulse width (%)	Grain size (Microns)	Hardness (VHN)
1	6	3	40	50	23.812	207.000
2	8	3	40	50	30.226	198.000
3	6	5	40	50	29.508	208.000
4	8	5	40	50	30.536	202.000
5	7	4	20	30	29.323	203.000
6	7	4	60	30	32.206	201.000
7	7	4	20	70	28.994	205.000
8	7	4	60	70	27.491	203.120
9	6	4	20	50	28.290	203.800
10	8	4	60	50	31.835	200.460
11	6	4	20	50	23.605	210.000
12	8	4	20	50	30.764	209.000
13	7	3	60	30	30.095	201.000
14	7	5	40	30	29.109	201.500
15	7	3	40	50	30.385	203.000
16	7	5	40	50	28.013	203.000
17	6	4	40	30	23.788	205.000
18	8	4	40	30	28.830	198.200
19	6	4	40	70	32.663	205.800
20	8	4	40	70	30.263	199.120
21	7	3	20	50	28.270	203.000
22	7	5	20	50	32.030	200.000
23	7	3	60	50	27.626	204.000
24	7	5	60	50	28.626	206.000
25	7	4	40	50	26.845	204.000
26	7	4	40	50	28.845	201.000
27	7	4	40	50	23.145	208.500

IOP Conf. Series: Materials Science and Engineering **283** (2017) 012007 doi:10.1088/1757-899X/283/1/012007

3. Recording output responses

3.1. Measurement of microstructure & grain size

Specimens for measurement of microstructure and grain size are prepared according to ASTM E3-1 standard. Three samples were prepared from each weld joint, leaving the edges. The samples are mounted on Bakelite and polished using 1/0, 2/0, 3/0 and 4/0 polishing papers. After polishing, disc polishing was done using Aluminum oxide. The polished specimens are etched with oxalic acid to reveal the microstructure. The etching time is increased to reveal the grain size of the welded specimen as per ASTM E407. Macrographs are taken using metallurgical microscope Axiovert 40MAT by Carl Zeissat x100 magnification. Figures 2 and 3 represent microstructure of heat-affected zone (HAZ) and fusion zone (FZ). Figures 3 and 4 represent grain size at HAZ and FZ.

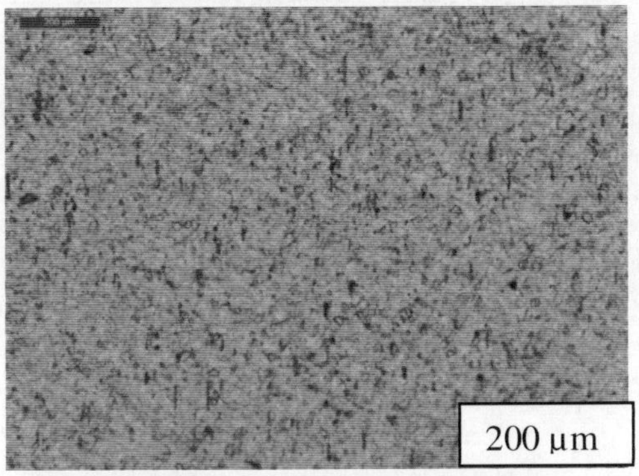

Figure 2. Microstructure of HAZ.

Figure 3. Microstructure of FZ.

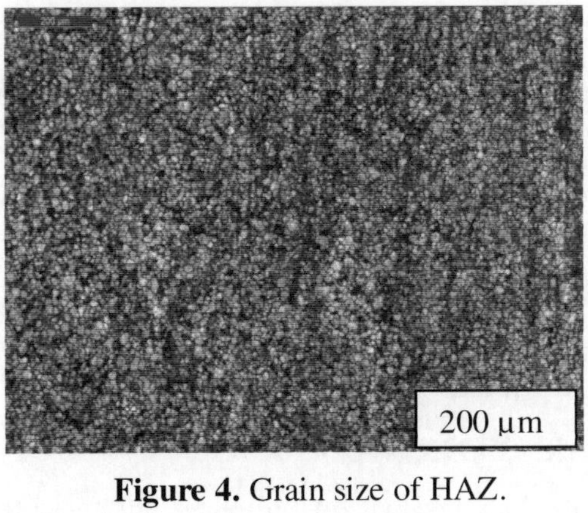

Figure 4. Grain size of HAZ.

Figure 5. Grain size of FZ.

As the grain size cannot be measured accurately, scanning electron microscope (SEM) is used to measure the exact grain size. The grain size represented in table 5 is average value of the measured grain size in FZ. Sample SEM image is presented in figure 6.

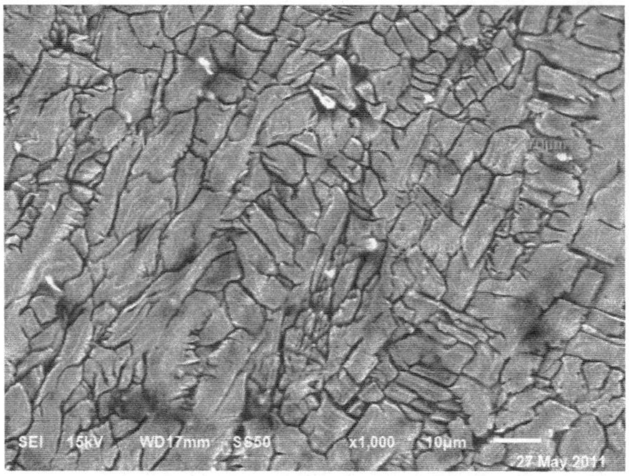

Figure 6. SEM image at FZ

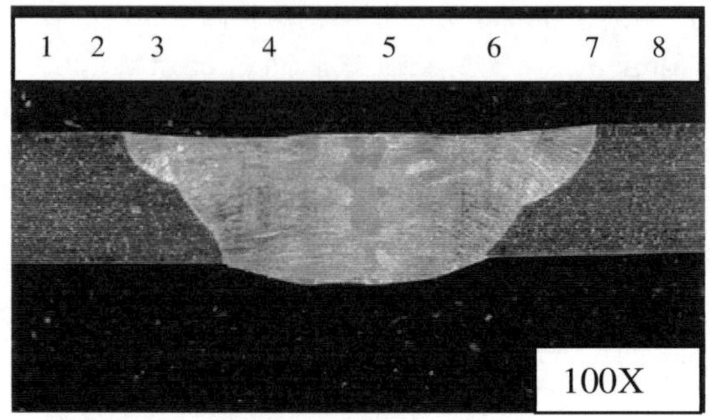

Figure 7. Location of hardness along the weld joint.

3.2. Measurement of hardness

Vickers hardness test was performed along HAZ and FZ of the welded joint at an interval of 0.3 mm. Using an MMT-X7 testing machine by Metsuzawa Ltd., Japan, Vickers's microhardness measurements were made by applying a load of 0.5 Kg as per ASTM E384. Figure 7 represents the indentation points. The values presented in table 5 are the average values at FZ.

4. Statistical analysis

Grain size and hardness values at FZ for all 7 samples were recorded and presented in table 5. In this study, the grain size and hardness measurements were made only for FZ.

4.1. Empirical mathematical modeling

In RSM design, mathematical models are developed using polynomial equations. The type of polynomial equation depends on the problem. A second order polynomial equation is adopted

$$Y = b_o + \sum b_i x_i + \sum b_{ii} x_i^2 + \sum\sum b_{ij} x_i x_j + \in \qquad \ldots\ldots..(1)$$

Empirical mathematical models are developed using MINITAB Ver.14 statistical software. The model is developed considering only significant coefficients at 95% confidence level.

$$\text{Grain size} = 69.9549 - 4.4302X_1 - 3.9241X_2 - 0.5701X_3 - 0.0126X_4 + 0.0866X_2X_3$$

$$\text{Hardness} = 243.566 - 43.377\,X_1 + 32.806X_2 - 1.046X_3 + 2.434X_4 + 2.877X_1^2 - 0.008X_3^2$$
$$-0.005X_4^2 + 0.171\,X_1X_3 - 0.058X_1X_4 - 0.199X_2X_3 - 0.370X_2X_4$$

The coded values of peak current, base current, pulse rate and pulse width are represented by X_1, X_2, X_3 and X_4 respectively.

4.2. ANOVA analysis of output responses

Analysis of variance (ANOVA) was performed at 95% confidence and presented in tables 6 and 7. From tables 6 and 7, it is revealed that the developed empirical mathematical models are found to be adequate at 95% confidence level and the coefficient of determination 'R^2' is 0.92, which shows a good agreement to existence between the experimental and predicted values. The variation in experimental and predicted values are shown in in figures 8 and 9 as scatter plots.

Table 6. ANOVA for grain size.

Source	DF	Seq SS	Adj SS	Adj MS	F	P
Regression	12	159.779	159.779	13.3149	9.62	0.000
Linear	4	92.160	3.586	0.8964	0.65	0.638
Square	4	27.776	2.139	0.5347	0.39	0.815
Interaction	4	39.843	39.843	9.9608	7.20	0.002
Residual Error	14	19.378	19.378	1.3842		
Lack-of-Fit	1	2.801	2.801	2.8010	2.20	0.162
Pure Error	13	16.577	16.577	1.2752		
Total	26	179.157				

Table 7. ANOVA test results for hardness.

Source	DF	Seq SS	Adj SS	Adj MS	F	P
Regression	12	252.513	252.513	21.043	15.31	0.000
Linear	4	106.073	81.911	20.478	14.90	0.000
Square	4	7.434	31.958	7.990	5.81	0.006
Interaction	4	139.006	139.006	34.752	25.29	0.000
Residual Error	14	19.241	19.241	1.374		
Lack-of-Fit	1	2.716	2.716	2.716	2.14	0.168
Pure Error	13	16.525	16.525	1.271		
Total	26	271.753				

In tables 6 and 7, SS is sum of squares, MS is mean square, DF is degree of freedom, F is Fisher's ratio, and P is probability ratio.

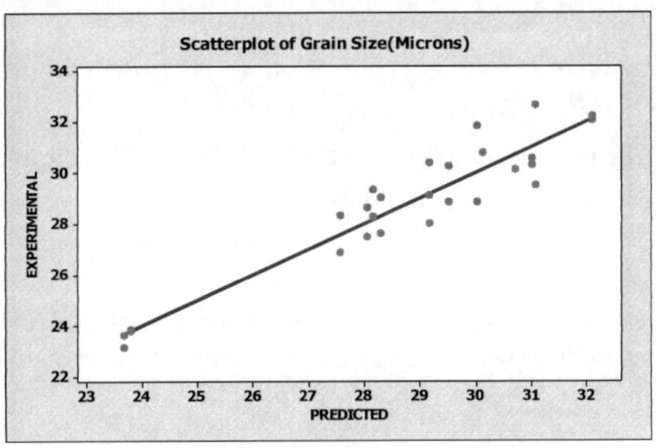

Figure 8. Scatter plot for grain size.

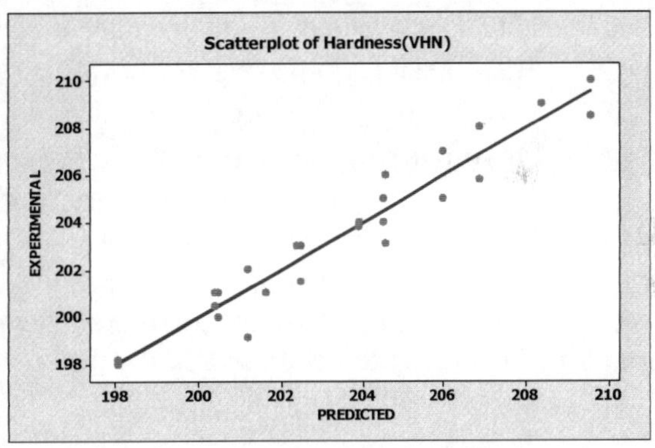

Figure 9. Scatter plot for hardness.

5. Results and discussion

Individual effect of welding variables on output responses can be studied using main plots as shown in figures 10 and 11.

5.1. Main effect plots for grain size and hardness

As peak current and base current increases, gain size increases. Because of change in cooling rate of the welded joint, when peak current is increased from 6 A, the grain size increases. The cooling rate is low during solidification and long-time interval available for grain coarsening. In contrast, the decrease in peak current leads to the decrease in the heat input, which leads to faster cooling rate and subsequently finer grain size in fusion zone. As pulse rate increases, the grain size increases, because of short on and off time of the welding current. As pulse width increases, grain size decreases, because of longer time gap between the peak current. According to the Hall-Petch equation, hardness and

IOP Conf. Series: Materials Science and Engineering **283** (2017) 012007 doi:10.1088/1757-899X/283/1/012007

strength of the weld joint are inversely proportional to grain size.

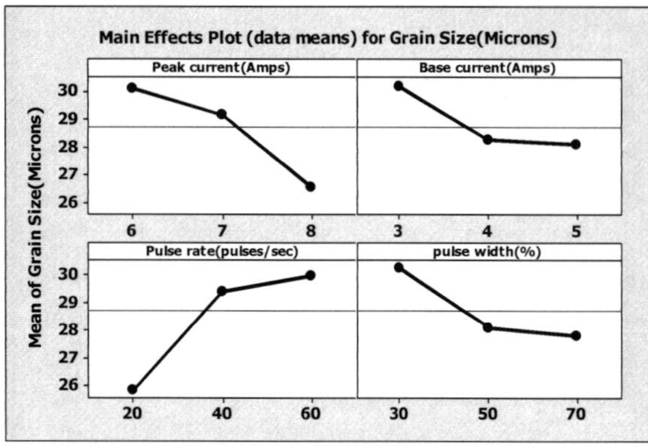

Figure 10. Main effect for grain size.

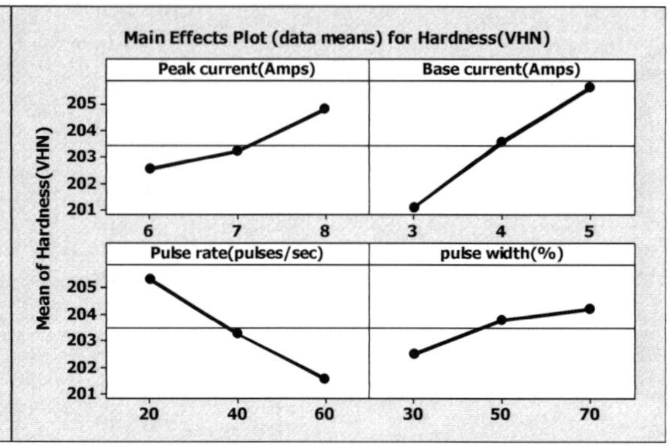

Figure 11. Main effect for hardness.

5.2. Contour plots for grain size and hardness

Figures 12(a) to 12(f) represent the contour plots for fusion zone grain size. From these plots, the interaction effect between the input process variables and output response can be observed as:

- Fusion zone grain size is more susceptible to change in peak current than base current (figure 12(a)), since the contour lines are more diverted towards peak current.
- Fusion zone grain size is susceptible to peak current than pulse rate (figure 12(b)), since the contour lines are more diverted towards peak current.
- Fusion zone grain size is more susceptible to peak current than pulse width (figure 12(c), since the contour lines are more diverted towards peak current.
- Fusion zone grain size is more susceptible to base current than pulse rate (figure 12(d)), since the contour lines are more diverted towards base current.
- Fusion zone grain size is more susceptible to base current than pulse width (figure 12(e)), since the contour lines are more diverted towards base current.
- Fusion zone grain size is more susceptible to pulse rate than pulse width (figure 12(f)), since the contour lines are more diverted towards pulse rate.

Among the welding variables considered, it is clear that peak current is the most influencing variables which affect the grain size in the fusion zone of the welded joints, followed by base current, pulse rate and pulse width.

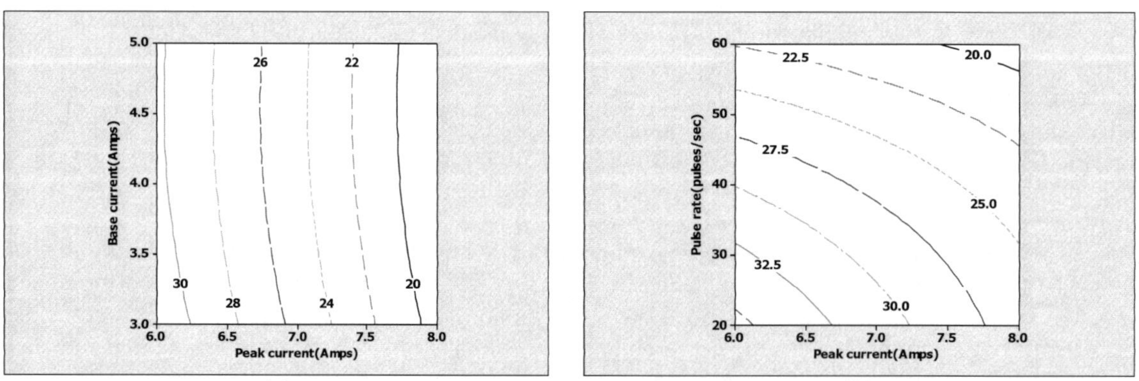

(a) (b)

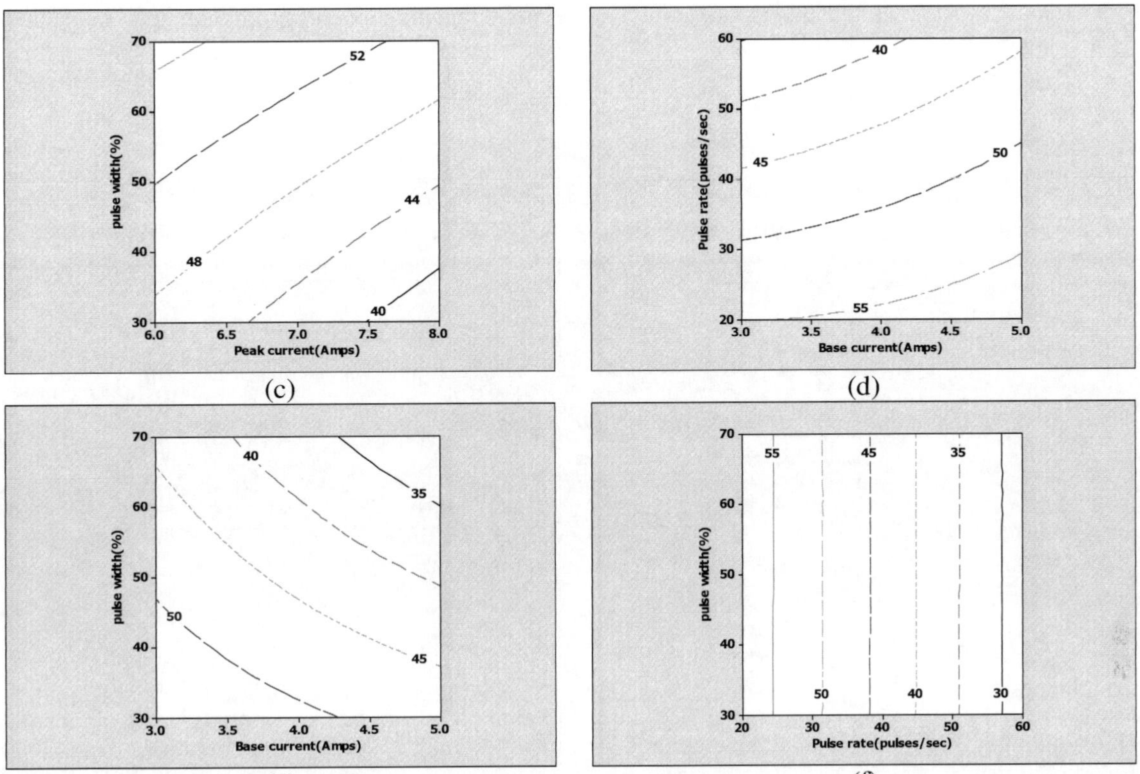

Figure 12. Contour plots for grain size: (a) peak current vs. base current, (b) peak current vs. pulse rate, (c) Peak current vs. pulse width, (d) Base current vs. pulse rate, (e) Base current vs. pulse width, (f) Pulse rate vs. pulse width.

Figures 13(a) to 13(f) represent the contour plots for fusion zone hardness. The effect between the input process variables and output response can be observed from contour plots as:

- Fusion zone hardness is more susceptible to change in peak current than base current (figure 13(a)), since the contour lines are more diverted towards peak current.
- Fusion zone hardness is susceptible to peak current than pulse rate (figure 13(b)), since the contour lines are more diverted towards peak current.
- Fusion zone hardness is more susceptible to peak current than pulse width (figure 13(c)), since the contour lines are more diverted towards peak current.
- Fusion zone hardness is more susceptible to base current than pulse rate (figure 13(d)), since the contour lines are more diverted towards base current.
- Fusion zone hardness is more susceptible to base current than pulse width (figure 13(e)), since the contour lines are more diverted towards base current.
- Fusion zone hardness is more susceptible to pulse rate than pulse width (figure 13(f)), since the contour lines are more diverted towards pulse rate.

Among the welding variables considered, it is understood that peak current is the most influencing variable, which affects the hardness in the fusion zone of the welded joints, followed by base current, pulse rate, and pulse width.

IOP Conf. Series: Materials Science and Engineering **283** (2017) 012007 doi:10.1088/1757-899X/283/1/012007

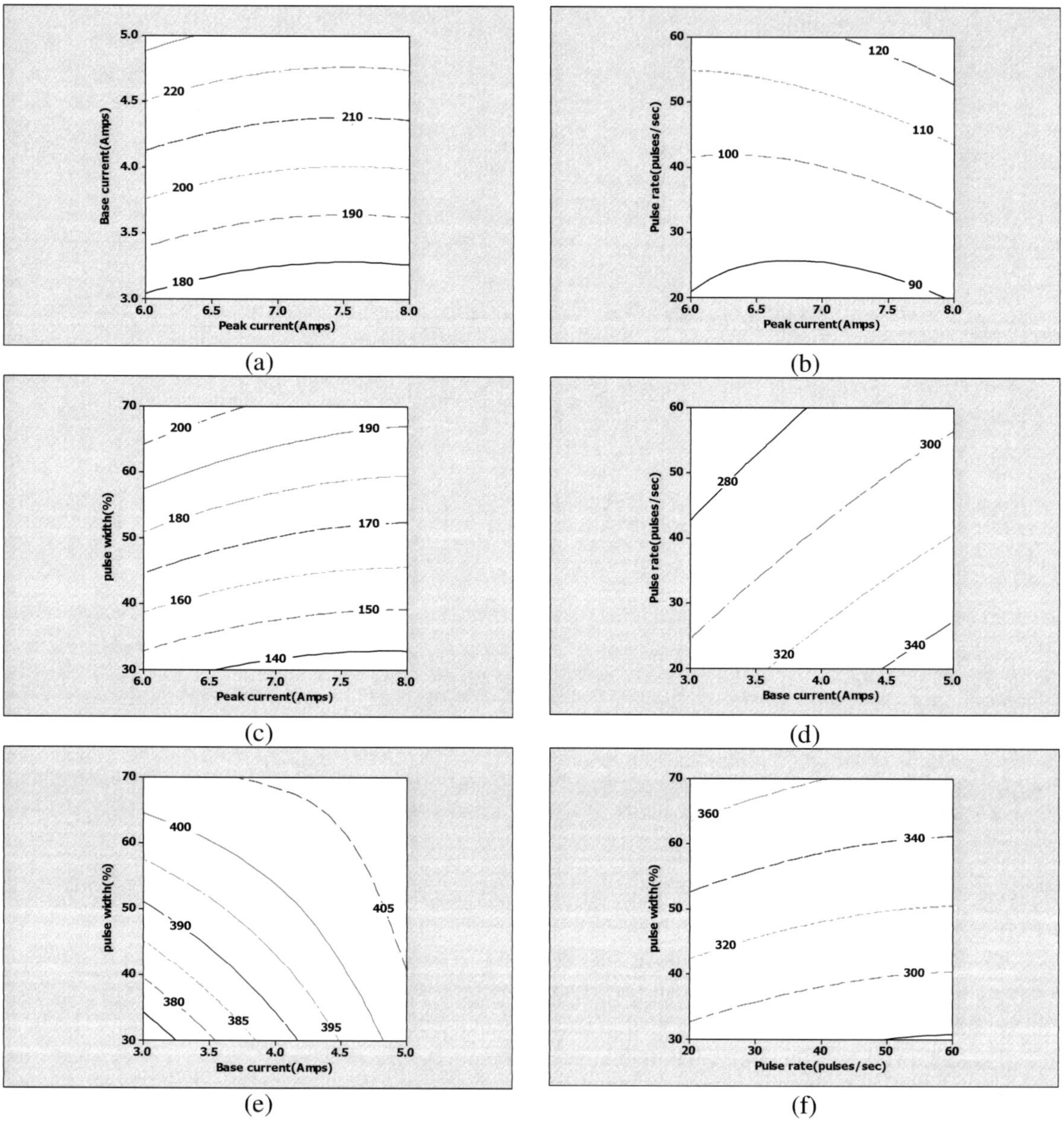

Figure 13. Contour plot for hardness: (a) peak current vs. base current, (b) peak current vs. pulse rate, (c) Peak current vs. pulse width, (d) Base current vs. pulse rate, (e) Base current vs. pulse width, (f) Pulse rate vs. pulse width.

5.3. Scatter plots for grain size and hardness

The minimum grain size is indicated by the nadir of the response surface, as shown in figures 14(a) to 14(f).

From the response figure 14(a), it is seen that at a peak current of 8A, the grain size is lower. Because of change in cooling rate, when peak current is increased from 6A, the grain size increases. Grain coarsening time has been improved because of slower cooling rates. From the surface plot, it is observed that at a peak current of 8A and base current of 4 A, the grain size is minimum.

From figure 14(b), it can be concluded that at a pulse rate of 60 pulse/second, grain size is lower. When the pulse rate decreases to 20 pulse/second, the grain size is increased. Because of finer grains

IOP Conf. Series: Materials Science and Engineering **283** (2017) 012007 doi:10.1088/1757-899X/283/1/012007

in fusion zone in strength of the welded joint is improved. Because of violent agitation of weld molten metal, at high pulse rate values, grain refinement in the weld region takes place. From the surface plot, it is observed that at a peak current of 8A and pulse rate of 60 pulse/second, grain size is minimum.

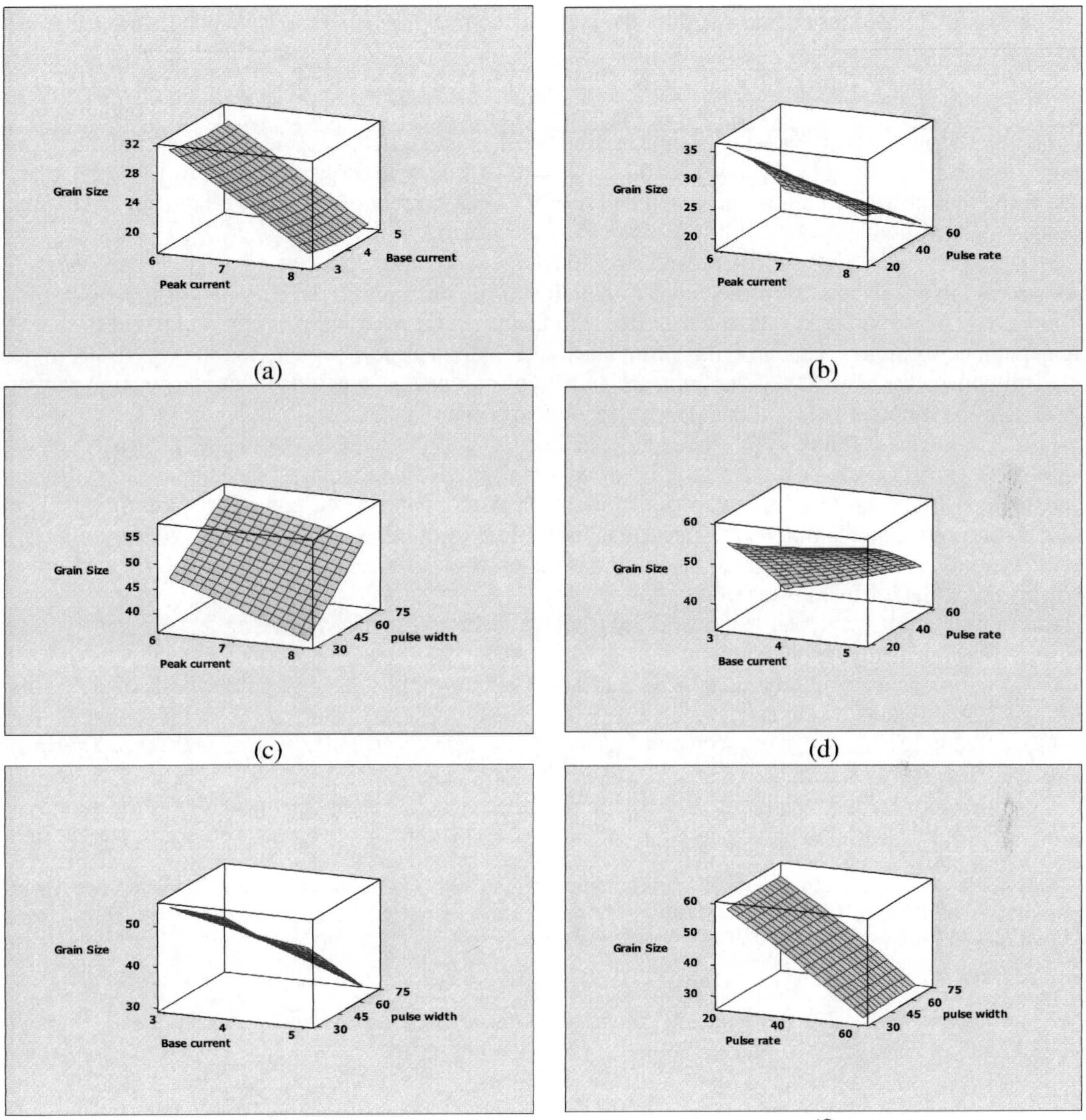

Figure 14. Surface plot for grain size: (a) peak current vs. base current, (b) peak current vs. pulse rate, (c) Peak current vs. pulse width, (d) Base current vs. pulse rate, (e) Base current vs. pulse width, (f) Pulse rate vs. pulse width.

From figure 14(c), it is seen that at a pulse width of 30%, the grain size is lower. Pulse width promotes the grain growth in the weld region. The time period of the pulse is improved, because of increase in pulse on time. From the surface plot, it is observed that at a peak current of 8A and pulse width of 30%, the grain size is minimum. From figure 14(d), it follows that when the base current is 4A and pulse rate of 60 pulse/second, grain size is minimum. From figure 14(e), it is observed that

IOP Conf. Series: Materials Science and Engineering **283** (2017) 012007 doi:10.1088/1757-899X/283/1/012007

when the base current is 4A and pulse width is 30%, the grain size is minimum. From figure 14(f), it is observed that with the pulse rate of 60 pulse/second and pulse width of 30%, grain size is minimum. The above observations reveal that for minimum grain size following is required: peak current of 8A, base current of 4A, pulse rate of 60 pulse/second and pulse width of 30%.

The maximum hardness is indicated by the apex of the response surface, as shown in figures 15(a) to 15(f).

From the response figure 15(a), it can be concluded that at the peak current of 7A, the hardness is lower. The hardness of the weld joint is improved because of fine equiaxed grains in fusion zone. Higher amperage and higher peak current lead to more heat and longer cooling time resulting in coarse grains, which is responsible for lower hardness. This is in line with the variation of fusion zone grain size. From the surface plot, it can be concluded that at a peak current of 7A and base current of 3A, the hardness is maximum.

From the response figure 15(b), it can be concluded that when the pulse rate is 20 pulse/second, the hardness is lower. Because of presence of fine grains in fusion zone, m the hardness of the weld joint is improved. At very low pulse rates, the effect of pulsing on the weld bead is less compared to that at high frequency pulsing. At high pulse rate values, the molten bead is agitated violently, resulting in grain refinement in the weld region. From the surface plot, it can be concluded that at a peak current of 7A and pulse rate of 20 pulse/second, the hardness is maximum.

From the response figure 15(c), it can be concluded that at the pulse width of 30%, the hardness is lower. The grain growth in weld region is influenced by pulse width. With the increase in pulse on time, there is an increase in the time period from the start of a pulse to the end of the base time. When pulse width is increased, the welding heat has more time to conduct into the fusion zone promoting grain coarsening. From the surface plot, it can be concluded that at a peak current of 7A and pulse width of 30%, the hardness is maximum.

From figure 15(d), it can be concluded that at a base current of 3A and pulse rate of 20 pulse/second, the hardness is maximum.

From figure 15(e), it can be concluded that at a base current of 3A and pulse width of 30%, the hardness is maximum.

From figure 15(f), it can be concluded that at a pulse rate of 60 pulse/second and pulse width of 30%, the hardness is maximum.

From the above observations, it may be inferred that to achieve maximum hardness the required values of welding variables are: peak current 7A, base current 3A, pulse rate 60 pulse/second and pulse width 30%.

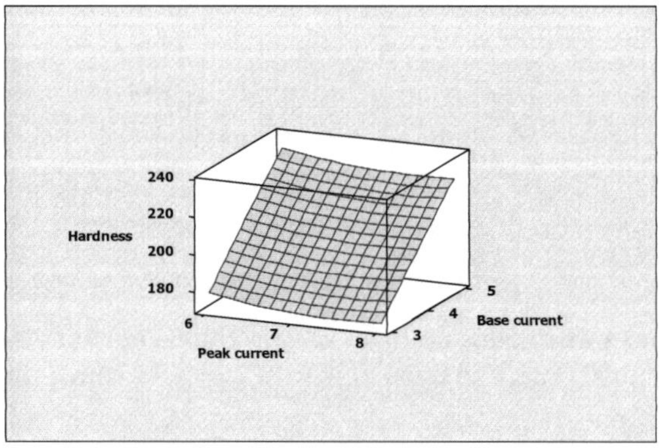

(a)

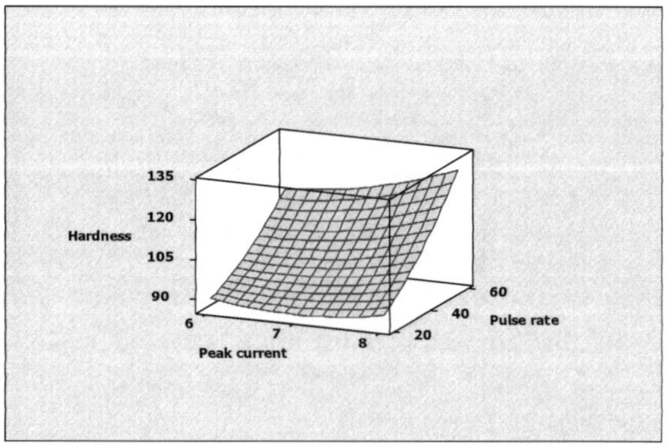

(b)

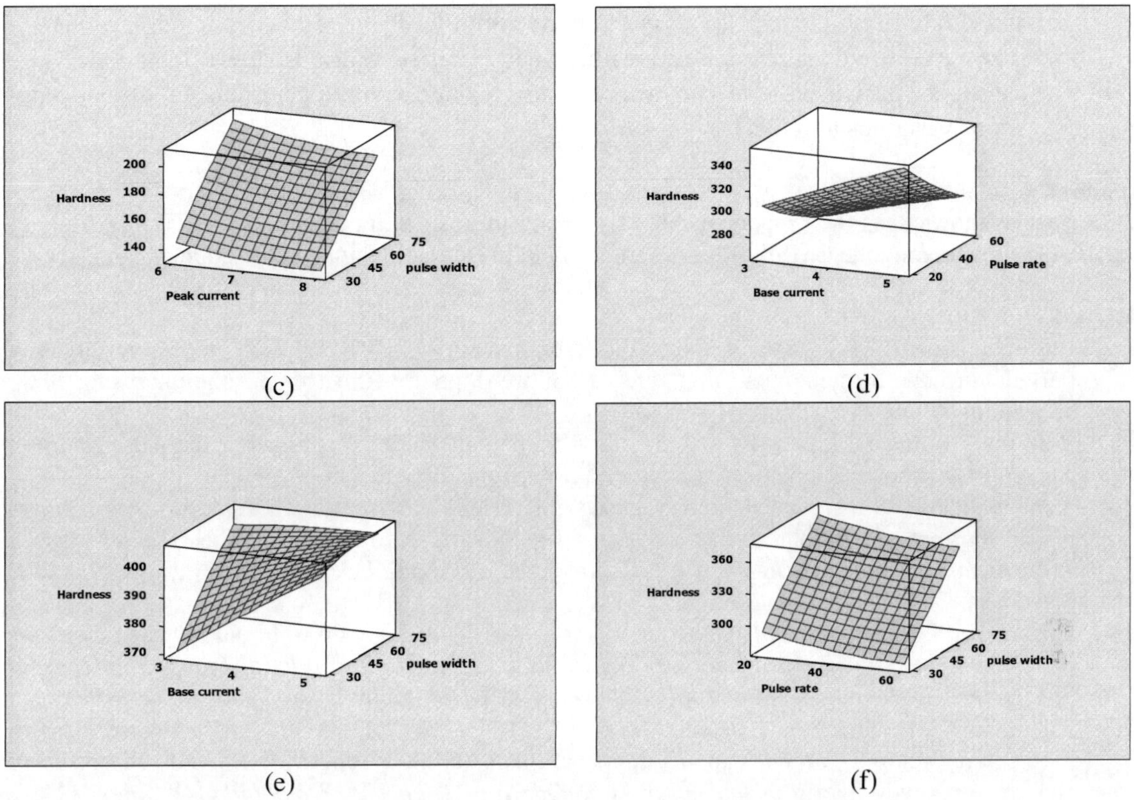

Figure 15. Scatter plot for hardness: (a) peak current vs. base current, (b) peak current vs. pulse rate, (c) peak current vs. pulse width, (d) base current vs. pulse rate, (e) base current vs. pulse width and (f) pulse rate vs. pulse width.

6. Conclusions

Based on the experiments performed and statistical analysis carried out, the following conclusions are drawn.

- The empirical mathematical models for grain size and hardness of the MPAW welded AISI 321 austenitic stainless sheets of 0.3 mm thickness were developed using the response surface method.
- Scatter plots are draw to understand the variation between experimental and predicted values and fond to be close to each other. The correlation coefficient R^2 is about 0.92.
- From the main effect plots, it is revealed that with the increase in peak current, base current and pulse width grain size decreases and hardness value increases. Whereas grain size increases and hardness decreases with an increase in the pulse rate.
- From the contour plots of grain size and hardness, it is revealed that peak current is the most influencing variables which affect the grain size and hardness in the fusion zone of the welded joints, followed by base current, pulse rate, and pulse width.
- From the surface plot, it is revealed that for minimum grain size the following is required: peak current of 8A, base current of 4A, pulse rate of 60 pulse/second, and pulse width of 30%.
- From the surface plot, it may be inferred that to achieve maximum hardness the required values of welding variables are: peak current of 7A, base current of 3A,

pulse rate of 60 pulse/second, and pulse width of 30%.

- The developed model is valid within the specified range of the selected welding variables; however, one can improve the accuracy of the models by considering more variables and their wider range.

Acknowledgments
The author acknowledges the support of AICTE, Government of India (F.No.10-18/RIFD/TG/Policy-1/2016-17), in attending the CMSE2017 conference held in Beijing, China, on October 24-27, 2017.

References

[1] Balasubramanian M, Jayabalan V and Balasubramanian V 2010 Effect of process variables of pulsed current tungsten inert gas welding on weld pool geometry of titanium welds *Acta Metall.Sin. (Engl. Lett.)* **23(4)** 312-20
[2] Balasubramanian B, Jayabalan V and Balasubramanian V 2006 Optimizing the pulsed current gas tungsten arc welding variables *J Mater Sci Technol* **22(6)** 821-5
[3] Kondapalli S P, Srinivasa R C and Nageswara R D 2011 Prediction of weld pool geometry in pulsed current micro plasma arc welding of SS304L stainless steel sheets *International Transaction Journal of Engineering, Management & Applied Sciences & Technologies* **2(3)** 325-36
[4] Kondapalli S P, Srinivasa R C and Nageswara R D 2011 A study on weld quality characteristics of pulsed current micro plasma arc welding of SS304L sheets *International Transaction Journal of Engineering, Management & Applied Sciences & Technologies* **2(4)** 437-46
[5] Kondapalli S P, Srinivasa R C and Nageswara R D 2011 Optimizing pulsed current micro plasma arc welding variables to maximize ultimate tensile strength of SS304L sheets using Hooke and Jeeves algorithm *Journal of Manufacturing Science & Production (DyGuter)* **11(1-3)** 39-48
[6] Kondapalli S P, Srinivasa R C and Nageswara R D 2012 Effect of process variables of pulsed current micro plasma arc welding on weld pool geometry of AISI 304L stainless steel sheets *Journal of Materials & Metallurgical Engineering* **2(1)** 37-48
[7] Kondapalli S P, Srinivasa R C and Nageswara R D 2012 Establishing empirical relations to predict grain size and hardness of pulsed current micro plasma arc welded SS 304L sheets *American Transactions on Engineering & Applied Sciences* **1(1)** 57-74
[8] Kondapalli S P, Srinivasa R C and Nageswara R D 2012 Effect of pulsed current micro plasma arc welding process variables on fusion zone grain size and ultimate tensile strength of SSS304L sheets *International Journal of Lean Thinking* **3(1)** 107-18
[9] Marimuthu K and Murugan N 2003 Prediction and optimization of weld bead geometry of plasma transferred arc hardfaced valve seat rings *Surf Eng* **19(2)** 143-9
[10] Gunaraj V and Murugan N 1999 Prediction and comparison of the area of the heat affected zone for the bead-on-plate and bead-on joint in SAW of pipes *J Mat Proc Tech* **95** 246-61
[11] Gunaraj V and Murugan N 1999 Application of response surface methodology for predicting weld quality in saw of pipes *J Mat Proc Tech* **88** 266-75
[12] Juang S C and Tarng Y S 2002 Process parameter selection for optimizing the weld pool geometry in the TIG welding of stainless steel *J Mat Proc Tech* **122** 33-7
[13] Allen T T, Richardson R W, Tagliable D P and Maul G P 2002 Statistical process design for robotic GMA welding of sheet metal *Weld J* **81(5)** 69–s-172-s
[14] Kim I-S, San J-S and Jeung Y-J 2001 Control and optimization of bead width for mutipass-welding in robotic arc welding processes *Aus Weld J* **46** 43-6
[15] Kim D, Kang M and Rhee S 2005 Determination of optimal welding conditions with a controlled random search procedure *Weld J* 125-s-130-s
[16] Kim D and Rhee S 2001 Optimization of arc welding variables using a genetic algorithm *Weld J* **80(7)** 184–s-9-s

IOP Conf. Series: Materials Science and Engineering **283** (2017) 012008 doi:10.1088/1757-899X/283/1/012008

Influence of Sb substitution on thermal and electrical characteristics of Ge-Sn-Se chalcogenide glass system

Surbhi Sharma[1*], Navjeet Sharma[2], Amit Sarin[3]
[1]Department of Applied Sciences, C. T. Institute of Engineering, Management, and Technology, Jalandhar, Punjab, India
[2]Department of Physics, DAV College, Jalandhar, Punjab, India
[3]Department of Physical Sciences, I.K.G. Punjab Technical University, Kapurthala, Punjab, India
Corresponding author: *physurbhi@gmail.com

Abstract. Present research work reports the systematic investigation of thermal and electrical characterizations of $Ge_{20}Sn_{10}Se_{70-x}Sb_x$ (x = 0, 3, 6, 9, 12, 15) glass system to probe the structural modifications. Bulk samples are prepared using melt quenching technique. Differential scanning calorimetric technique (DSC) is used at the constant heating rate of 10^0C under nonisothermal conditions. Different kinetic parameters viz. glass transition temperature T_g, Hruby parameter H_r, crystallization temperature T_c ,etc., have been calculated. Glasses under study shows good thermal stability, hence, can find practical applications especially in optical fiber technology. Further, in the current study, an attempt has been made to observe the variation in conductivity with respect to the increase in temperature for amorphous $Ge_{20}Sn_{10}Se_{70-x}Sb_x$ (x = 0, 3, 6, 9, 12, 15) using two probe method and hence the band structure and corresponding conduction mechanism is studied.

1. Introduction

Chalcogenide based thin films have attracted great attention of scientific community owing to their unique features like interesting electron transport mechanism, good thermal, mechanical and chemical properties [1-2]. Amorphous nano-scale structures have various advantages over crystalline nano- structures. It is well known that synthesis of crystalline nano-structures is controlled by deposition of atoms one by one. On the other hand in amorphous chalcogenide thin films atomic level handling is meaningless owing to its inherent property of disorderness. Thus chalcogenide films exhibit various varieties of structures due to lack of bonding constraints those arises in crystalline unit cells [3].

In Ge-Se system, Ge atoms acts as bond modifiers, which increase average bond strength by crosslinking with Se. This make the system more rigid, thus, increasing the glass transition temperature and resistivity [4]. The addition of third impurity introduces modifications into the local atomic structure, which further influence physical properties of the system. Various studies reveal that addition of Sn to Ge-Se matrix weakens the system and makes it at higher wavelength [5].

Much attention has been focused on quaternary chalcogenide glass (C.G) systems. Thus, study [6] revealed that the addition of Sb to Ge-Sn-Se matrix enhanced the glass-forming ability and increased the glass transition temperature T_g.

Thermal characterization of glasses involves the study of glass transition temperature, heat capacity and crystallization kinetics of chalcogenides, which, in turn, helps one to determine the thermal stability, effective working limits, and their practical applications.

In the present work, we have attempted to study the effect of Sb concentration on the thermal properties Ge-Sn-Se chalcogenide glass system. Differential Scanning Calorimetry (DSC) is very useful technique for the study of crystallization in glass system. This technique is particularly important due to the fact that (i) it is easy to carry out (ii) it is quite sensitive (iii) it requires little apriori sample preparation, etc. [7]. In the present study, non-isothermal method for calorimetric measurements is used.

Electrical properties of chalcogenide glasses C.Gs are important to understand the transport mechanism and hence the phenomenon of conductivity in glasses. Electrical study offer vital information regarding electronic band structure, conduction mechanism in chalcogenide materials. Study of electrical properties of C.Gs is still under extensive study. In the current study, an attempt has been made to observe the variation in conductivity with respect to the increase in temperature for amorphous $Ge_{20}Sn_{10}Se_{70-x}Sb_x$ (x = 0, 3, 6, 9, 12, 15) using two probe method and hence to predict the band structure and corresponding conduction mechanism.

2. Experimental techniques

Bulk samples of chalcogenide glass system $Ge_{20}Sn_{10}Se_{70-x}Sb_x$ (x = 0, 3, 6, 9, 12, 15) were prepared using conventional melt quenching technique. The constituent elements of 5N purity were weighed according to their atomic percentages and were sealed in evacuated quartz ampoules of 8 mm inner diameter. The ampoules were then heated to 1000 °C at slow rate of 2-3 °C per min. and the peak temperature was maintained for 48 hours. During heating, ampoules were shaken regularly to ensure the homogeneous mixing of constituent elements. The samples were quenched in ice cold water. The samples were recovered from the ampoules by dissolving the ampoules in solution of $HF+H_2O_2$.

Thin films of $Ge_{20}Sn_{10}Se_{70-x}Sb_x$ (x = 0, 3, 6, 9, 12, 15) were then produced using thermal evaporation process (Hind High Vacuum Coating Unit (model no. 12A4D)). Thin films were deposited onto a pre-cleaned glass susbstrate at the rate of 5-8 nm/sec, in a vacuum better than 10^{-5} torr. For cleaning of substrate three steps were used (i) washing with soap solution (ii) cleaning with acetone (vapor cleaning) (iii) cleaning with methanol. Washing with soap solution involved scrubbing substrates with soap and rinsing with distilled water. This removes dirt, oil and grease from the substrate. Cleaning with acetone removes organic impurities whereas methanol removes inorganic impurities from it. After cleaning, substrate was dried by placing in oven at temperature ~ 110^0C. Then substrates were placed in deposition chamber.

In order to confirm the amorphous nature of specimen so produced, the X-Ray diffraction technique was used. Absence of prominent peaks confirms the amorphous nature of specimens. Figure 1 shows the XRD diffractograms of $Ge_{20}Sn_{10}Se_{70-x}Sb_x$ (x = 0, 3, 6, 9, 12, 15).

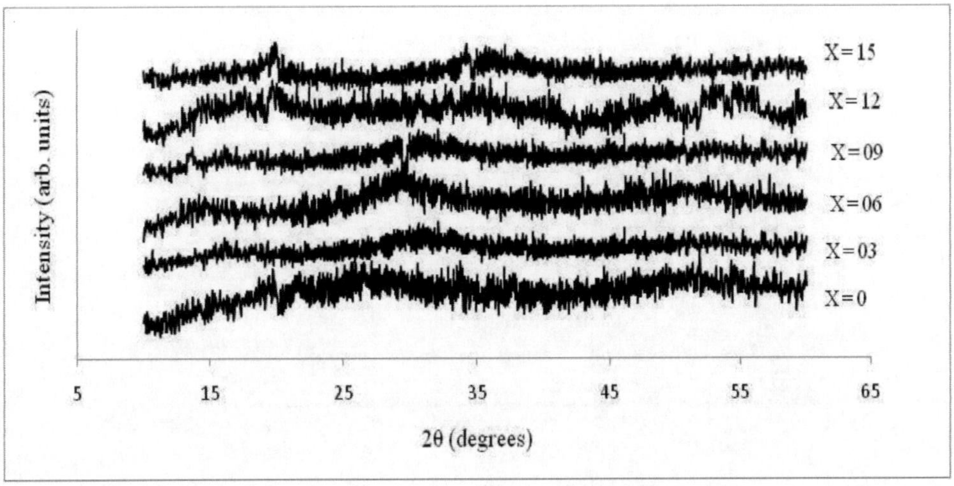

Figure 1. XRD diffractograms of $Ge_{20}Sn_{10}Se_{70-x}Sb_x$ (x = 0, 3, 6, 9, 12, 15)

In the present study, FESEM-EDAX was carried out to verify the compositions of as prepared thin films. FESEM was conducted using an Hitachi SU 8010 model with field emission tungsten-based gun. The EDAX analysis of the samples was performed by using Bruker EDX flash detector model 6130. For FESEM platinum coating of the specimen was done using sputtering.

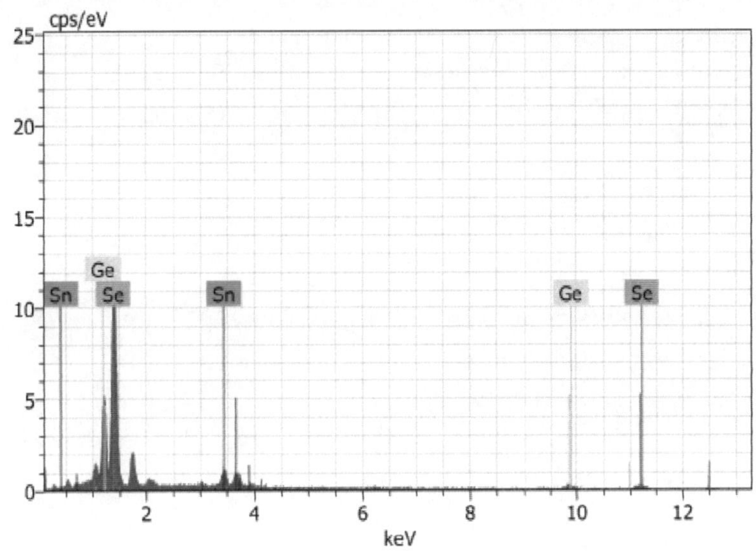

Figure 2. EDAX results for $Ge_{20}Sn_{10}Se_{70}$

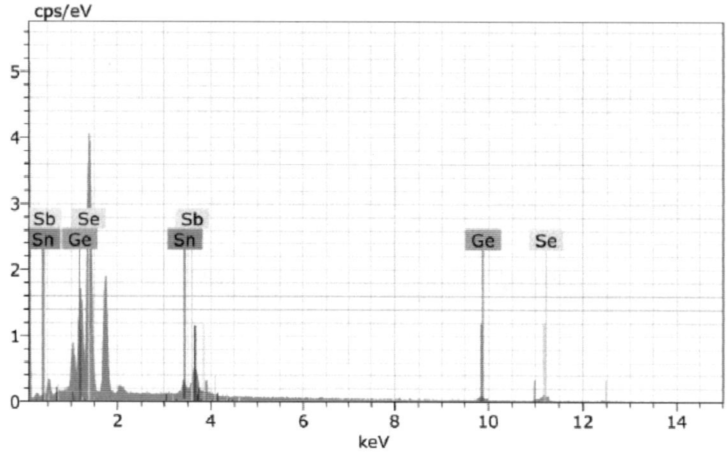

Figure 3. EDAX results for $Ge_{20}Sn_{10}Se_{64}Sb_6$

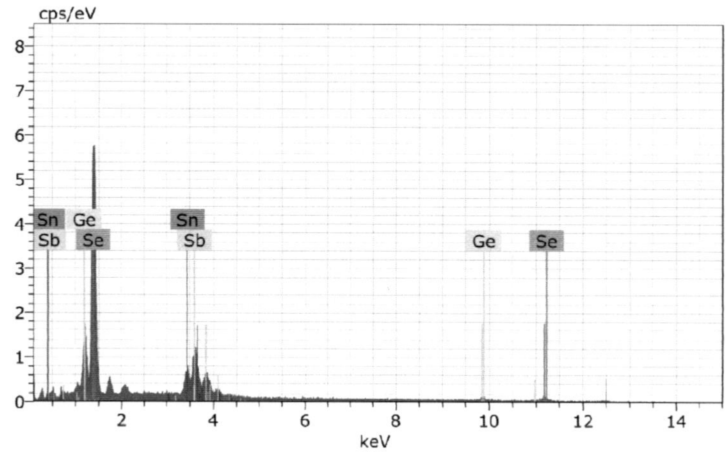

Figure 4. EDAX results for $Ge_{20}Sn_{10}Se_{58}Sb_{12}$

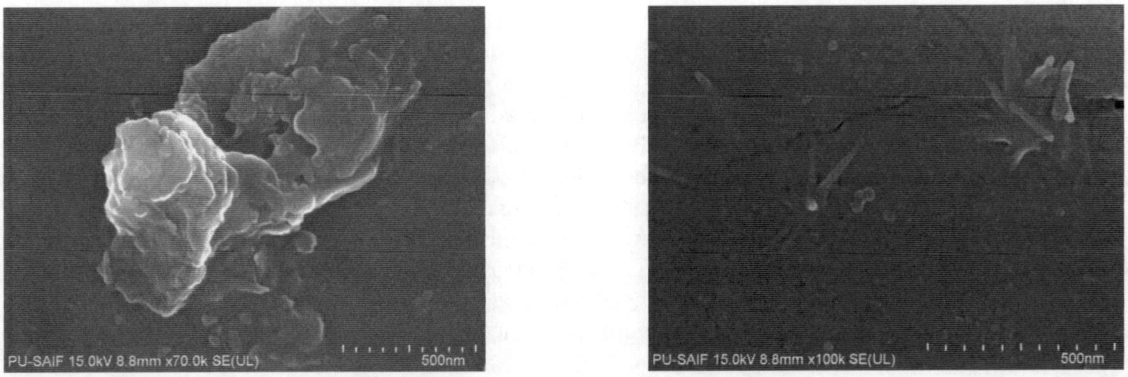

Figure 5. FESEM images of $Ge_{20}Sn_{10}Se_{70}$

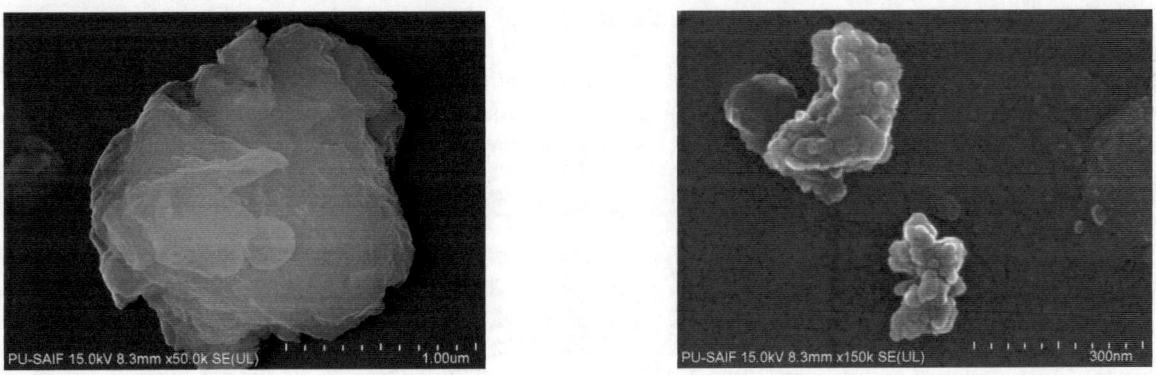

Figure 6. FESEM images of $Ge_{20}Sn_{10}Se_{64}Sb_6$

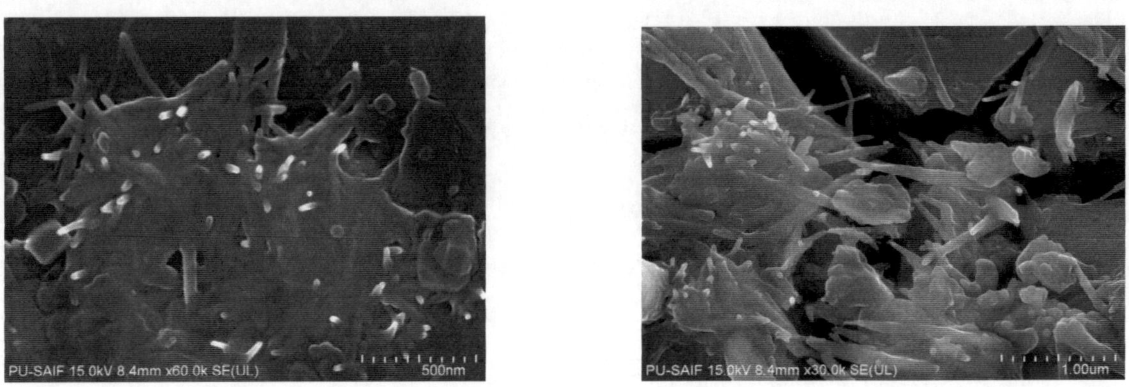

Figure 7. FESEM images of $Ge_{20}Sn_{10}Se_{58}Sb_{12}$

IOP Conf. Series: Materials Science and Engineering **283** (2017) 012008 doi:10.1088/1757-899X/283/1/012008

In the present study, the differential scanning calorimetric DSC technique was used under nonisothermal conditions for thermal characterization. DSC instrument was first calibrated with alumina powder as standard material before recording thermal history of the sample. DSC scan was produced under the constant flow of N_2 (200 ml/min) to avoid oxidization of the sample, which could lead to some unwanted features in the scan. About 10mg of the sample of $Ge_{20}Sn_{10}Se_{70-x}Sb_x$ (x = 0, 3, 6, 9, 12, 15) was taken in alumina pan with 10mg of alumina powder as the reference material. Scan is produced in the temperature range of 70^{0}C to 500^{0}C with the constant heating rate of 10^{0}C/min.

For electrical characterization of thin films two probe methods was used by authors. Measurements were taken in the temperature range of 289-333K. Electrodes were designed using silver paste with ~99.99% purity. For I-V characteristics, D.C voltage ranging 0-500V was applied across the ohmic contacts, and the corresponding current was noted through picometer, whereas Keithley (model 6487) was used by the authors for electrical characterization.

3. Results and discussion

3.1 Thermal characterization of $Ge_{20}Sn_{10}Se_{70-x}Sb_x$ (x = 0, 3, 6, 9, 12, 15) chalcogenide glasses

DSC thermograms of all the samples in the temperature range 70 °C – 500 °C are shown in figure 8. Single glass transition temperature observed in thermograms for all the samples indicates that the samples were compositionally homogenous. From DSC curves, glass transition temperature has been calculated from the intersection of tangents from base line and slope of the step. The values of glass transition temperature (T_g), onset crystallization temperature (T_c) and peak crystallization temperature (T_p) calculated from thermograms are summarized in table 1.

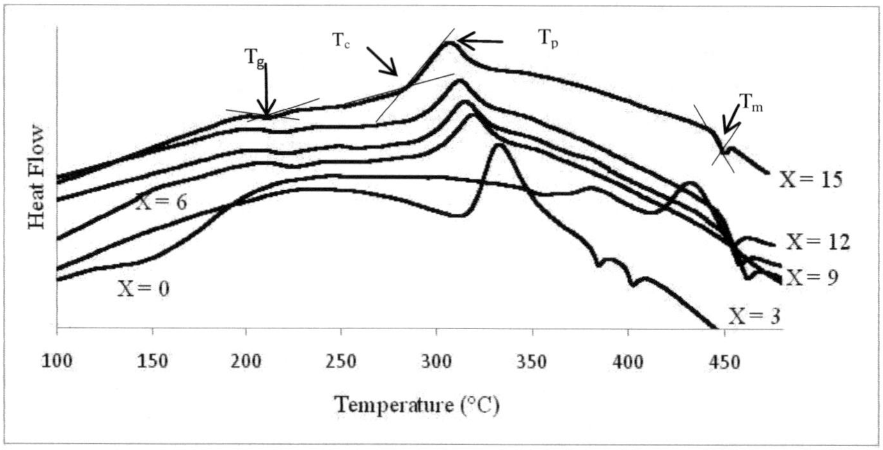

Figure 8. DSC thermograms of $Ge_{20}Sn_{10}Se_{70-x}Sb_x$ (x = 0, 3, 6, 9, 12, 15) chalcogenide glasses.

The values of glass transition temperature T_g, crystallization temperature T_c, peak temperature T_p, melting temperature T_m, Thermal stability temperature T_c-T_g and Hruby parameter H_r has been calculated for $Ge_{20}Sn_{10}Se_{70-x}Sb_x$ (x = 0, 3, 6, 9, 12, 15) chalcogenide glasses are given in table 1.

Table 1. Glass transition temperature (T_g), onset crystallization temperature (T_c) and peak crystallisation temperature (T_p), Hruby parameter (H_r), reduced glass transition temperature (T_{rg}) for $Ge_{20}Sn_{10}Se_{70-x}Sb_x$ (x = 0, 3, 6, 9, 12, 15) chalcogenide glasses.

Composition	T_g (^{0}C)	T_c (^{0}C)	T_p (^{0}C)	T_m (^{0}C)	T_c-T_g (^{0}C)	H_r	T_{rg}
$Ge_{20}Sn_{10}Se_{70}$	209.63	369.91	432.10	-	160.28	-	-
$Ge_{20}Sn_{10}Se_{67}Sb_3$	216.41	316.31	332.56	394.42	99.90	1.28	0.55
$Ge_{20}Sn_{10}Se_{64}Sb_6$	219.72	298.13	319.11	461.92	78.41	0.48	0.48
$Ge_{20}Sn_{10}Se_{61}Sb_9$	214.33	295.92	314.92	457.76	81.59	0.50	0.47
$Ge_{20}Sn_{10}Se_{58}Sb_{12}$	210.92	297.41	311.23	454.15	86.49	0.55	0.46
$Ge_{20}Sn_{10}Se_{55}Sb_{15}$	206.47	283.62	305.56	451.12	77.15	0.46	0.46

It can be observed from table 1 that glass transition temperature T_g increases up to 6 at.% of Sb, after which a decrease in T_g is observed. This trend can be explained on the basis of CON (chemically ordered network) model. Calculating the hetropolar bond energies using the Pauling relation [8], we get

$$\mathrm{E(A-B)} = [E(A-A) \times E(B-B)\,]^{\frac{1}{2}} + 30(\chi_A - \chi_B)^2 \quad (1)$$

where E(A-A) and E(B-B) are energies of homonuclear bonds and χ_A and χ_B are electronegativity values for atoms involved, the values of bond energies are given in table 3. According to CON model bonds are formed in the sequence of decreasing energy. Thus Ge-Se stronger bonds will form first followed by Sn-Se, Se-Sb and so on. It is evident from table 2 that upto 6 at.% of Sb stronger bonds viz. Ge-Se, Se-Sn, Se-Sb dominates over weaker homopolar Se-Se, Sb-Sb bonds leading to the increase in mean bond energy and hence glass transition temperature. But with further increase in Sb atoms, concentration of weaker Sb-Sb, Sn-Sb, Sn-Sn bonds increases resulting into decrease in glass transition temperature T_g.

Table 2. Values of distribution of bonds for $Ge_{20}Sn_{10}Se_{70-x}Sb_x$ (x = 0, 3, 6, 9, 12, 15).

	Distribution of bonds						
	Ge-Se	Se-Sn	Se-Sb	Se-Se	Sn-Sn	Sn-Sb	Sb-Sb
$Ge_{20}Sn_{10}Se_{70}$	0.57	0.28	-	0.14	-	-	-
$Ge_{20}Sn_{10}Se_{67}Sb_3$	0.60	0.30	0.06	0.04	-	-	-
$Ge_{20}Sn_{10}Se_{64}Sb_6$	0.60	0.31	0.05	-	-	-	0.03
$Ge_{20}Sn_{10}Se_{61}Sb_9$	0.59	0.30	-	0.01	-	-	0.10
$Ge_{20}Sn_{10}Se_{58}Sb_{12}$	0.59	0.26	-	-	0.01	-	0.13
$Ge_{20}Sn_{10}Se_{55}Sb_{15}$	0.58	0.22	-	-	0.02	0.02	0.15

IOP Conf. Series: Materials Science and Engineering **283** (2017) 012008 doi:10.1088/1757-899X/283/1/012008

Table 3. Values of bond energy for Ge-Sn-Se-Sb System [9]

Bonds	Bond Energy (Kcal/mol)
Ge-Se	49.42
Sn-Se	49.23
Se-Sb	43.96
Se-Se	44.00
Sn-Sn	34.20
Sn-Sb	32.38
Sb-Sb	30.20

Thermal stability of glasses is very important from their practical application point of view, which reflects the resistance to devitrification after the glass formation. Glass forming ability GFA is the ease with which the melt can be cooled to form glass. Nasciemento et al [10] suggested that slower is the rate of crystallization higher will be the glass formation ability [10]. Although thermal stability and glass forming ability GFA are two important and independent characteristics of glasses but number of studies are made to relate these two parameters [11-15]. Glass transition temperature T_g represents the rigidity of glassy structure. Thus, T_g can be associated with thermal stability of glass but GFA cannot be determined with T_g alone. Sakka and Meckenzie [16] used the parameter of reduced glass transition temperature $T_{rg}= T_g/T_m$, as glass stability parameter. For glass formation, T_{rg} has range i.e.$1/2 \leq T_{rg} \leq 2/3$. In the present work, T_{rg} is approximately 0.5 i.e. lies in the lower accepted range. Turnbull [17] proposed that for a given composition glass forming ability decreases with the decrease in T_{rg}. So, there is a need to search for another parameter that can relate both GFA and T_g.

Dietzel 18] suggested parameter $\Delta T=T_c-T_g$ which is the strong indicator for both thermal stability and GFA of glasses, as calculated in table 1. Unstable glasses generally have small difference between crystallization temperature and glass transition temperature and soon these glasses tend to crystallize. Higher difference between T_c and T_g retards the nucleation process, thus GFA increases. [19]. Hruby [20] introduced a new parameter, which combines both growth and nucleation aspects of phase transformation:

$$H_r = \left(\frac{T_c - T_g}{T_m - T_c}\right) \tag{2}$$

Glass formation will be difficult for $H_r \leq 0.1$ and easy for $H_r \geq 0.4$ [20]. In this paper, glasses under study have good glass-forming ability as indicated in table 1. Hence as prepared glasses can find practical applications viz for optical fibres, phase-change memory, etc.

3.2 *Improvement of electrical properties of ternary Ge-Sn-Se chalcogenide thin films by the Sb additions*

In the present study of amorphous $Ge_{20}Sn_{10}Se_{70-x}Sb_x$ (x = 0, 3, 6, 9, 12, 15), the whole mechanism is divided into two parts (i) deviation from ohmic to non-ohmic behavior of I-V curve at higher voltage (ii) dependence of conductivity on composition.

Efforts have been made to evaluate DC conductivity at room temperature via relation

$$\sigma_{RT} = \frac{L}{RA} \tag{3}$$

where L is sample thickness, A is cross-sectional area, and R is sample resistance. The values of R for all samples at room temperature are derived from the ohmic (linear) part of I-V curve.

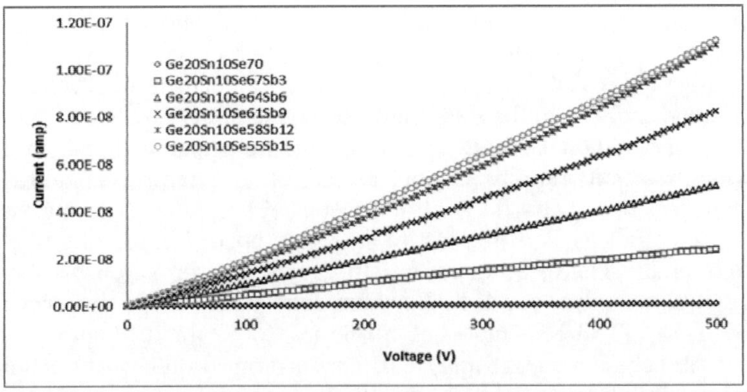

Figure 9. Current (A) vs. voltage (V) for $Ge_{20}Sn_{10}Se_{70-x}Sb_x$ (x = 0, 3, 6, 9, 12, 15).

Figure 9 shows the variation of current with respect to voltage. The non-ohmic nature of samples is shown in figure 3 for a specific range of voltages, where the ohmic part for 0-5V is observed, after which the graph deviates from its linear path at high voltage. Linear portion of the graph reveals that the charge carriers are generated due to thermal excitations. However, the deviation from the linear path reflects the non-ohmic aspect of films. Space charge limited conduction (SCLC) model implies that higher voltages applied to the sample causes electrodes to inject the non-equilibrium density of charge carriers. The latter populate above the Fermi level. The larger population of these charges, as compared to thermally generated charge carriers, causes a deviation from the linearity [21].

Table 4. Values of activation energy E_a, optical band gap E_g[9], conductivity at room temperature σ_{RT}, resistivity ρ_0, pre-exponential function σ_0 ,cohesive energy C.E* for $Ge_{20}Sn_{10}Se_{70-x}Sb_x$ (x =0,3, 6, 9, 12, 15).

S. No.	Composition	Conductivity σ_{RT} (mho m^{-1})	Resistivity ρ_{RT} (ohm m)	Activation energy E_a (eV)	E_g^{*} (eV) Optical band gap (Tauc)	$\ln\sigma_0$	Pre-exponential function σ_0 (mho m^{-1})	C.E*[9] (Kcal/mol)
1	$Ge_{20}Sn_{10}Se_{70}$	1.88×10^{-4}	5308.54	0.75	1.66	-1.57	2.08×10^{-1}	48.59
2	$Ge_{20}Sn_{10}Se_{67}Sb_3$	2.06×10^{-2}	48.49	0.44	1.54	-9.74	5.88×10^{-5}	48.79
3	$Ge_{20}Sn_{10}Se_{64}Sb_6$	$1.12x10^{-2}$	89.31	0.37	1.49	-13.05	2.16×10^{-6}	47.54
4	$Ge_{20}Sn_{10}Se_{61}Sb_9$	9.00×10^{-3}	111.11	0.33	1.45	-13.13	1.98×10^{-6}	47.39
5	$Ge_{20}Sn_{10}Se_{58}Sb_{12}$	8.22×10^{-3}	121.68	0.29	1.42	-13.15	1.94×10^{-6}	46.60
6	$Ge_{20}Sn_{10}Se_{55}Sb_{15}$	7.97×10^{-3}	125.46	0.28	1.38	-14.81	3.71×10^{-7}	45.64

From table 4, it is observed that there is an overall decrease in conductivity at room temperature σ_{RT} with the increase in concentration of Sb in Ge-Sn-Se thin films. Conductivity at room temperature σ_{RT} shows an abrupt increase for 3 at% of Sb. This observation is in agreement with cohesive energy as shown in the table 4. Similar trend is observed for glass transition temperature T_g as studied in previous section of our work which shows the system is rigid for this composition and with the further increase in Sb concentration the system losses its rigidity. Overall decrease in σ_{RT} can be explained by the fact that Se has tendency to form polymerized network. Scottmiller et al. [2] reported that, in glassy state, 60% of Se atoms are bonded as polymeric chains, while rest of atoms are bonded in the ring-like structure [22]. The addition of Sb to the ternary Ge-Sn-Se glass system causes the formation of stronger Se-Sb (43.96 Kcal/mol) bonds leading to an increase in C.E. However, further increase in Sb concentration leads to the formation of weaker Sn-Sb (32.38 Kcal/mol) and Sb-Sb (30.20 Kcal/mol) bonds at the cost of stronger bonds, which reduces the cohesive energy and, thus, conductivity. In other words, conductivity at room temperature decreases with Sb concentration due to the increase in concentration of defect states associated with impurity atoms [23]. Observations are in agreement with the results of Jain and Saxena [24]. The decrease in σ_{RT} can be correlated with the decrease in mean free path owing to the short range order of glasses. The temperature dependence of DC conductivity has been determined for the temperature range of (289-333K) for as prepared thin films of amorphous $Ge_{20}Sn_{10}Se_{70-x}Sb_x$ (x = 0, 3, 6, 9, 12, 15). At higher temperature, dependence of DC conductivity with temperature is given by the Arrhenius formula

$$\sigma(T) = \sigma_0 \exp\left(\frac{-E_a}{k_b T}\right) \tag{4}$$

where σ_0 is the pre-exponential factor which includes charge carrier mobility and density of states. Pre-exponential factor σ_0 indicates whether the conduction mechanism takes place in extended states above mobility edge or by hopping in localized states. E_a represents the activation energy. According to Mott and Davis (1979) for conduction in localized states, σ_0 is in the range of 10^3 to 10^5 $\Omega^{-1}m^{-1}$ [25]. Value of E_a has been determined by fitting a slope for the plot between ln σ and 1000/T, as shown in figure 4. Activation energy E_a is the smallest energy required by electrons to overcome the potential barriers, thus influencing the structural relaxation. Thus, E_a implies the energy differences either (E_C-E_F) or (E_F - E_V) for conduction in localized states.

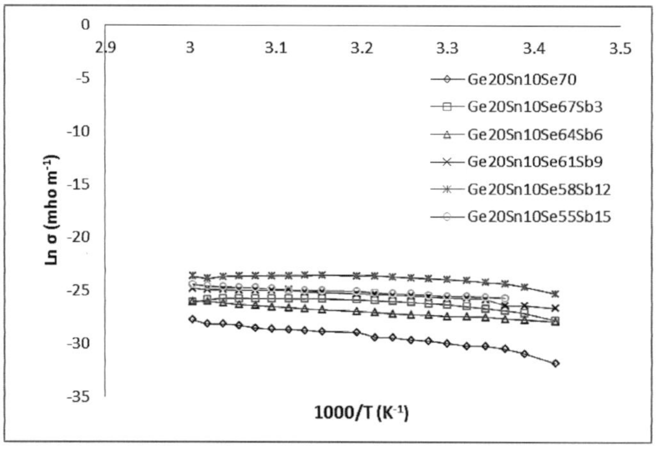

Figure 10. lnσ versus 1000/T (K^{-1}) for various compositions of Sb in $Ge_{20}Sn_{10}Se_{70-x}Sb_x$ (x = 0, 3, 6, 9, 12, 15) chalcogenide thin films.

E_a is generally found to be half of the optical band gap E_g ($E_g \sim 2E_a$) [26] implying that the Fermi level lies near the mid gap. In the present study, it is found that values of E_a is less than the half of E_g thus Fermi level in the present study is not pinned to the middle of band gap.

It is observed that E_a and σ_0 decrease with the concentration of Sb. It is known that Sb forms ionic bonds and, thus, makes the system ionic in nature. This weakens the network and deviates the structure towards fragility causing the decrease in C.E and conductivity [27].

Also decrease in activation energy can be attributed to the formation of weaker homopolar bonds, which, in turn, reduce the average single bond energy [28]. Such small values of E_a and pre-exponential function σ_0 imply that the conductivity is due to thermally generated charge carrier motion in the band tails of localized states, which finding is in accordance with the results of earlier studies [29-30].

4. Conclusions

This study confirmed that glass transition temperature T_g increases with Sb concentration up to 6 at.% Sb, after which a decrease in T_g is observed in $Ge_{20}Sn_{10}Se_{70-x}Sb_x$ (x = 0, 3, 6, 9, 12, 15) chalcogenide glass system. DSC thermographs reveal a higher difference between glass transition temperature T_g and crystallization temperature T_c, which implies that the glassy system is quite stable and shows resistance to nucleation process. Further glass forming ability and stability is verified using Hruby parameter and reduced glass transition temperature. Glasses under study show good thermal stability and, hence, can find practical applications, especially, in optical fiber technology field.

An overall conductivity deterioration was observed at room temperature σ_{RT} with the concentration of Sb in Ge-Sn-Se thin films. Conductivity at room temperature σ_{RT} shows an abrupt increase for 3 at% of Sb. Although an overall decrease in σ_{RT} has been observed, it can be correlated with the mean free path decrease due to the short range order of glasses.

References

[1] Singh A K 2012 *J. Opto-Elect. Rev.*, **20** **(3)** 226
[2] Saxena M, Kukreti A K, Gupta S and Aggarwal M K 2013 MIT Int. *J. Elect. Comm. Engg.* **3** **(1)** 29-35
[3] Khan Z H, Khan S A, Salah N, Ghamdi A A and Habib S 2011 *Phil. Mag. Lett.,* **91** **(3)** 207.
[4] Sharda S, Sharma N, Sharma P and Sharma V 2012 *Chalco. Let.* **9** **(9)** 389.
[5] Busnach I H, Dror J and Croitoru N 1990 *J. Mat. Res.* **5** **(6)** 1215
[6] Asokan S and Prasanth B 2009 *J. Non-Cryst. Sol.* **355** 164.
[7] Jain P K and Saxena N S 2009 *J. Non-Oxide and Photonic Glasses* **1(1)** 43
[8] Pauling L 1960 *The Nature of the Chemical Bond*, Ed 3rd (New York: Cornell University Press)
[9] Sharma N, Sharma S, Sarin A and Kumar R 2016 *J. Opt. Mat.* **51** 56
[10] Nascimento M L F, Souza L A, Ferreira E B and Zanotto E D 2005 *J. Non-Cryst. Sol.* **351** 3296
[11] Mousa M A I, 2011 *Phys. B* **406** 482
[12] Moynihan CT, Easteal A J, Wilder J and Tucker J 1974 J. Phys. Chem. **78** 2673
[13] Senapati U, Varshneya A K 1995 *J. Non-Cryst. Sol.* **185** 289
[14] Omar A L, Mousa M A 2011 J. Alloys and Comp. **509** 5090
[15] Adel A S, Mousa, M A I, Omar A L and Ma'rouf K A 2011 Curr. App. Phys. **11** 492
[16] Sakka S and Mackenzie J D 1971 *J. Non-Cryst. Sol.* **6** 145
[17] Turnbull D 1969 Contemp. Phys., 10, pp. 473–488.
[18] Dietzel A 1968 *Glass Science and Technology* **22** 41
[19] Mehta N 2006 *J. Sci. Ind. Res.* **65** 777
[20] Hruby A 1972 Czech J. Phys B **22** 1187
[21] Singh S P, Kumar A and Kumar S 2006 *Mat. Lett.* **60** Issues 13–14 1640–1645

IOP Conf. Series: Materials Science and Engineering **283** (2017) 012008 doi:10.1088/1757-899X/283/1/012008

[22] Schottmiller J, Tabak M, Lucovsky G and Ward A, J. *Non-Cryst. Sol.* **4** 80
[23] Okano S, Suzuki M, Imura K, Fukada N and Hiraki A 1983 J. *Non – Cryst. Sol.* **59-60** 969
[24] Jain P K and Saxena N S 2009 *J. Non-Oxide and Photonic Glasses* **1(1)** 43
[25] Mott N F and Davis E A 1979 *Electronic processes in non-crystalline materials*, 2nd ed. (Oxford: Claesndon press) p 328
[26] Elliott S R 1990 *Physics of Amorphous Materials*, 2nd ed. (London: Longman)
[27] Nidhi A V, Modgil V and Rangra V S 2013 *New J. Glass and Ceramics* **3** 91
[28] Hassanien A S and Akl A A 2016 *J. Non-Cryst. Sol.* **432** 471
[29] Fayek S A 2000 Mat. Chem. Phys. **62** 95
[30] Thangaraj R and Chander R 2012 *Thin Sol. Films* **520** 1757

Research on rolling element bearing fault diagnosis based on genetic algorithm matching pursuit

R W Rong[1] and T F Ming[2*]

[1]Department of Technology Information Hubei Police University Wu Han, 430034, China
[2]College of Power Engineering Naval University of Engineering Wu Han, 430033, China
*E-mail: hgming@126.com

Abstract. In order to solve the problem of slow computation speed, matching pursuit algorithm is applied to rolling bearing fault diagnosis, and the improvement are conducted from two aspects that are the construction of dictionary and the way to search for atoms. To be specific, Gabor function which can reflect time-frequency localization characteristic well is used to construct the dictionary, and the genetic algorithm to improve the searching speed. A time-frequency analysis method based on genetic algorithm matching pursuit (GAMP) algorithm is proposed. The way to set property parameters for the improvement of the decomposition results is studied. Simulation and experimental results illustrate that the weak fault feature of rolling bearing can be extracted effectively by this proposed method, at the same time, the computation speed increases obviously.

1. Introduction

Rolling element bearings are one of the most extensively used elements in rotary machine. They also play a key role and are easily damaged components in the whole system. It may cause serious economic loss and safety accident when failures occur in rolling bearing. In order to ensure the safety of the machine and reduce the cost of production, it is meaningful to do research on rolling bearing fault diagnosis.

The most common rolling bearing fault detecting method is vibration monitoring method. Vibration signal generated by normal rolling bearing is constructed by the low frequency components which are caused by the rotation of shaft, the changing of the stiffness and the variation of loading. Once local fault occurred, the impulse components will be excited when rolling elements pass the fault position. These components are high frequency impulse components and these impulses lead to the vibration signal non-stable. However, the fault feature is weak to incipient defects and the mixed faults have a high proportion in practical engineering [1]. The key to the diagnosis of rolling bearing is to extract out the fault feature from vibration signals effectively by property methods.

Recently, a lot of meaningful researches have been done on the analysis and processing of non-stationary signal. Lots of relative theories and methods have proposed, such as, Fourier transform [2], wavelet transform [3], Wigner-Ville distribution [4], and so on. However, all the above methods have fixed basis functions which have great effect on the expression of vibration signals [5]. In order to enhance the flexibility of the expression of vibration signals and make the decomposition process self-adaptive, the idea that over-complete redundant atom dictionary can be used to obtain signal sparse decomposition results is proposed and the MP algorithm is introduced by Mallet [6]. MP algorithm soon got widely used. MP algorithm is applied to analyzing bearing vibration signal by Liu, and the bearing faults are analyzed by time-frequency distribution [7]. Wang regards the feature signal waveform as the basis atom of dictionary, and successfully applied to the pattern recognition of bearing fault [8]. The MP algorithm is used in the complexity analysis of vibration signal by Tang, realized the feature extraction of rolling bearing [9]. However, because the dictionary is very enormous, the vast majority of computation time is spent on the search of optimum atoms, so the decomposition speed is very slow. Therefore, it is important to select property dictionary and improve the search method to increase the computation speed.

Because the process of searching for optimum atom is an optimization problem, so we can use genetic algorithm to solve the problem. In this paper, because the Gabor expression is constructed by the functions which can reflect the local time-frequency character well, the Gabor function is used to construct the dictionary [10]. Combined the genetic algorithm with MP algorithm, the decomposition speed increased obviously, a time-frequency analysis method based on GAMP is proposed. The effectiveness and practicable of this method is verified by simulation and experimental.

2. The basic principle and algorithm fundamental of Gabor dictionary

According to the signal processing theory, signal $f(t)$ can be represented as some basis functions [11]:

$$f(t) = \sum_{n=0}^{M-1} a_{\gamma n} g_{\gamma n}(t) \tag{1}$$

where M is the order of decomposition, $a_{\gamma n}$ are the Gabor coefficients, and $g_{\gamma n}(t)$ are the Gabor atoms, also called basis functions. A Gabor atom consists of a cosine-modulated Gaussian window function:

$$g_{\gamma n}(t) = \frac{1}{\sqrt{s}} g(\frac{t-u}{s}) \cos(vt + w) \tag{2}$$

where $g(t) = e^{-\pi t^2}$ is the Gaussian window function and $\gamma = (s, u, v, w)$ are the time-frequency parameters. The function $g_\gamma(t)$ is centered at u and its energy is mostly concentrated in a neighborhood of u whose size is proportional to s. The space of time-frequency parameters can be discretized as $\gamma = (a^j, pa^j\Delta u, ka^{-j}\Delta v, i\Delta w)$, with $a = 2$, $\Delta u = 1/2$, $\Delta v = \pi$, $\Delta w = \pi/6$, $0 < j \le \log_2 N$, $0 \le p < N2^{-j+1}$, $0 \le k < 2^{j+1}$, $0 \le i \le 12$ to form the so called Gabor dictionary [10]. Here N is the number of samples.

- Fundamental of MP algorithm

MP algorithm is a greedy algorithm that decomposes the signal according to atom dictionary. It spends the majority of time searching for optimum atom. The signal component which matches the optimum atom well is extracted from original signal when the optimum atom is selected. This procedure is repeated each time on the following signal component that is obtained. The procedure will not stop until the energy of the residual signal is less than given threshold value.

$\mathbf{D} = \{g_\gamma, \gamma = 1, 2, \mathrm{K}, K\}$ is the dictionary. Let $\mathbf{H} = \mathbf{R}^N$, for K ? N, be a Hilbert space. The atoms of the dictionary are the unit vector of a Hilbert space. In order to select an optimum atom from $\mathbf{D}$, the original signal $\mathbf{f}$ is projected on a vector $\mathbf{g}_{\gamma 0} \in \mathbf{D}$. It must satisfy the following condition:

$$\left|\langle \mathbf{f}, \mathbf{g}_{\gamma 0} \rangle\right| = \sup\left|\langle \mathbf{f}, \mathbf{g}_\gamma \rangle\right| \tag{3}$$

The original signal $\mathbf{f}$ is decomposed into

$$\mathbf{f} = \left|\left\langle \mathbf{f}, \mathbf{g}_{\gamma 0} \right\rangle\right| \mathbf{g}_{\gamma 0} + \mathbf{R}^1\mathbf{f} \tag{4}$$

where $\mathbf{R}^1\mathbf{f}$ is the residual vector after approximating $\mathbf{f}$ in the direction of $\mathbf{g}_{\gamma 0}$. Clearly $\mathbf{g}_{\gamma 0}$ is orthogonal to $\mathbf{R}^1\mathbf{f}$, hence

$$\left\|\mathbf{f}^2\right\| = \left|\left\langle \mathbf{g}_{\gamma 0}, \mathbf{f} \right\rangle\right|^2 + \left\|\mathbf{R}^1\mathbf{f}\right\|^2 \tag{5}$$

To minimize $\left\|\mathbf{Rf}\right\|$, we must choose $\mathbf{g}_{\gamma 0} \in \mathbf{H}$ so that $\left|\left\langle \mathbf{f}, \mathbf{g}_{\gamma 0} \right\rangle\right|$ is maximum. In infinite dimensional space, it is only possible to find a vector $\mathbf{g}_{\gamma 0}$ that is almost the best in the sense that

$$\left|\left\langle \mathbf{f}, \mathbf{g}_{\gamma 0} \right\rangle\right| \geq \alpha \sup \left|\left\langle \mathbf{f}, \mathbf{g}_{\gamma} \right\rangle\right| \tag{6}$$

where α is an optimality factor that satisfies $0 < \alpha \leq 1$. In finite dimensional space, $\left\langle \mathbf{f}, \mathbf{g}_{\gamma} \right\rangle$ has maximum. In this case, we usually choose $\alpha = 1$.

MP algorithm is an iterative algorithm that decomposition the $\mathbf{Rf}$ by projecting it on a vector of $\mathbf{D}$ that matches $\mathbf{Rf}$ almost at best, as it was done for $\mathbf{f}$. After compute $n+1$th iteration, the result is obtained as follow:

$$\mathbf{R}^n\mathbf{f} = \left|\left\langle \mathbf{R}^n\mathbf{f}, \mathbf{g}_{\gamma n} \right\rangle\right| + R^{n+1}\mathbf{f} \tag{7}$$

Because $R^{n+1}\mathbf{f}$ is orthogonal to $\mathbf{g}_{\gamma n}$, hence

$$\left|\left\langle \mathbf{R}^n\mathbf{f}, \mathbf{g}_{\gamma n} \right\rangle\right| = \sup \left|\left\langle \mathbf{R}^n\mathbf{f}, \mathbf{g}_{\gamma} \right\rangle\right| \tag{8}$$

Let us carry this decomposition up to the order M . $\mathbf{f}$ is decomposed into the concatenated sum

$$\mathbf{f} = \sum_{n=0}^{M-1} \left\langle \mathbf{R}^n\mathbf{f}, \mathbf{g}_{\gamma n} \right\rangle \mathbf{g}_{\gamma n} + \mathbf{R}^M\mathbf{f} \tag{9}$$

Similarly, energy $\left\|\mathbf{f}\right\|^2$ is decomposed into a concatenated sum

$$\left\|\mathbf{f}\right\|^2 = \sum_{n=0}^{M-1} \left|\left\langle \mathbf{R}^n\mathbf{f}, \mathbf{g}_{\gamma n} \right\rangle\right|^2 + \left\|\mathbf{R}^M\mathbf{f}\right\|^2 \tag{10}$$

The M th estimate of vector $\mathbf{Rf}$ is obtained by the above equation, and the error *is* $\mathbf{R}^M\mathbf{f}$.

- Time-frequency analysis method based on GAMP algorithm

The genetic algorithm is an iterated self-adaptive probability search method which is based on natural selection and natural genetic principle. It achieves the optimal goal according to Darwin's theory of evolution and Mendel's theory of gene. The process of searching for optimum atom when using MP algorithm is an optimized problem. Therefore, we can combine genetic algorithm with MP algorithm.

The steps of the time-frequency analysis method based on GAMP are similar to MP algorithm. The absolute value of the inner product of atom and signal is regarded as the fitness function of genetic algorithm, and the parameters of the atom is regarded as the chromosome. Thereby, the optimum atom is found out from dictionary by genetic algorithm. The process is shown in figure 1.

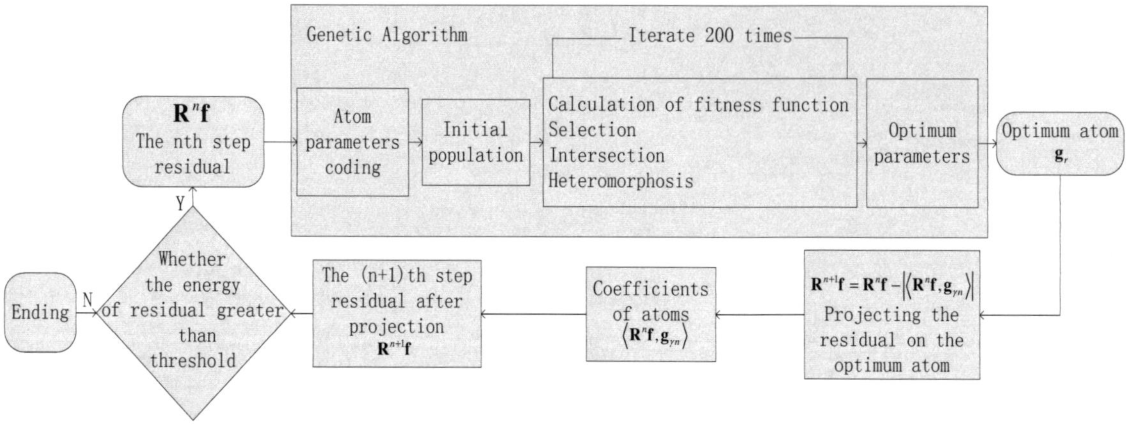

Figure 1. The process of the time-frequency analysis method based on GAMP.

Genetic algorithm possesses favorable whole situation optimum performance. The translation parameter, scale parameter and modulating parameter are the parameters waiting for optimization. The fitness function is established by the inner product of atom and signal. Therefore, the model of the time-frequency analysis method based on GAMP is established.

- Time-frequency energy distribution

Wigner distribution possesses well time-frequency resolution. However, it exists cross terms when applied to complicated signal. It is hard to reflect the time-frequency character well.

$$Wf(t,\omega) = \sum_{n=0}^{+\infty} \left|\left\langle \mathbf{R}^n\mathbf{f}, \mathbf{g}_{\gamma n}\right\rangle\right|^2 W_{g_{\gamma n}}(t,\omega) + \sum_{n=0}^{+\infty}\sum_{m=0,\, m\neq n}^{+\infty} \left\langle \mathbf{R}^n\mathbf{f}, \mathbf{g}_{\gamma n}\right\rangle \mathrm{g}\overline{\left\langle \mathbf{R}^m\mathbf{f}, \mathbf{g}_{\gamma m}\right\rangle} W[\mathbf{g}_{\gamma n}, \mathbf{g}_{\gamma m}](t,\omega) \tag{11}$$

The second term in equation (11) is the cross term. We can separately get the Wigner distribution of the optimum atoms, then we can overlap the result. In this way, the effect of cross terms is eliminated, and the energy distribution with well time-frequency resolution is obtained.

3. Simulation studies

In order to validate the correctness and validity of the rolling bearing fault signal extraction method based on GAMP algorithm, according to the fact that inner race prone to failure, the simulation signal of rolling bearing surface fault was studied and analyzed [12].

3.1. Simulated signal

The vibration signal model [13] for rolling bearing inner ring single damage is shown in equation (12).

$$\begin{cases} x(t) = \sum_{i=1}^{N} A_i s(t - iT - \tau_i) + B(t) + n(t) \\ A_i = A_0 \cos(2\pi f_r t + \varphi_A) + C_A \\ B(t) = B_0 \cos(2\pi f_m t + \varphi_B) \\ s(t) = e^{-Rt} \sin(2\pi f_n t + \varphi_w) \end{cases} \tag{12}$$

Where A_i is the amplitude modulation. f_r is rotation frequency of the shaft. $B(t)$ is the background harmonic component. f_m is the frequency of background signal. $s(t)$ is exponential decay pulse. T is the interval of two adjacent shock. τ_i is cycle delay of the ith impulse due to slip. $n(t)$ is white noise. A_0, B_0, is constant, $C_A > A_0$. R is the attenuation coefficient determined by the system. f_n is the natural frequency of the system.

3.2. Verification and analysis

The sampling frequency is 16384 Hz, f_r is 52 Hz, fault frequency is 180 Hz, and the duration of the signal is 62.5 ms. In early stage of bearing failures, the feature information is usually buried by strong background noise coming from gear meshing vibration, mass unbalance, misalignment, etc. The hybrid signal waveform obtained from equation (12) is shown in figure 2(a), the time-domain waveform contains significant harmonic components, with the influence of noise, the fault impact cannot be seen clearly. The time-frequency distribution of the hybrid signal is shown in figure 2(b), the energy distribution of each frequency component can't be distinguished obviously.

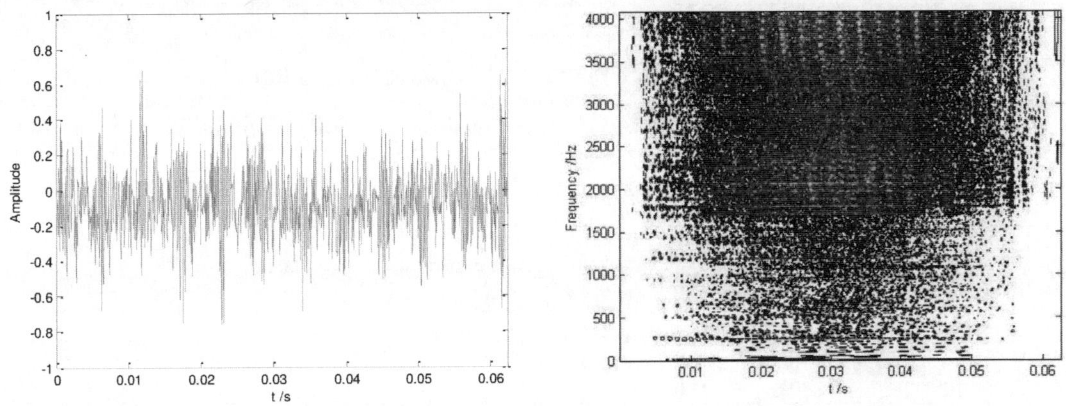

Figure 2. Simulation signal: (a) time-domain waveform, (b) Wigner distribution.

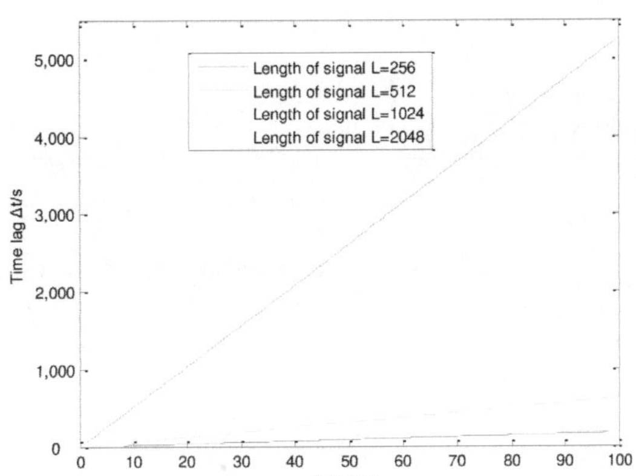

Figure 3. Comparison of the decomposition speed.

The GAMP algorithm and MP algorithm are applied to vibration signal separately. We find that both of them possess similar reconstruction property, but there is a big difference in decomposition speed. The comparison of two methods in decomposition speed is shown in figure 3. The value of vertical axis means the time lag of two methods. It is obvious that the calculating speed of GAMP is many times quicker than MP method, and the advantage of GAMP method reflects more and more obviously when it regards to the longer time signal.

The GAMP algorithm is applied to simulation signal. Setting the parameters according to table 1, then the time-frequency distribution by decomposing 60 times is obtained as shown in figure 4(a). There exists some small-scale atoms, which are regularly spaced in frequency above 3 KHz, and the time interval between the atoms equals to the inner fault period of rolling bearing (about 5.6 ms). These are the impact components of simulation signal. In the extracted signal, the fault impact and amplitude modulation can be seen obviously. According to the above results, the inner fault can be judged easily. Because the resonance components of vibration signal usually appear in high frequency with short time period, we can improve the ability to detect the fault feature by the small scale atoms of high frequency components. According to the fact that the frequency of small scale atoms is greater than 3 KHz, we can set the scale parameter $j = 4$ to utilize the high frequency atoms. The extracted signal is shown in figure 4(b). In the extracted signal, the fault impact related to the bearing fault can be seen clearly, and the time interval between the adjoin impact equals to inner fault period of bearing (about 5.6 ms). At the same time, the amplitude modulation can be seen obviously. According to the above analysis, the rolling bearing fault feature can be extracted effectively by GAMP method when the parameters are set properly.

Table 1. The related parameters of genetic algorithm.

Population size	Maximum generation	Hybrid method	Hybrid probability	Mutation probability
80	200	Single hybridization	0.9	0.1

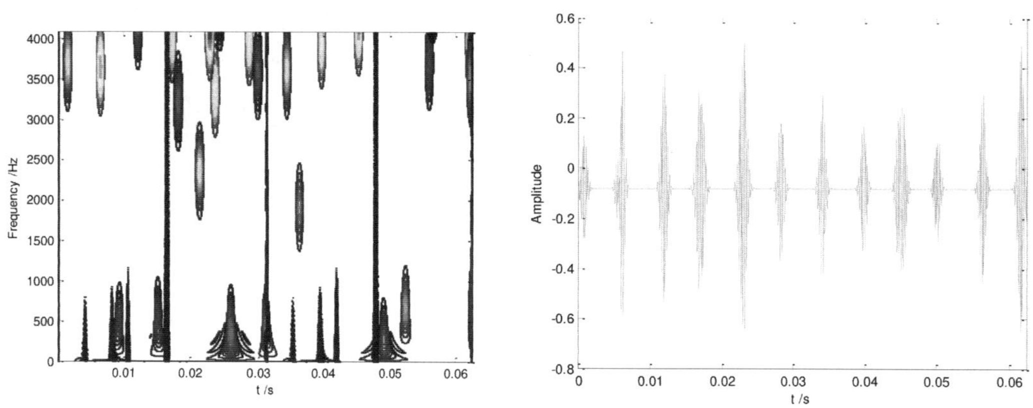

(a) (b)

Figure 4. Extracted signal: (a) time-frequency distribution, (b) time-domain waveform.

4. Experimental studies

4.1. Experimental apparatus

The rolling bearing vibration signals are measured from rolling bearing fault simulation platform in laboratory. The type of rolling bearing is 6010. The structure parameters are as follows: the bearing pitch diameter is D=65 mm, the rolling element diameter is d=9 mm, the number of rolling element is Z=13, the contact angle is $\alpha = 0^o$. The rolling bearing faults include inner fault, outer fault, rolling element fault and cage fault. However, the inner fault is the most hard to be detected, and we analyze the vibration signal with inner fault.

The rev of the shaft is 2947 r/min. The sampling frequency is 65536 Hz. According to the geometric parameters of bearing and the relationship between character frequency and rotation frequency, we can obtain the inner fault frequency, which is 363 Hz.

4.2. Verification and analysis

The time-domain waveform of rolling bearing vibration signal is shown in figure 5(a). Due to the effect of attenuation and machine's vibration, it is hard to resolve the fault period impact. The time-frequency distribution is shown in figure 5(b). We cannot resolve the energy distribution at each frequency component of measured signal.

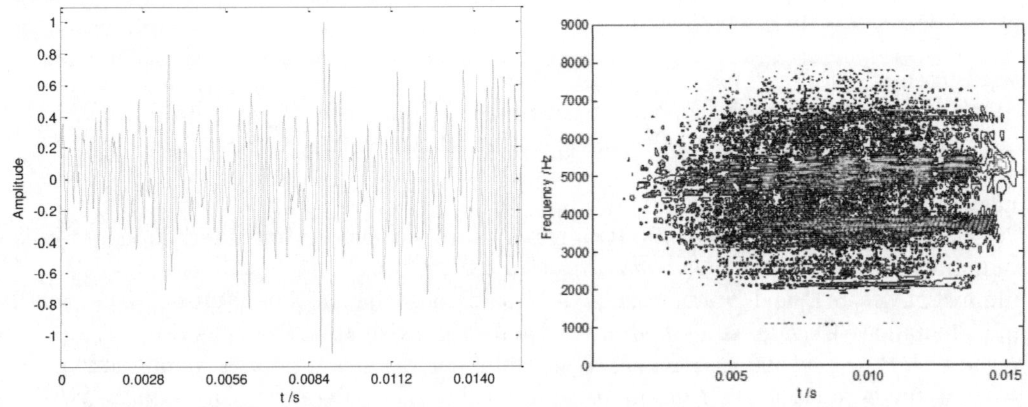

Figure 5. Measured signal: (a) time-domain waveform, (b) Wigner distribution.

The GAMP algorithm is applied to the measured signal. The time-frequency distribution by decompose 60 times is obtained as shown in figure 6(a). There are some small scale atoms with regularly and same spaced in time between 4 KHz and 7 KHz, and the time interval between the atoms equals to the inner fault period of rolling bearing (about 5.6 ms). These are the impact components of measured signal. In the extracted signal, the fault impact and amplitude modulation can be seen obviously. According to the above results, the inner fault can be judged easily. We can set the scale parameter $j = 5$ to utilize the high frequency atoms. The extracted signal is shown in figure 6(b). In the extracted signal, the fault impact related to the bearing fault can be seen clearly, and the time interval between the adjoin impact equals to inner fault period of bearing (about 2.75 ms). At the same time, the amplitude modulation can be seen obviously. According to the above analysis, the rolling bearing fault feature can be extracted effectively by GAMP method when the parameters are set properly.

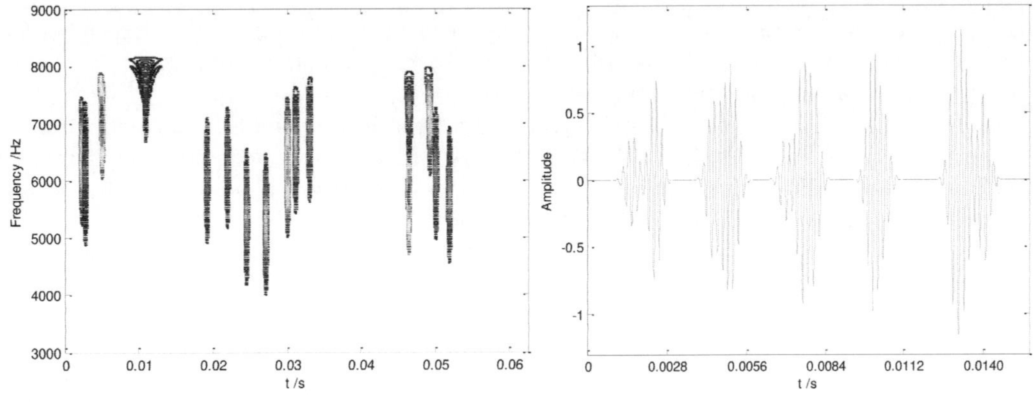

Figure 6. Reconstructed signal: (a) time-frequency distribution, (b) time-domain waveform.

5. Conclusions

In this paper, based on the structure principle of Gabor dictionary and the principle of MP algorithm, searching the optimum atoms from redundancy dictionary is the reason why the decomposition speed is slowly pointed out, at the same time, the searching process which is an optimization process is illustrated. Then the principle why GAMP algorithm can improve the calculate speed is analyzed. According to time-frequency energy distribution, setting the decomposition parameters, the fault feature can be extracted out effectively. Simulation and experimental results illustrate that the weak fault signal of rolling bearing can be extracted effectively by this proposed method, at the same time, the computation speed increases obviously.

Acknowledgments

The author would like to thank the support of government (Funding ID 9140A27020413JB11076) and National Natural Science Foundation of China (Funding ID 50979109).

References

[1] Cui L L, Wang J and Lee S 2014 Matching pursuit of an adaptive impulse dictionary for bearing fault diagnosis *Journal of Sound and Vibration* **333** 2840-62
[2] Klein R, Ingman D and Braun S 2001 Non-stationary signals: Phase-energy approach-theory and simulations *Mechanical Systems and Signal Processing* **15(6)** 1061-89
[3] Wang S S, Huang W and Zhu Z 2011 Transient modeling and parameter identification based on wavelet and correlation filtering for rotating machine fault diagnosis *Mechanical Systems and Signal Processing* **25(4)** 1299-320
[4] He Q B and Wang X X 2013 Time-frequency manifold correlation matching for periodic fault identification in rotating machines *Journal of Sound and Vibration* **332** 2611-26
[5] Kang C H, Cui L L, Wang J, *et al* 2012 Research on the composite dictionary multi-atoms matching algorithm based on the signal character *Journal of Mechanical Engineering* **48(12)** 1-6
[6] Mallat S and Zhang Z 1993 Matching pursuit with time-frequency dictionaries *IEEE Trans on Signal Processing* **41(12)** 3397-415
[7] Liu B, Ling S F and Gribonval R 2002 Bearing failure detection using matching pursuit *NDT & E International* **35** 255-62
[8] Wang G D, Li M, Yang J H, *et al* 2010 Fault pattern recognition of rolling bearings based on characteristic waveform sparse matching *Journal of University of Science and Technology Beijing* **32(3)** 390-6
[9] Tang H F, Chen J and Dong G M 2010 Research on application of matching pursuit based on complexity analysis method for fault diagnosis of rolling element bearings *Journal of Vibration Engineering* **23(5)** 541-5
[10] Lobo A and Loizou P 2003 Voiced/unvoiced speech discrimination in noise using Gabor atomic decomposition//Proceedings of ICASSP'03 *IEEE International Conference on Acoustics, Speech, and Signal Processing* (Hongkong) 820-3
[11] Qian S and Chen D 1996 *Joint Time-Frequency Analysis* (Englewood Cliffs: Prentice Hall)
[12] Jiang R L, Chen J, Dong G M, *et al* 2012 The weak fault diagnosis and condition monitoring of rolling element bearing using minimum entropy deconvolution and envelop spectrum *Journal of Mechanical Engineering Science* **227(5)** 1116-29
[13] Ming Y, Chen J and Dong G M 2011 Weak fault extraction of rolling bearing based on cyclic Wiener filter and envelop spectrum *Mechanical Systems and Signal Processing* **25** 1773-85

IOP Conf. Series: Materials Science and Engineering **283** (2017) 012010 doi:10.1088/1757-899X/283/1/012010

Mechanical responses of filled thermoplastic elastomers

M Wang[1*], D C Shan[2] and Y Liao[2]

[1]School of Nuclear Science and Technology, University of South China, Hengyang, Hunan 421001, China
[2]School of Chemistry and Chemical Engineering, University of South China, Hengyang, Hunan 421001, China
*E-mail: wmeng020321@163.com

Abstract. In this paper, several mechanical responses of thermoplastic polyurethane (TPU) filled with nano-scale attapulgite (AT) particles, including cyclic loading-unloading behavior, dynamic mechanical behavior and stress relaxation have been investigated. With the addition of AT, it is noticed that the strength, modulus and stress relaxation time of TPU increased significantly compared to that of unfilled materials. It is also observed that, whether filled or unfilled TPU, pronounced inelastic mechanical features such as Mullins effect, residual strain and hysteresis, occurred mainly during the first load loop, but became more notable with AT increase.

1. Introduction

TPU is one of the most thermoplastic block polymers composed of soft and hard segments. Because of its good elasticity and easy of melt-processing, TPU can be fabricated by diverse method and displays many excellent properties, such as high elasticity, wear resistance, good resistance to oils, and good weather stability [1-3]. However, relative to chemical crosslinked thermoset polymer materials, TPU does not show high mechanical properties and desired heat resistance. Therefore, in recent years, many efforts have been devoted to improve the polyurethane matrix and reduce costs by introducing various fillers, including inorganic clay, metallic oxide and carbon nanoparticles [4-6]. Attapulgite (AT), a natural fibrillar clay with many hydroxyl groups on the surfaces, it is known to be very effective for material enhancement [7,8]. In this paper, we firstly prepared the TPU/AT composites by solution blending method, and then discussed the effect of AT on mechanical properties. In particular, the inelastic mechanical behaviors of materials were mainly studied and analyzed because of its importance in practical application [9].

2. Materials and methods

2.1. Materials

AT and TPU were purchased from Jiangshu XuYi MinChem Co., Ltd. and BASF chemical company, respectively. TPU/AT composites were prepared by solution blending and the detailed process has been reported in our earlier paper [10]. The elastic composites with AT contents of 0, 1, 2, and 5% were marked as TPU/AT0, TPU/AT1, TPU/AT2 and TPU/AT5.

2.2. Experimental methods

SEM tests were conducted on a JEOL JSM-5600LV instrument. The surface of samples was all coated with a gold layer for the observation of cross-section images.

FTIR were conducted on a Nicolet 8700 spectrometer. The spectra for all samples were obtained at a resolution of 4 cm^{-1}, and the resulting spectra were the average of 32 scans to meet an adequate signal-to-noise.

Cyclic mechanical tests were made on thin-sheet samples approximately 50×2×0.2 mm^3 on an elastic testing machine with a 500 mm/min rate. The initial tensile modulus (E) was determined by first loading slope $d\sigma / d\varepsilon$ at $\varepsilon = 0.1$, and residual strain (ε_r) was irreversible strain after experiencing first cyclic deformation .

DMA measurements were performed from 30-90℃ at frequency of 10 Hz and at heating rate of 2° C/min using a TA Q800 instruments.

Stress relaxation behaviors were tested when the elastomers were stretched to 300% strain with a 1000 mm/min rate.

3. Results and discussion

SEM photographs of the TPU/AT0 and TPU/AT1 composite are showed in figure 1. It is found that the fractured section of TPU/AT1 sample shows many AT nano-fibrils with irregular distribution, while that of pure TPU is smooth. Furthermore, there is not a clear interface between AT and the TPU matrix and only a few AT marked in red circle are pulled out from matrix, which implies that AT is uniformly dispersed and randomly oriented in TPU.

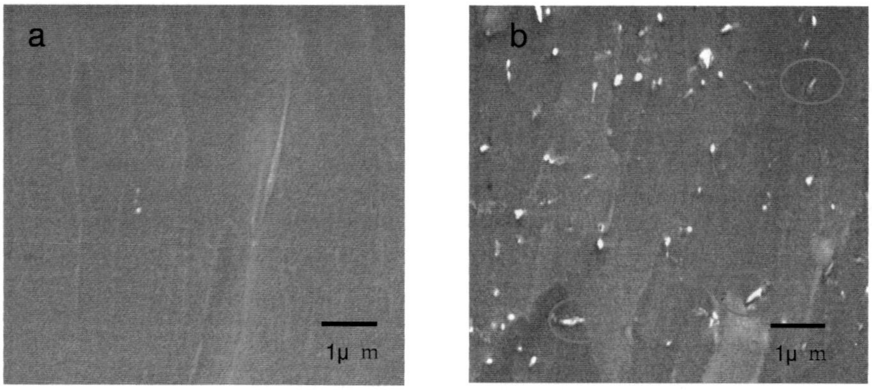

Figure 1. SEM photographs of elastomers. (a) TPU/AT0; (b) TPU/AT1.

Figure 2 presents the load-strain curve of TPU composites at 500 mm/min cyclic tensile rate. It is clear that filled TPUs have discernibly higher strength and initial tensile modulus than those of pure TPU in first loading process. When subjected to cyclic loading (cycle number, N=1-3), all materials also exhibit a significant stress softening behavior (e.g. Mullins effect) [11]. There is a distinct stress reduction after the first load loop, while the stress achieves a stable state in the subsequent cycles. At the same time, there is also pronounced hysteresis in the first cycle, which exhibits a large hysteresis loop with a distinct residual strain. In the following cycles, the loading and unloading paths are both less independent of cycle numbers and the corresponding hysteresis loops become quite small than that of the first loading-unloading cycle.

The stress relaxation of TPU/AT composites can also reflect the inelastic mechanical behaviors caused by large viscous dissipation. Figure 3 shows the curves of tensile stress relaxation, where the relative relaxation value, σ_R is plotted against time, t. σ_R is expressed by

$$\sigma_R = \sigma_t / \sigma_0 \times 100\%$$

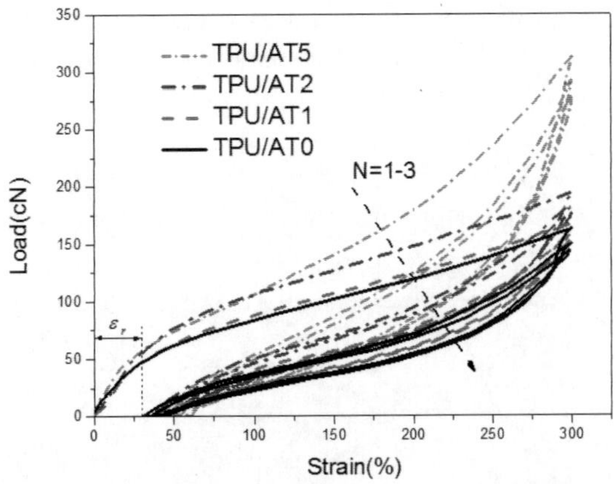

Figure 2. The load-strain curve of TPU/AT composites during thrice cyclic deformation.

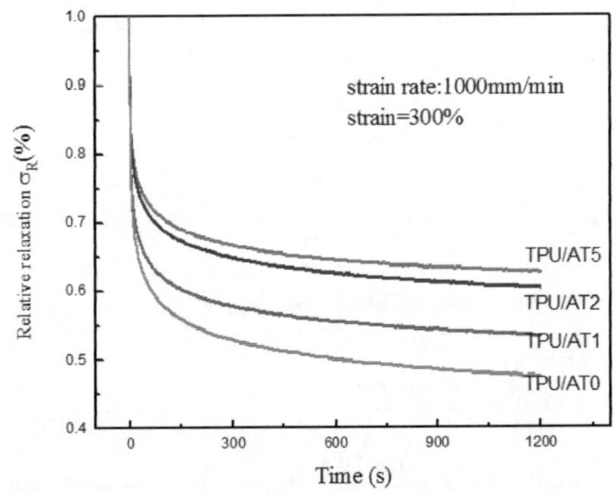

Figure 3. Impact of AT concentration on stress relaxation of TPU/AT composites.

where σ_0 is instantaneous stress of the initial holding period, σ_t is stress with time during the holding period. It can be seen that there are two phenomenon happening with increasing AT content, (1) the overall relative strain is higher (strain=300%) at any time during loading, and (2) the stress decreases gradually during the strain hold period and shows a wide relaxation time spectrum. However, the relaxation behavior is strongly dependent on AT content, decreasing with the increase of AT content, which means that there may be a larger viscous friction between AT and TPU matrix.

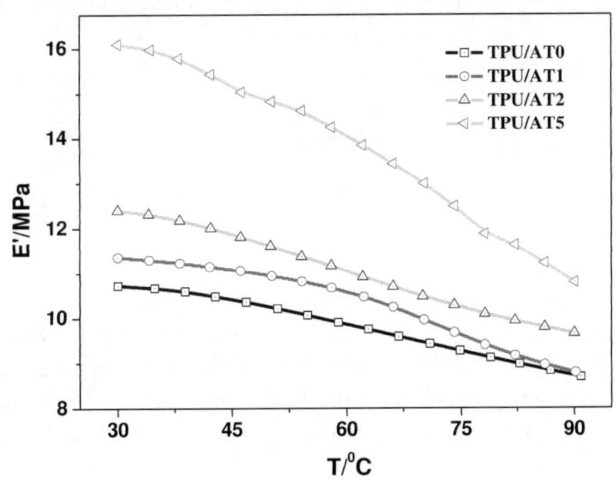

Figure 4. Impact of AT concentration on the storage modulus of TPU/AT composites.

Figure 4 shows that the dynamic storage modulus (E') of TPU/AT composites has a significant temperature dependence. In the range of measuring temperature (30-90℃), the magnitude of E' increases consistently with the content of AT in the composites, which can reflect the physical interaction formed between AT and TPU matrix. However, the dynamic storage modulus is getting smaller and smaller as the temperature increases and becomes more sensitive to temperature with the increase of AT content. It is clear that the physical bond between AT and TPU can be destroyed at high temperature.

The FTIR spectra of the TPU/AT composites are displayed in figure 5(a). Distinct bands related to the amine (–NH) and carboxyl (C=O) groups were observed at 3300 cm^{-1} and 1725 cm^{-1}, respectively. Figure 5(b) exhibits the carboxyl spectra of the TPU/AT composites that are expanded from figure 5(a).

In this group curves, it is noticed that the carboxyl peaks move distinctly to lower wave number than that of the unfilled polymer, and it is a clear and convincing evidence of the H-bond forming between the –OH of AT particle and C=O of TPU matrix [12], and the corresponding schematic diagram of H-bond is shown in figure 6.

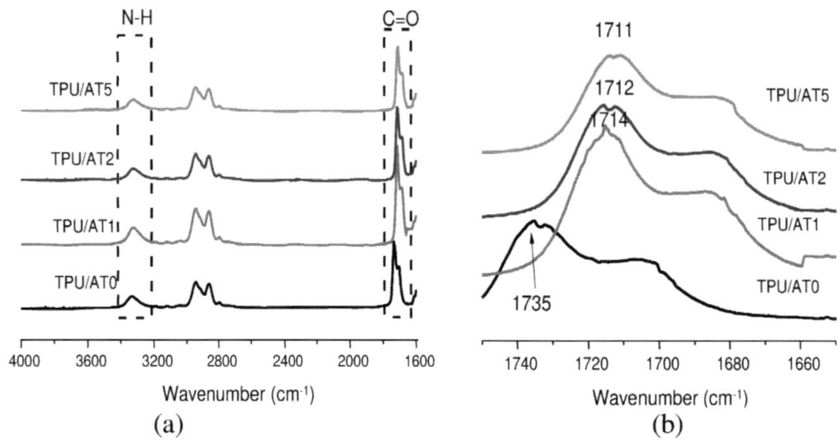

Figure 5. (a) Infrared spectrum of the TPU/AT composites; (b) C=O stretching region of the composites.

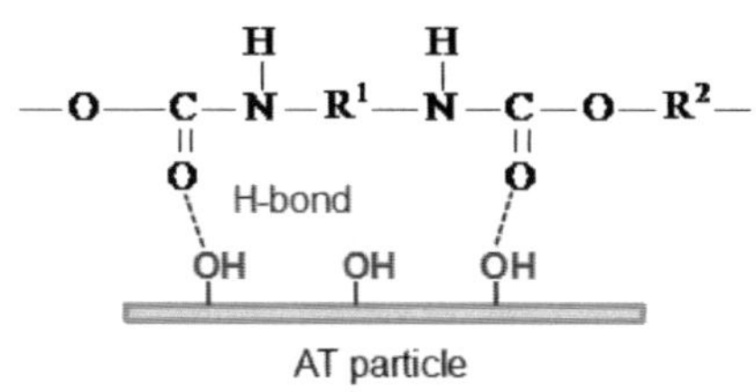

Figure 6. Hydrogen bond between TPU matrix and AT.

4. Conclusions

The strength, modulus, and relaxation time of thermoplastic polyurethane elastomers filled with AT nano-fibrils are higher than those of unfilled elastomer. In addition, for both filled and unfilled elastic materials, pronounced inelastic mechanical behaviors are also observed during the loading-unloading cycles. These results strongly suggest that AT enhances the viscous friction of the inter-molecular force, because of physical cross-link forming between AT and the TPU matrix by the hydrogen bond.

Acknowledgments
This work was supported by the National Natural Foundation of China (51403099) and National Key R & D Program of China (SQ2017ZY060105, 2017YFD0200705).

References

[1] Mattia J and Painter P 2007 *Macromolecules* **40** 1546-54
[2] Avanzini A and Gallina D 2011 *J. Eng. Mater. Technol.* **133** 021005-1-021005-9
[3] Jia R P, Wang C F, Zheng K S, He X Y and Huang M S 2015 *J. Reinf. Plast. Comp.* **34** 1090-8
[4] Liff S M, Kumar N and Mckinley G H 2007 *Nat. Mater.* **6** 76-83
[5] Dai X H, Xu J, Guo X L, Lu Y L, Shen D Y, Zhao N, Luo X D and Zhang X L 2004 *Macromolecules* **37** 5615-23
[6] Wang M, Duan X D, Xu Y X and Duan X F 2016 *ACS Nano* **10** 7231-47
[7] Pan H X and Chen D J 2007 *Eur. Polym. J.* **43** 3766-72

[8] Zilg C, Thomann R, Mulhaupt R and Finter J 1999 *Adv. Mater* **11** 49-52
[9] Yi J, Boyce M C, Lee G F and Balizer E 2006 *Polymer* **47** 319-29
[10] Wang M, Shan D C, Liao Y and Xia L S 2017 *Polym. Bull.* doi: 10.1007/s00289-017-2051-x
[11] Merckel Y, Diani J, Brieu M and Berghezan D 2011 *Mat. Sci. Eng A* **528** 8651-9
[12] Buckley C P, Prisacariu C and Martin C 2010 *Polymer* **51** 3213-24

IOP Conf. Series: Materials Science and Engineering **283** (2017) 012011 doi:10.1088/1757-899X/283/1/012011

Research on stratified evolution of composite materials under four-point bending loading

M J Hao[a,1,*], Q J You [b,2], J C Zheng [b,2], Z Yue [a,2] and Z P Xie [a,2]

[a] Academy of Safety and Environmental Engineering, Capital University of Economics and Business, Beijing, China

[b] Urban Systems Engineering Research Center, Beijing, China

*E-mail: 18810598300@163.com

Abstract: In order to explore the effect of stratified evolution and delamination on the load capacity and service life of the composite materials under the four-point bending loading, the artificial tectonic defects of the different positions were set up. The four-point bending test was carried out, and the whole process was recorded by acoustic emission, and the damage degree of the composite layer was judged by the impact accumulation of the specimen - time-amplitude history chart, load-time-relative energy history chart, acoustic emission impact signal positioning map. The results show that the stratified defects near the surface of the specimen accelerate the process of material failure and expansion. The location of the delamination defects changes the bending performance of the composites to a great extent. The closer the stratification defects are to the surface of the specimen, the greater the damage, the worse the service capacity of the specimen.

1. Introduction

Glass fiber reinforced composite materials have excellent plasticity, fatigue resistance, and durability. Therefore, they are widely used in aerospace, shipbuilding, military equipment, sports equipment, water conservancy, infrastructure, and many other areas [1-5]. However, during their operation, sub-layer damage may enhance the composite material stratification, which issue becomes more and more topical.

Some scholars and experts have made a study of composite materials. Huget et al [6] used the acoustic emission (AE) method to monitor the AE signal of polyester / GF composites during static tensile loading. By analyzing the data on duration, the two types of failure modes of fracture and debonding were found and the neural network was used to identify the damage. Alander et al [7] performed a three-point load loading test on composite specimens and found that the damage of the specimen occurred at 19% -32% of the failure load. Wu [8] used the Weibull function to describe the composite layer by the material damage mode and found that the tensile strength of the material with the number of layers is gradually decreased. The tensile strength values of the two-, three- and four-layer fiber reinforced composites were lower that of single-layer ones by 10, 20, and 25%, respectively. However, there is scarce research on the stratified damage evolution behavior of glass fiber reinforced composites, which is critical to the bearing capacity and service life of composite materials.

The acoustic emission can monitor and record the development and evolution of damage in real time, and can provide real-time or continuous information with defects such as load, time, temperature and other external variables to detect the activity of these defects under external structural stress [9-14].

In this paper, four-point bending loading was realized via a universal testing machine, and composite specimens with different tiered defects were loaded and tested. The whole process was monitored by the AE method, and the damage behavior and degree of composite layered defects was assessed by analyzing the amplitude, impact accumulation, and relative energy. It is an important basis for the safety and efficient application of composite materials and the reliability of the composite materials [15,16].

2. Four-point bending test of specimens with different internal defects

2.1 Experimental equipment and materials

Full-wave acoustic transmitter, computer control electronic universal tensile testing machine, vacuum oven, glass template, vacuum bag plastic film, release cloth, diversion network, glass fiber unidirectional cloth(ECW600-1270， 600g/m^2), teflon film, epoxy resin, epoxy resin curing agent, sealing tape, resin tube.

2.2 Preparation of test pieces

The glass fiber unidirectional cloth was cut into a rectangular shape of 280 × 200mm, tiled on a flat die, and ten layers were laid in the same direction. The polytetrafluoroethylene film with width of 25 mm was placed in the middle of the different layers to obtain different depth of stratification defects. In the experiment, polytetrafluoroethylene film was placed on the first, second, fifth and sixth layers, respectively.

After the laying of the epoxy resin, the epoxy resin was vacuum-filled and allowed to stand at room temperature for 48 hours. Then, the temperature of the vacuum oven was set at 80℃, and the cured material was heated for 12 hours, and then cooled to the temperature of the room, and the laminates with thickness of about 3.8 mm were obtained. Finally, 160 x 25 mm composite specimens were prepared. The schematic of specimens is shown in figure 1.

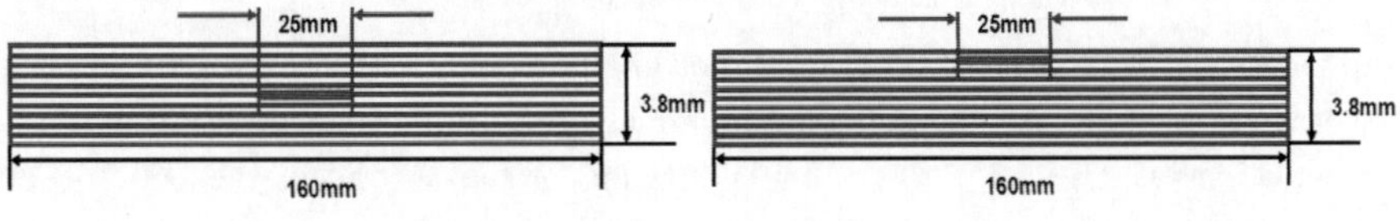

Figure 1. Sketch map of specimen

2.3 Mechanical loading and acoustic emission monitoring

In the experiment, high vacuum grease was used to couple the two VS150-RIC sensors to the specimen, and the tape was tight on both sides of the specimen 25mm. The position of each sensor was set in the same way as the artificial, the centerline of the tiered defect is 60 mm apart, the bands are 100 kHz to 450 kHz, the amplifier gain is 34 dB, the center frequency is 150 kHz, the sample frequency is 5 MHz, and the threshold is set to 40dB. To ensure the reliability of the experimental data, the number of valid specimens for each type is at least five. The loading of the sensor, specimen and fixture is shown in figure 2.

IOP Conf. Series: Materials Science and Engineering **283** (2017) 012011 doi:10.1088/1757-899X/283/1/012011

Figure 2. Schematic diagram of specimen loading

3. Results and discussion

3.1 Mechanical properties and failure characteristics of composite materials with artificial prefabricated layers

The load-deflection curve of the specimen used in the experiment is shown in figure 3. The delamination defect of specimen a lies in the first two layers, and the stratification defects of specimen *b* are located between the fifth and sixth layers. When the loading is over, the load of specimen *a* is 1.29 kN and the load of specimen *b* is 1.38 kN.

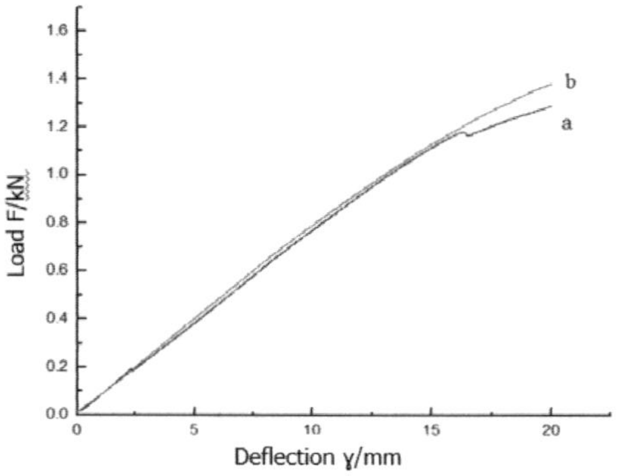

Figure 3 Load deflection curves of composite specimens

From curve *a* , we can see that the whole loading process can be divided into two stages, the first stage of the load and the deflection curve as a whole linear relationship, but with a slight fluctuation, which is due to the initial loading of the specimen, the bottom of the specimen occurred slightly fiber broken and so on. In the second stage, the load and the deflection show obvious nonlinear relationship, which shows that there is obvious damage occurs, with the increase in deflection, damage continues to accumulate, when the load reaches a certain degree, the specimen a prefabricated crack cracking and fiber/matrix interface layered expansion, the sudden expansion of the crack led to a rapid decline in load, the curve is still stable and increased. Until the application of the load process cut off, respectively, to the left extension of 1 cm, extended to the right by 0.8 cm. This indicates that the specimen stiffness and strength decreased significantly.

It can be seen from curve *b,* that the relationship between load and deflection is a good linear characteristic, indicating that when the stratification defect is located between the fifth and sixth layers, the mechanical properties of the specimen are not significantly affected, and the bending resistance of the specimen is good.

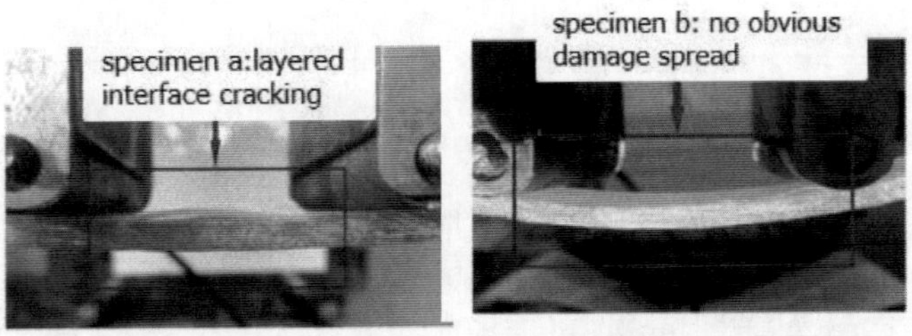

Figure 4. Specimen failure site map

Figure 4 is for the stop loading, the two types of specimen site map, we can see that specimen *a* experienced a layer interface cracking, while specimen *b* had no obvious damage. Figure 5 shows the failure characteristics of the specimen, the black box that the manual placement of layered defects, arrows indicate the direction of damage to the expansion of the direction, we can see that in the load phase, the specimen a has significant damage expansion, and concentrated in the specimen near the surface area, the main form of damage is prefabricated layered interface cracking, interlayer cracking and fiber fracture, and the specimen *b* did not experience any significant damage.

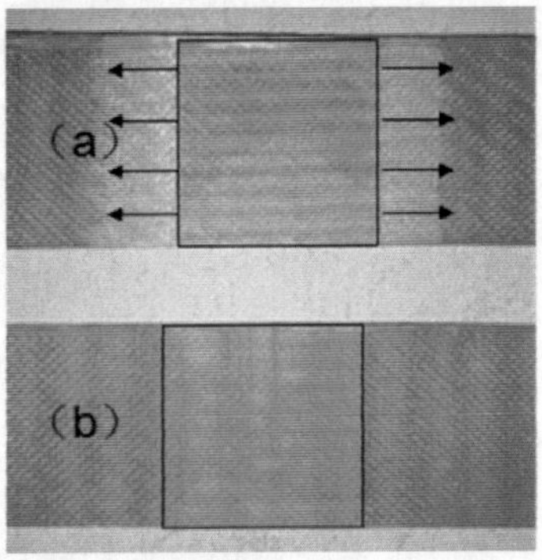

Figure 5. Specimen failure characteristics

By comparing the above two specimens, it can be found that the closer the stratification defect is to the surface of the specimen, the worse is the interface carrying capacity and the easier is the specimen damaged. The delamination defect is located at the center of the specimen interface, which has little effect on the load capacity of the specimen interface. This indicates that the stratum is in the part of the specimen, which is closely related to the load capacity of the specimen. The closer the stratification defect is to the surface of the specimen, the lower the load capacity of the specimen.

3.2 Acoustic emission response of composites with artificial stratification

Figure 6 is the impact accumulation-time-amplitude plot of the test piece in the test of the acoustic emission during the four-point bending load, impact accumulation and amplitude change reflect the intensity and frequency of acoustic emission, so as to study the damage behavior of composite specimens.

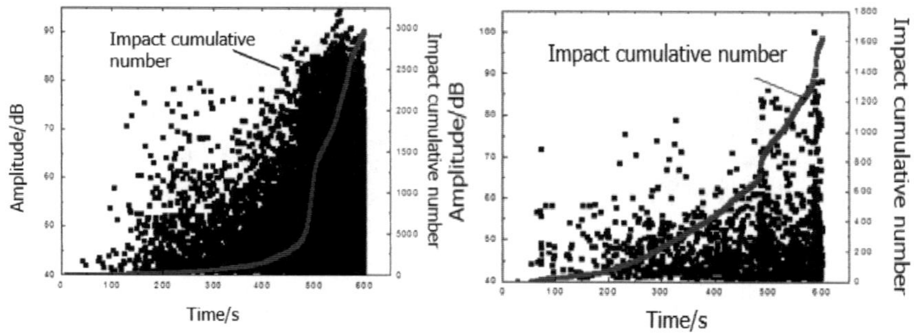

Figure 6. Impact accumulation/time/amplitude history diagram for specimen *a* (a) and specimen *b* (b)

It is seen in figure 6(a) that specimen *a* underwent three processes during the test: the initial stage, the surface rupture stage and the stratified evolution damage stage. At the beginning of the loading of the specimen, the load on the specimen is small, the deformation is very subtle, the signal collision cumulative curve is very gentle, only a few low frequency signals appear; with the increase of load, the specimen has a certain degree of deformation, structural damage, the signal increases, the amplitude increases, amplitude in the 50 dB ~ 70 dB floating, in the stratified interface cracking stage; load continues to increase, the specimen bending deformation Large-scale and high-amplitude acoustic emission signals appear, the amplitude range increases to 70 dB ~ 90 dB, and the damage of stratified damage is intensified, and the cumulative number of acoustic impact signals is increased abruptly. The final specimen to reach the bending limit, was destroyed failure.

It can be seen from figure 6(b) that, in the initial stage, the specimen bending deformation is small, the cumulative increase in the number of crash, only a small amount of low-amplitude signal appears, in the evolutionary stage, the load increases, the bending curve of the specimen becomes larger, the signal is higher than 70dB, and the load is about 480s,the cumulative number of impacts increases abruptly and rises linearly, and the signal is higher than 80dB, indicating that the specimen under the action of deformation increased, began to damage.

Figure 6(a) and (b) shows that the two experimental phenomena are in accordance with the bending theory, and relative b specimen, a specimen in the stratification evolution failure stage amplitude, impact number and other acoustic emission signals are more abundant, The value is higher, the impact number is more, the change is very obvious. The more the position of the layered defect is closer to the surface of the specimen, the higher the amplitude signal, the greater the impact on the interface carrying capacity, the worse the bending resistance of the specimen.

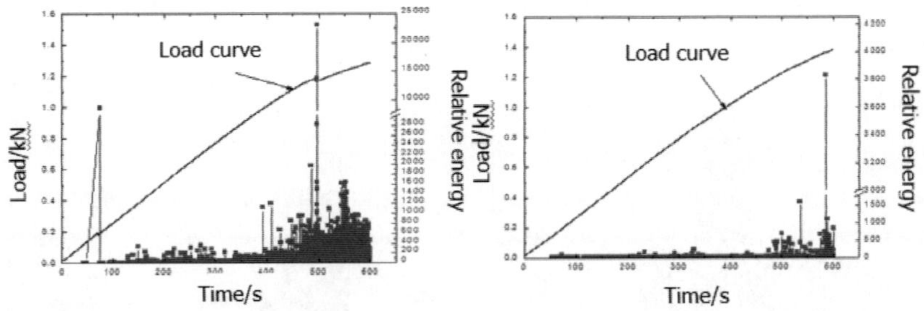

Figure 7. Load/time/ relative energy diagram for specimen *a* (a) and *b* (b)

Figure 7 depicts the load-acoustic emission relative energy-time history of the four-point bending tests of two composite material specimens. It can be seen from figure 7(a) that the relative energy of the acoustic emission signal is close to zero at the initial stage of loading, load and bending deformation is small; with the increase of load, the bending deformation of the specimen increases and the structure is obviously damaged. In the prefabricated stratified cracking stage, the relative energy reaches the high-energy signal level of 8609.98, indicating that the prefabricated stratification has completely cracked and destroyed; the load continues to increase, Deformation continues to grow, damage continues to accumulate, the relative energy compared to the previous stage as a whole increase, the final specimen fiber / matrix interface layered expansion damage, and produce the highest relative energy signal. It can be seen from the figure (b) that the relative energy of the acoustic emission signal is very low at the initial stage, the load and the bending deformation are very low. Under the load, the bending deformation of the specimen is only slightly damaged. When the experiment is carried out for 480s, the acoustic emission signal becomes relatively active, the relative energy is from 240 to 1600, the bending deformation of the specimen is aggravated, and the damage is seriously serious. When the load is about 600s, the relative energy signal is close to 3818.99. This phenomenon coincides with the trend of the crash-time-amplitude plot of the specimen b after 480 s.

Comparing the four-point bending loading experimental results of two specimens, these are in accordance with the bending theory. The relative AE energy of specimen *a* is 23119.90, and that of specimen *b* is much lower, namely 3818.99. The magnitude in the cumulative time amplitude diagram of specimen *a* is significantly larger than that of specimen *b*. It can be seen that the closer the layered defect is to the surface of the specimen, the better the bending resistance and the bearing capacity of the specimen.

Figure 8 depicts the AE impact signal positioning map for the composite material specimen bending failure process: In figure 8(a), bending deformation increases with the load, the AE signal of specimen *a* is mainly concentrated in the central part of the right side, while that on the left is relatively small. In figure 8(b), the acoustic emission positioning signal not only appears in the center of the specimen, but also a certain number appears in the adjacent area, which corresponds to the failure characteristics of the composite in figure 5. Comparison of the two maps shows that the number of AE targets for specimen *a* and *b* is 13865 and 1869, respectively, which difference is due to a poor bending resistance, prefabricated crack cracking and fiber/body interface stratification expansion of specimen *a*. Specimen *b* has better bending resistance, which drops only due to the occurrence of fiber damage and other defects. This indicates that a smaller distance between the stratified defect and the specimen reduces the bending ability of the composite material, which exacerbates its damage evolution and failure.

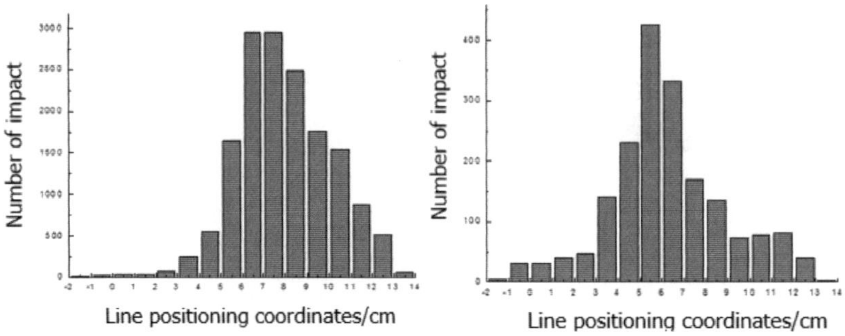

Figure 8. AE impact signal location diagrams of specimen *a* (a) and specimen *b* (b)

4. Conclusions

Through the four-point bending load of composite specimens containing artificial defects, the delamination evolution behavior is studied, and the following conclusions are obtained:

(a) Delamination defects located at the specimen surface, in the experiment process, there will be set manually in layered interface cracking, fiber / matrix interface crack and fiber breaking damage types; no significant defects in the specimen breakage phenomenon near the middle position.

(b) The experimental results of the two kinds of specimens are in line with the bending theory. The defects at the surface, the load deflection curves show obvious nonlinear characteristics, the specimen showed defects in a middle position for good linear characteristics, that delamination location has a great influence on the bending properties of composite materials.

(c) When the defect is located in the surface layer, the evolution stage is shortened obviously, and the specimen is locally damaged prematurely. When the defect is located in the middle, no obvious damage or damage occurs during the loading process. It shows that the delamination near the specimen surface accelerates the failure process of material expansion.

(d) In four-point bending loading test, with the load and bending deformation increasing, delamination defects in the specimen position, relative to the defects in the specimen surface, its amplitude is smaller, the impact cumulative number is less, the highest relative low energy is lower, specimens from the interface crack to layer failure evolution amplitude has been 40 ~ 90 dB defect on the surface, the extent of damage is more serious. Therefore, the smaller the distance of the delamination to the surface of the specimen is, the worse the bending resistance of the specimen is, the more likely it will be damaged or fractured, and the more severe will be the damage.

(e) In the four-point bending test process, with the damage increasing, the relative energy, amplitude, cumulative hits etc. acoustic emission signals constantly enrich and change. Therefore, to study the four-point bending loads with delamination defects in the application of acoustic emission monitoring experiment process of glass fiber reinforced composite delamination damage evolution behavior is correct and effective, which will promote the further development of research on the history of composite materials.

Acknowledgments

This project was supported by Beijing Municipal Natural Science Foundation of Beijing Science and Technology Research Institute joint project (L150005), Beijing Natural Science Foundation (No.4172023), and Beijing financial project (No. PXM2017_178215_000001). We also thank the reviewers for their insightful and valuable hints.

References:

[1] Li Y J, Zhou W and Liu R, et al. 2014 Study on acoustic emission monitoring of intermolecular cracks in wind turbine blades. *Journal of Hebei University*, **34** (2): 220-221.

[2] Liao Z L and Wu Y. 2008 Performance and application of epoxy resin/glass fiber composites, *Application of Engineering Plastics*, 2008, **36** (9): 47-57.

[3] Zhang L, Sun Q and Wang H C et al. 2010 Experimental study on mechanical properties of glass fiber reinforced epoxy resin composites, *Power Construction*, **31** (9): 118- 121.

[4] Mei D. 2010 Study on mechanical properties of glass fiber reinforced resin matrix composites, Master's degree thesis of Wuhan University of Technology.

[5] Zhang F P. 2000 Study on Layered Damage Behavior of Fiber Reinforced Laminated Composite, Department of Doctoral Dissertations, Northeastern University.

[6] Guo Y, Li D G and Chen Y X et al. 2013 Application of acoustic emission in fiber reinforced polymer matrix composites, *Journal of Engineering Plastics*, 2013, **41** (4): 107-110.

[7] Alander P, Lassila L V, Tezvergil A, et al. 2004 Acoustic emission analysis of fiber-reinforced composite in flexural testing, *Dental Materials,* (20): 305-312.

[8] Wu Y B. 2014 Study on damage evolution of fiber reinforced composites based on Weibull probability distribution, *Fiberglass / Composites*, (8): 49-53.

[9] Cai H P, Wang J P and Zhao X X, et al. 2013 Mechanical properties of composite wrapped pipes under bending load, *Glass Fiber Reinforced Plastic/Composite*, (8): 31-34.

[10] Liu H X, Zhang H and Yan Y C. 2002 Application and research development of fiber composites, *The Acoustic Emission Technology in Composite Materials*, (4): 50-52.

[11] Liu H X, Zhang H and Yan Y C 2004 Glass fiber / epoxy resin composite material damage and fracture of acoustic emission in the process of new technology, *New technology characteristics,* (03): 43~44.

[12] Zhou W, Sun S R and Feng Y N, et al. 2013 Wind turbine blade composites tensile damage acoustic emission behavior, *Journal of Composites*, **30** (2): 240-246.

[13] Zhou W, Zhang H B and Ma L H, et al. 2010 Research progress on non-destructive testing of structural defects of wind turbine blade composite, *Plastics Technology*, **38** (2): 84-86.

[14] Zhou W, Zhang X X and Wei Z H, et al. 2011 Wind turbine blade composite compressive damage acoustic emission monitoring, *Engineering plastics applications*, **39** (11): 61-64.

[15] Zhou W, Zhang X X and Han J, et al. 2014 Study on acoustic emission monitoring of composite single - joint adhesive joint, *Engineering plastics applications*, **42** (3): 69-72.

[16] Li Y J, Zhou W and Liu Ran, et al. 2015 Composite II type layered damage evolution acoustic emission monitoring, *FRP / Composite Materials*, **1**: 012.

IOP Conf. Series: Materials Science and Engineering **283** (2017) 012012 doi:10.1088/1757-899X/283/1/012012

The stochastic model of pitting corrosion of metals

Y H You[1*], B R Wang[1] and H Y Hu[1]

[1]School of Physics, Beijing Institute of Technology, Beijing, 100081, China
*E-mail: huhy@bit.edu.cn

Abstract. Considering that pitting corrosion is a stochastic process, the purpose of this paper is to use stochastic theory to investigate the growth process of pitting corrosion of metals. Based on the mechanism of pit growth and nonequilibrium statistical theory, the time evolution equations of pit depth and corrosion rate are obtained, and the probability density function can be derived by solving Fock-Plank Equation. Subsequently, the failure probability or reliability of metals is calculated based on the weakest link model of pitting corrosion. Finally, the stochastic model is used on the aircraft aluminum alloy LD2. Furthermore, this methodology can be applied to analyze the reliability and predict the pitting life of metals.

1. Introduction

Pitting corrosion is one of the common and highly destructive localized corrosion types, which is likely to cause catastrophic damage without any obvious signs in metals and alloys. Accordingly, it is more dangerous than uniform corrosion [1, 2]. Due to the inhomogeneity of the component, heat-treatment temperature, the surface state of metal [3-5] and temperature, pH, etc., of the corrosive environment [6, 7], the pitting corrosion of metal can be dealt with the stochastic theory. Shekari E *et al* [8] used the non-homogenous Markov process to investigate the model of maximum pit depth, and calculated the burst pressure of a defected pipe. Caleyo *et al* [9-11] established the stochastic model of pitting corrosion by Monte Carlo simulation method and obtained the probability distribution of pitting corrosion depth and rate in underground steel pipeline. In this work, considering that pitting corrosion of metal can be regarded as an uniform environment superimposed by inhomogeneous fluctuation, Xing's nonequilibrium statistical theory [12,13] is used to investigate the pitting corrosion.

2. The stochastic equation of pitting corrosion

According to the nonequilibrium statistical theory, the corrosion rate should obey the following generalized Langevin equation [14]

$$v = \frac{\mathrm{d}a}{\mathrm{d}t} = K(a,t) + F(a,t) = K(a,t) + \beta(a) f(t) \tag{1}$$

where a is the pit depth; $K(a,t)$ is the transportation rate, the deterministic part of the corrosion rate; $F(a,t) = \beta(a) f(t)$ is the fluctuation rate, the stochastic part of the corrosion rate; $\beta(a)$ is the fluctuation amplification function; $f(t)$ is the fluctuation function, which satisfies Gaussian distribution. The stochastic nature of pitting corrosion allows one to regard the pitting corrosion as a Markov process [15].

The Fock-Plank equation [14], which is equivalent to (1), can be derived as follows

$$\frac{\partial P(a_0,a;t)}{\partial t}=-\frac{\partial}{\partial a}\left\{\left[K(a,t)+\frac{Q\beta(a)}{2}\frac{\partial\beta(a)}{\partial a}\right]P(a_0,a;t)\right\}+\frac{Q}{2}\frac{\partial^2}{\partial a^2}\left[\beta^2(a)P(a_0,a;t)\right] \tag{2}$$

This is a stochastic partial differential equation, which describes the growth process of pitting corrosion, where Q is the fluctuation growth coefficient, $P(a_0,a;t)\mathrm{d}a$ is the probability that the initial pit depth a_0 grows into the depth between a and $(a+\mathrm{d}a)$ at time *t*, it is obvious that $P(a_0,a;t)\mathrm{d}a$ satisfies normalization condition.

3. Pit depth and probability density function

3.1. Pit depth and corrosion rate

Normally, pitting corrosion can be regarded as two physical processes: the metastable pitting (the pit generation process) and the stable pitting (the pit growth process) [16,17].

Metastable pitting consists of the metastable pit nucleation, growth and the process of the metastable pit growth to stable pit [18]. Frankel *et al* [19] considered the corrosion current density of metastable pit to be a constant, the period of pitting corrosion from occurring to nucleation can be defined as the pit initiation time t_d, for the corrosion current density, there is

$$I_\mathrm{corr}=i_\mathrm{a} \qquad 0<t<t_\mathrm{d} \tag{3}$$

where i_a is a constant, which is independent of the corrosion time *t*.

Stable pitting follows the growth of metastable pit. The passive film of the metal surface is destroyed at some point, aggressive anionic species (such as Cl^-, Br^- or $ClO4^-$ and so on) flow into the pit and promote the transition from metal atoms to ions. On the one hand, the ions form corrosion current by diffusion. On the other hand, corrosion products and environment hinder the diffusion of ions. This process can be equivalent to the charging process of RC circuit, and the corrosion current density of stable pitting (after the pit initiation time) can be proposed as

$$I_\mathrm{corr}=i_\mathrm{a}\exp\left[-\lambda(t-t_\mathrm{d})\right] \qquad t\geq t_\mathrm{d} \tag{4}$$

where λ is the attenuation parameter of corrosion current density. Equation (4) is similar to the current density formula in the anodic dissolution model which is proposed by Ford [20,21]. Substituting equations (3) and (4) into Faraday formula [22], the transportation rate $K(a,t)$ of the equation (1) can be derived as follows

$$K(a,t)=\begin{cases}\dfrac{Mi_\mathrm{a}}{ZF\rho} & 0<t<t_\mathrm{d}\\[2ex] \dfrac{Mi_\mathrm{a}}{ZF\rho}\exp\left[-\lambda(t-t_\mathrm{d})\right] & t\geq t_\mathrm{d}\end{cases} \tag{5}$$

where M is the atomic weight of metal cation, Z is the valence of metal cation, ρ is the metal density, and F is the Faraday constant. Integrated equation (5), the time evolution of pit depth can be expressed as

$$a=\frac{Mi_\mathrm{a}t_\mathrm{d}}{ZF\rho}+\frac{Mi_\mathrm{a}}{ZF\rho\lambda}\left\{1-\exp\left[-\lambda(t-t_\mathrm{d})\right]\right\} \qquad t\geq t_\mathrm{d} \tag{6}$$

If $t\to\infty$, equation (6) becomes

$$a_\mathrm{m}\approx\frac{Mi_\mathrm{a}t_\mathrm{d}}{ZF\rho}+\frac{Mi_\mathrm{a}}{ZF\rho\lambda} \tag{7}$$

where a_m is regarded as the maximum pit depth.

3.2. *Probability density function of pitting corrosion*

According to equations (5) and (6), the transportation rate can be written as

$$K(a,t)=\frac{Mi_a t_d \lambda}{ZF\rho}+\frac{Mi_a}{ZF\rho}-a\lambda \qquad t\geq t_d \tag{8}$$

Let

$$A=\frac{Mi_a}{ZF\rho}, \qquad \beta(a)=1+t_d\lambda-\frac{a\lambda}{A} \tag{9}$$

The transportation rate $K(a,t)$ in equation (8) can be regarded as an average value $\bar{v}$. In fact, due to the inhomogeneous fluctuations of the material and corrosive environment, the atomic weight of metal cation, the corrosion current density, the valence of metal cation, and the metal density should be $M\pm\Delta M$, $i_a\pm\Delta i_a$, $Z\pm\Delta Z$, $\rho\pm\Delta\rho$. The corrosion rate of equation (1) should be written as

$$v=\frac{(M\pm\Delta M)(i_a\pm\Delta i_a)}{(Z\pm\Delta Z)(\rho\pm\Delta\rho)F}\beta\approx\frac{Mi_a}{ZF\rho}(1\pm\eta)\beta=\bar{v}\pm\eta\bar{v} \tag{10}$$

where η is the sum of relative deviations of the four physical quantities M, i_a, Z, ρ

$$\eta=\frac{\Delta M}{M}+\frac{\Delta i_a}{i_a}+\frac{\Delta Z}{Z}+\frac{\Delta\rho}{\rho} \tag{11}$$

The fluctuation growth coefficient has been obtained in the [14]

$$Q=\eta^2A^2\tau \tag{12}$$

where τ is the pitting life of the metal.

Substituting equations (8), (9), (12) into equation (2), the probability density function $P(a_0,a;t)$ of pitting corrosion can be expressed by the following equation

$$P(a_0,a;t)\mathrm{d}a=\frac{1}{\sqrt{2\pi Qt\beta(a)^2}}\exp\left\{-\frac{U(a)^2}{2Qt}\right\}\mathrm{d}a \tag{13}$$

where $U(a)$ is expressed as

$$U(a)=\frac{A}{\lambda}\ln\left(\frac{t_d\lambda A+A-a_0\lambda}{t_d\lambda A+A-a\lambda}\right)-At \tag{14}$$

It can be seen that $P(a_0,a;t)$ varying with the pit depth can be fitted to the logarithmic normal distribution function.

It is assumed that $P(t,a)\mathrm{d}t$ is the probability of finding the pit depth a in the range of time $(t,t+\mathrm{d}t)$ over the period of time $t\in(0,\infty)$

$$P(t,a)\mathrm{d}t=\frac{A\exp[-\lambda(t-t_d)]}{\sqrt{2\pi Qt\beta(a)^2}}\exp\left\{-\frac{U(a)^2}{2Qt}\right\}\mathrm{d}t \tag{15}$$

The initial and boundary conditions of $P(t,a)$ are given by

$$P(t=t_{\rm d},a)=\delta(a-a_0),\qquad P(t\to\infty,a)=0 \tag{16}$$

It indicates that pits of any depth cannot exist in an infinite time. That is to say, the pitting life of the metal is always limited.

4. Failure probability and reliability

Generally, there are a large number of pits in the metal. Let S denote the surface area of material and N denote the number of pits per unit area of material, so the total number of pits G is SN. The material will be perforated due to pitting corrosion when the pit depth reaches the material thickness, on the basis of the weakest link model of pitting corrosion, the failure probability of material $P_{\rm f}(t)$ during the time interval $0-t$ is

$$P_{\rm f}(t)=1-\left[1-\int_0^t P(t,a_{\rm m}){\rm d}t\right]^{G-1}\approx 1-\exp\left[-(G-1)\int_0^t P(t,a_{\rm m}){\rm d}t\right] \tag{17}$$

where $a_{\rm m}$ is the maximum pit depth, it can be assumed as the thickness of the material.

According to the [13], the integral $\int_0^t P(t,a_{\rm m}){\rm d}t$ in equation (17) can be approximately expressed as

$$\int_0^t P(t,a_{\rm m}){\rm d}t\approx\frac{1}{\sqrt{\pi}}\int_X^\infty \exp(-X^2){\rm d}X\approx\frac{1}{\theta X^{10}} \tag{18}$$

where θ is a constant, and X is $U(a_{\rm m})/\sqrt{2Qt}$.

Substituting equation (18) into equation (17), $P_{\rm f}(t)$ can be approximately expressed as

$$P_{\rm f}(t)\approx 1-\exp\left\{-\frac{(G-1)}{\theta X^{10}}\right\} \tag{19}$$

Therefore, the reliability of metal $R(t)$ is the probability that the material can work safely without perforation within time interval $0-t$ in the corrosive environment, which can be deduced by the equation (19).

$$R(t)=1-P_{\rm f}(t)\approx\exp\left\{-\frac{(G-1)}{\theta X^{10}}\right\} \tag{20}$$

5. Application in the aircraft aluminum alloy LD2

5.1. Calculation of pit depth and corrosion rate

The theory is used on the pitting corrosion of aircraft aluminum alloy LD2, according to the accelerated corrosion test in [23, 24], the parameters are selected as follows: $\rho=2.7\times10^3$ kg m^{-3}, $M=27.18\times10^{-3}$ kg mol^{-1}, $F=96500$ C mol^{-1}, $t_{\rm d}=2\times3.1536\times10^7$ s=2 yr, $i_{\rm a}=0.009$ A m^{-2}, $Z=3$, $\lambda=5.71\times10^{-9}$ s$^{-1}=0.18$ yr^{-1}. Then, the curves of pit depth and corrosion rate over corrosion time can be drawn by using the equations (6) and (10), as shown in figures 1 and 2 (η is 0.2).

It can be seen from figure 1, the pit depth a increases with corrosion time t, and the maximum pit depth $a_{\rm m}$ is around 74 μm. When the corrosion time takes $5/\lambda$, the pit depth a reaches $0.993a_{\rm m}$, so the approximation of pitting life τ of aluminum alloy is 20 years. Figure 2 shows that the corrosion rate v decreases with corrosion time t. In addition, the theoretical simulation results present the general trend of pit depth and corrosion rate versus corrosion time, which provides the basis for calculation of probability density function.

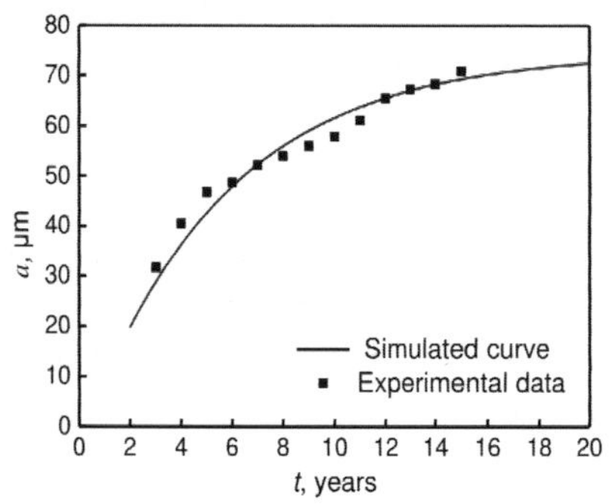

Figure 1. Comparison between the theoretical results and the experimental data of pit depth versus corrosion time.

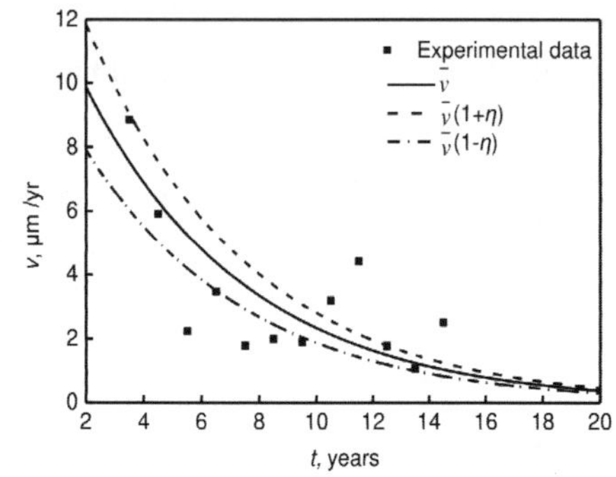

Figure 2. Comparison between the theoretical results and the experimental data of corrosion rate versus corrosion time.

5.2. Calculation of probability density function

According to equations (13) and (15), the curves of probability density function $P(a_0,a;t)$ and $P(t,a)$ are plotted respectively, as shown in figures 3 and 4.

Figure 3 shows that the distribution of the probability density function $P(a_0,a;t)$ tends to be high in the middle and lower at both ends, but it's not a Gaussian distribution. The maximum probability density $P(a_0,a;t)$ is corresponding to the most probable pit depth a_p, which can be used to predict the most probable pit depth a_p in different corrosion time ranges. For instance, when the corrosion time is 4 years, the pit depth a is the most likely to be 45-55 μm. As the corrosion time increases, the peak value of curve shifts to the right, and the shape of the curve becomes narrower. That is to say, pits grow up continuously, and most of the pit depths are concentrated around the most probable pit depth when the corrosion time t is larger.

Figure 4 shows that the maximum probability density function $P(t,a)$ is corresponding to the most probable corrosion time t_p, which can be used to predict the most probable corrosion time range of different pit depths. For instance, pit depth a of 50 μm is the most likely to occur at the time range 4-5 years.

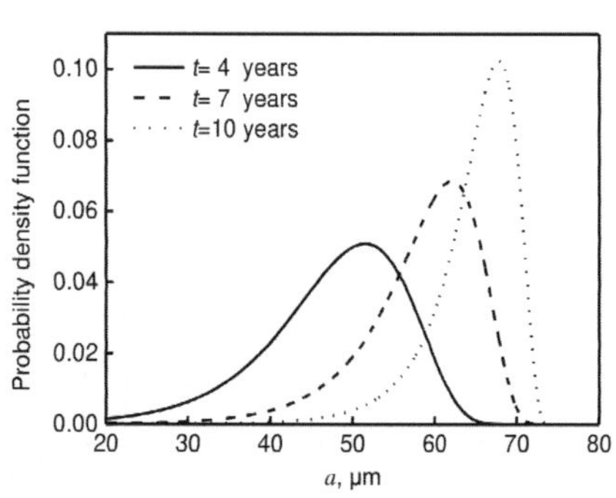

Figure 3. Probability density function $P(a_0,a;t)$ at different corrosion times.

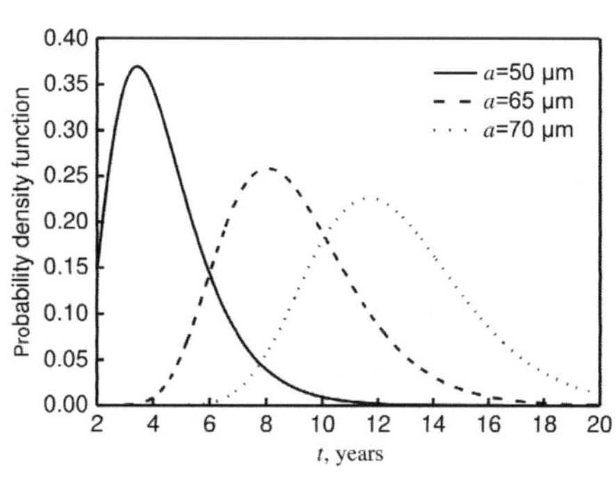

Figure 4. Probability density function $P(t,a)$ at different pit depths.

5.3. Calculation of failure probability and reliability

Substituting the surface area S=1 mm^2 and the pit density N of aluminum alloy into the equations (17) and (20), therefore, the failure probability P_f and the reliability R of aluminum alloy can be calculated, as shown in figure 5.

As can be seen from figure 5(a), at the same corrosion time, the greater the pit density, the lower the reliability, and the easier it is to perforate the aluminum alloy, so improving corrosion resistance and the reliability of the material can reduce pit density, and an example is the nanocrystalline material which has been studied by Pan C *et al* [25]. As shown in figure 5(b), when the corrosion time t is less than 12 years, the reliability is close to 1, denoting that aluminum alloy will not be perforated due to pitting corrosion. But when the corrosion time increases to the limit value (about 20 years), the failure probability P_f basically reaches 100%, while the reliability R is reduced to zero.

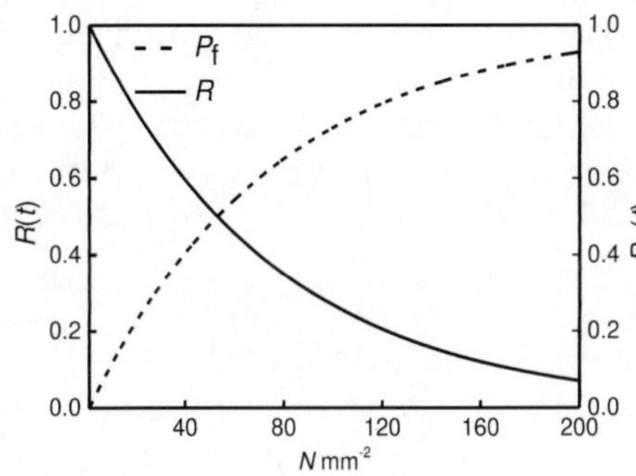

(a)

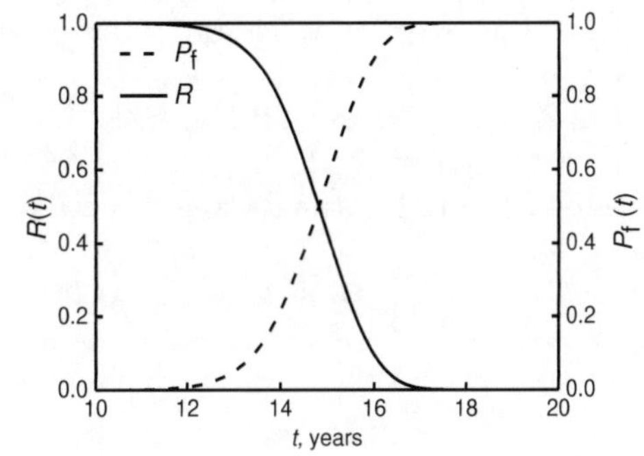

(b)

Figure 5. Time curves of the failure probability and reliability. (a) versus pit density (t=15 year) and (b) versus corrosion time (N=67 mm^{-2}).

6. Conclusions

In conclusion, the pit depth, corrosion rate and the maximum pit depth of metal with corrosion time are calculated. Establishing and solving the Fock-Plank Equation of pitting corrosion, the probability density functions $P(a_0,a;t)$ and $P(t,a)$ are calculated, which can be used to predict the most probable pit depth in different time ranges and the most probable corrosion time range of different pit depths. Moreover, the failure probability and reliability of metal are estimated by using the weakest link model of pitting corrosion, which is dependent on the pit density N, corroded area S, corrosion time t, etc., the safety period of pitting corrosion of metal can be deduced.

Acknowledgments

This work was supported by the National Natural Science Foundation of China (No. 11272053).

References

[1] Deyab M A 2016 *Electrochim. Acta* **202** 262-8
[2] Caines S, Khan F and Shirokoff J 2013 *J. Loss Prev. Process Ind.* **26** 1466-83
[3] Galit K L and Eli A 2016 *Adv. Eng. Mater.* **18** 269-76
[4] Liang C H, Guo J W and Hu X Q 2007 *J Wuhan Univ Technol Mater Sci Ed.* **22** 12-6
[5] Zhao Q H, Liu W, Zhu Y C, Zhang B L, Li S Z and Lu M X 2017 *Acta Metall. Sin. (Engl. Lett.)* **30** 164-75
[6] Ezuber H M 2014 *Mater. Des.* **59** 339-43
[7] Deyab M A 2009 *J. Solid State Electrochem.* **13** 1737-42
[8] Shekari E, Khan F and Ahmed S 2015 *Int. J. Press. Vessels Pip.* **137** 13-21
[9] Valor A, Caleyo F, Alfonso L, Rivas D and Hallen J M 2007 *Corros. Sci.* **49** 559-79
[10] Caleyo F, Velázquez J C, Valor A and Hallen J M 2009 *Corros. Sci.* **51** 1925-34
[11] Velázquez J C, Caleyo F, Valor A and Hallen J M 2009 *Corrosion* **65** 332-42
[12] Xing X S 1986 *Scientia Sinica (Series A)* **10** 1051-62

IOP Conf. Series: Materials Science and Engineering **283** (2017) 012012 doi:10.1088/1757-899X/283/1/012012

[13] Xing X S 1986 *Scientia Sinica (Series A)* **12** 1318-33
[14] Xing X S 1991 *Prog. Mater. Sci.* **5** 22-8
[15] Valor A, Caleyo F, Alfonso L and Velázquez J C 2013 *Math. Probl. Eng.* doi: 108386
[16] Hong H P 1999 *Corrosion* **55** 10-6
[17] Liu L, Li Y and Wang F H 2014 *Acta. Metall. Sin.* **50** 212-8
[18] Burstein G T, Pistorius P C and Mattin S P 1993 *Corros. Sci.* **35** 57-62
[19] Frankel G S, Stockert L, Hunkeler F and Boehni H 1987 *Corrosion* **43** 429-36
[20] Ford F P 1978 *Metal Science* **12** 326-34
[21] Ford F P 1979 *Corrosion* **35** 281-7
[22] Xiao J M and Cao C N 2002 *Corrosion Mechanism of Materials* (Beijing: Chemical Industry Press)
[23] Liu Z G, Cai Z J, Bian R P and Zhao W Y 2012 *Equipment Environmental Engineering* **9** 66-71
[24] Li X D, Zhu W F, Mu Z T and Liu Z G 2013 *Equipment Environmental Engineering* **10** 8-12
[25] Pan C, Liu L, Li Y and Wang F H 2013 *Corros. Sci.* **73** 32-43

IOP Conf. Series: Materials Science and Engineering **283** (2017) 012013 doi:10.1088/1757-899X/283/1/012013

Transparent superhydrophobic coating application to the interlocking clay block for the microbial growth mitigation

O Nimittrakoolchai[1*], M Suwan[1] and S Supothina[1,2]

[1]National Metal and Materials Technology Center, National Science and Technology Development Agency (NSTDA), 114 Thailand Science Park, Phahonyothin Rd., Klong Luang, Pathum Thani 12120, Thailand

*E-mail: sitthis@mtec.or.th

Abstract. Interlocking clay block is of widespread use in many countries due to its less labor-intensive design. The block comprises of considerable amount of pore structures; therefore water can easily penetrate into it causing deterioration by natural weathering as well as the growth of microbes on its surface. In this present work, a transparent, superhydrophobic coating was applied onto the block's surface to prevent water absorption. The coating which is a poly(methyl hydrogensiloxane) base containing fumed SiO_2 nanoparticle was designed to possess surface topographic feature similar to that of the lotus leaf. After coating the block with the superhydrophobic compound, its water absorption decreased from 11.5 to 4.8% and porosity substantially decreased from 22.5 to 9.2%, while bulk and apparent densities did not change. Accelerated weathering test revealed that the coating had good stability. The coated samples exhibited growth inhibition of algae.

1. Introduction

An interlocking clay block is widely employed in buildings and other artefacts essentially due to its simplicity and less labor-intensive of construction as no mortar is required for joining as in the case of a conventional clay brick [1]. The block can be fabricated in a variety of shapes and designs simply by changing the mold of a pressing machine. Nowadays, the interlocking clay block has been utilized not only for structural purpose but also for decoration and gardening purposes which require natural-like appearance. In the tropical zone, there is plenty of raw material called 'laterite' or 'red earth' which is one of fine starting materials for production of the interlocking clay block. As the laterite's color is red, the block made from this material has natural appearance which, for a certain artefact, do not require painting.

The interlocking clay block is typically fabricated by uniaxial dry pressing of the feed consisting of red earth, sand and cement after which it is dried in open air for a few days. Since most of raw materials are coarse grains and there is no heat treatment for densification, the finished block comprises of both micro- and nanopores, which can draw significant amount of water into it, causing a rapid deterioration by natural weathering. In addition, when utilized in humid environment it is observed that microbes such as lichen and algae can grow rapidly, especially, during a rainy season because of high moisture content at the block's surface, leading to a loss of aesthetic appearance as a result of such biological fouling. This is a major drawback that can limit the utilization of the block in humid environment. To protect the material from natural weathering and more importantly to reduce the biological fouling, the surface must be coated with a protective layer that inhibits water penetration. The most common protective coating is partially fluorinated polymeric compounds containing acrylic, urethane or silicone resin. For instance, it was found that perfluoropolyethers provide a very satisfactory protecting performance and compatibility with the stone heritage [2]. An influence of side-chain fluorination on the performance of acrylic-based copolymers has been investigated for stone protection. The presence of fluorine was found to increase water repellency, thus providing protective efficiency [3]. Similarly, siloxane branches of a polydimethylsiloxane-grafted polymer can enhance hydrophobic property [4]. A fluoropolymer has also been utilized to coat glazed tiles to enhance water repellency and to reduce soil attachment, thus providing a better cleanability [5]. Environmental friendly, solvent-free hybrid organic-inorganic methacrylic-based coatings have been introduced with a performance in terms of hydrophobicity and color change comparable to those of the commercial products [6]. Recently, oligoamide grafted with perfluoropolyether blocks which are environmental friendly has been synthesized. This coating material can be used as a protective coating for low porous stone materials such as stone artworks [7]. For large porous stones, it has been reported that coating application performed by brushing resulted in better water-repellent property due to the formation of a superficial film compared to capillary absorption [8]. In order to evaluate their protective performance, weathering stability of waterproof polymers have been evaluated under artificial weathering. It was found that all the polymers under investigation underwent irreversible modifications with reduction of their conservative properties, color change and impossibility of their complete removal [9]. The coatings of these compounds usually provide water contact angle in a range of 110-140 degrees, and some of them have glossy or darken appearance which may be undesirable for certain artefacts especially those requiring natural appearance.

In this present work, a very thin super water-repellent coating (water contact angle $\geq$ 150 degrees) with high transparency was applied onto the interlocking clay block to minimize water penetration as well as reduce moisture absorption. The coating's transparency provided the advantage that the coated block preserved its naturally aesthetic appearance. The coating comprising of a siloxane compound, fluorosilane and oxide nanoparticle was designed by mimicking surface feature of a lotus leaf that exhibited a "lotus effect" [10,11]. Physical and chemical properties of the blocks such as chemical composition, apparent porosity, water absorption, bulk and apparent densities of the bare and coated blocks were also measured. Growth inhibition of microbes was evaluated according to the ASTM standard, as well as by means of natural weathering test. In addition, stability of the coating on glass substrate was evaluated by subjecting to an accelerated weathering test. Surface topography and transparency of the coating on the glass substrate were also characterized.

2. Experimental procedure

2.1. Fabrication of interlocking clay blocks

The interlocking clay blocks of 10 cm x 10 cm x 10 cm in size were fabricated by pressing of raw material in a steel mold. The raw material consisting of 52 wt.% red earth, 31 wt.% sand, 11 wt.% cement and 6 wt.% water was mixed thoroughly, fed into the mold and pressed at 2.0-2.3 MPa using a hand-operated hydraulic machine. Finally, the resulting interlocking blocks were dried in open air for at least 5 days.

2.2. Transparent superhydrophobic coating

To reduce water absorption, the blocks were coated with a water-repellent coating as follows. The coating solution was prepared by dissolving 1 g of poly(methyl hydrogensiloxane) (PMHS, Dow corning®) in hexane followed by an addition of 0.05 ml of 100 ppm Pt-catalyst which was prepared by dissolving platinum(0)-1,3-divinyl-1,1,3,3-tetramethyldisiloxane complex (Aldrich) in hexane. Then, 0.025 g of a fumed SiO_2 (Aerosil 200, JJ Degussa) nanoparticles, and 4.95 ml of trichloro (1*H*, 1*H*, 2*H*, 2*H*-perfluorooctyl) silane ($CF_3(CF_2)_5CH_2CH_2SiCl_3$, 97%, Aldrich) solution prepared by dissolving 0.4 ml of the fluorosilane in 2-propanol were added respectively into the PMHS solution under a continuous stirring. The resulting mixture was stirred at room temperature for at least 10 min. The water-repellent coating was applied onto the interlocking blocks by hand painting with special attention on the amount of the coating precursor applied. The coated blocks were dried at room temperature for 15 min before they were re-coated for 2 more times to obtain uniform coating.

2.3. Characterizations

Physical property analysis of the clay block: A specific surface area (SSA) of the block was measured by means of nitrogen adsorption using an Autosorb-1 (Quantachrome). Surface morphology was observed by using a field-emission scanning electron microscope (FE-SEM, JSM 6301F). Apparent porosity ($\%P$), water absorption ($\%A$), bulk density (ρ_b) and apparent density (ρ_a) were measured by means of water displacement according to the standard test method (ASTM C67-03a) as follows. The block was cut into 2 cm x 2 cm x 2 cm pieces and dried in an oven at 110°C for 24 h. After cooling to room temperature in a desiccator, the weight of the oven-dried specimen was measured. Then, the specimen was immersed in a beaker filled with de-ionized water maintained at 24±1°C for 24 h, and the specimen's weight when it was submerged in water was measured. After submersion in water, surface water was wiped off with a damp cloth and the specimen was weighed again. The apparent porosity, water absorption, bulk density and apparent density were calculated from the equations below:

$$\%P = \frac{W_{WS} - W_D}{W_{WS} - W_{SS}} \times 100, \quad (1)$$

$$\%A = \frac{W_{WS} - W_D}{W_D} \times 100, \quad (2)$$

$$\rho_b = \frac{W_D \cdot \rho_L}{W_{WS} - W_{SS}}, \quad (3)$$

$$\rho_a = \frac{W_D \cdot \rho_L}{W_D - W_{SS}}, \quad (4)$$

where W_{WS} is the weight of water-saturated specimen, W_D is the weight of oven-dried specimen, W_{SS} is the weight of water-saturated specimen when it is submerged in water, and ρ_L is the density of water.

Characterization of the coating: Surface morphology of the uncoated and coated blocks was observed by using a field-emission scanning electron microscope (FE-SEM, JSM 6301F). Surface topography, transparency of the coating deposited onto the glass slide were characterized by using an atomic force microscopy (AFM, SPA 400) performed in a non-contact mode and a UV-Vis-NIR spectrophotometer (Perkin-Elmer Lambda 900), respectively. Thickness of the coating was determined by making a scratch on the sample, and then the z-height AFM scan was performed.

Accelerated weathering and algae inhibition tests: The accelerated weathering test was performed in a Ci3000 Xenon Weather-Ometer (Atlas Material Testing Solutions) following the ASTM standard (G 155-04a) by using the following operating condition: Xenon arc lamp irradiation: 0.35 W/m·nm at 340 nm; black panel temperature: 63°C; relative humidity: 50%. The specimens were mounted on the holders which were placed on the specimen rack. The testing was conducted for a total time span of 325 h during which the specimens were taken out for water contact angle measurement every 50 or 100 h interval. Growth inhibition of algae was tested following the ASTM D5589-97 standard.

3. Results and discussion

3.1. Physical analysis of the clay block

Physical properties of the uncoated and coated blocks are summarized in table 1. Since the block did not subject to high-temperature treatment, the uncoated block consists of high porosity of 22.50%, and low bulk and apparent densities of 2.53 and 1.96, respectively, leading to high water absorption of 11.50%. The specific surface area is 23.80 m^2g^{-1}.

Table 1. Physical properties of the uncoated and coated blocks.

Properties	Uncoated blocks	Coated blocks
SSA (m^2g^{-1})	23.80	20.84
%*P*	22.50	9.20
%*A*	11.50	4.80
ρ_a (gcm^{-3})	1.96	1.92
ρ_b (gcm^{-3})	2.53	2.11

3.2. Surface treatment with transparent, superhydrophobic coating

Most of the coating precursor was easily absorbed into the pores of the blocks during the first painting, and was drawn at a lesser extent during the subsequent painting cycles. After the block surface was modified with a water-repellent coating, the water absorption decreased from 11.50 to 4.80% and the porosity substantially decreased from 22.50 to 9.20%, while the bulk and apparent densities did not change. This result indicated that the water-repellent coating did not form a continuous film on the block surface, but part of it penetrated into the block and wetted the pores. It was anticipated that the extent of water absorption of the coated block in a condition of use would be much lower since water will form spherical droplet on its surface and roll off on a vertical surface very easily.

Surface morphology analysis of the uncoated and coated blocks (figure 1) revealed that surface of the coated block was still porous and there was merely continuous coating observed. Both the coated and uncoated blocks had similar surface morphology, indicating that the coating was very thin and most of the water-repellent compound penetrated into the pores and wetted the pore's interior. Wetting of the pore's interior by the water-repellent compound would prevent water absorption by the capillary force exerted by the pore structure. Thickness of the coating determined by performing z-height AFM scan was only 0.27±0.08 μ. This hand-painted coating was less uniform compared to the coating on a glass slide performed by using a dip coater.

Figure 1. SEM images of (a) uncoated block and (b) coated block.

Surface topography analysis of the water-repellent coating on the glass slide (figure 2) revealed that the surface was roughened, and had surface roughness of 95.7±5 nm which was much higher than surface roughness of bare glass slide (3.6±0.3 nm). The film consisted of sub-micrometer-sized protrusions created by SiO_2 agglomerates which consisted of many nanosized SiO_2 primary particles. The SiO_2 nanoparticle was added as a surface roughness enhancer to improve degree of hydrophobicity as described by Wenzel's and Cassie-Baxter's models [12,13]. This surface had water contact angle of 160.2±1.0 degrees, exhibiting a superhydrophobic property. With no addition of the

SiO_2 nanoparticle, the contact angle of only 103.9±0.4 degrees was obtained.

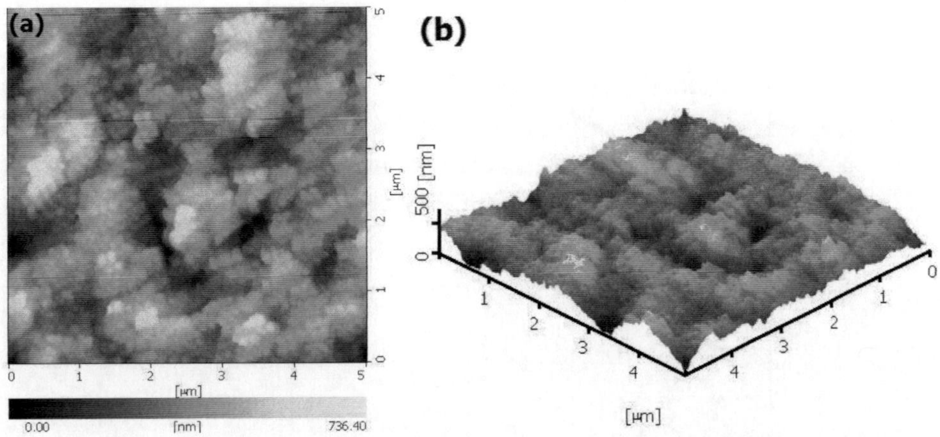

Figure 2. AFM images of the water-repellent coating on a glass slide.

Figure 3 shows SEM images of the superhydrophobic films on the glass slide. For the film without the addition of SiO_2 oxide filler (figure 3(a)), the surface is observed as a flat surface comprising of sub-micrometer to micrometer white spots. It is believed that these white spots took place during the drying step which resulted in rapid evaporation of the solvents employed. It is clearly seen that the film containing the SiO_2 filler (figure 3(b)) composed of the SiO_2 particles covering most of the surface. These particles formed loose agglomerates as seen in the inserted image. It was anticipated that the coating on the block would have more or less similar topographic feature to the coating on the glass slide. The transmittance of the coating (figure 4) was in a range of 60-80% in a visible region. Although, the transmittance of the modified glass was lower than that of the bare glass, the film was sufficiently transparent which can be employed as a water-repellent coating on the clay block without altering its aesthetic appearance.

Figure 5 shows photographs of the uncoated and coated blocks after water was dropped on their surfaces. The water was instantly drawn by the uncoated block. In contrast, it formed a droplet on the surface of the coated block, and rolled off when the block was tilted. The coated block had water contact angle of 151.4±2.1 degrees. This water-repellent property is similar to that observed at the surface of many plant leaves and insects [10,11,14]

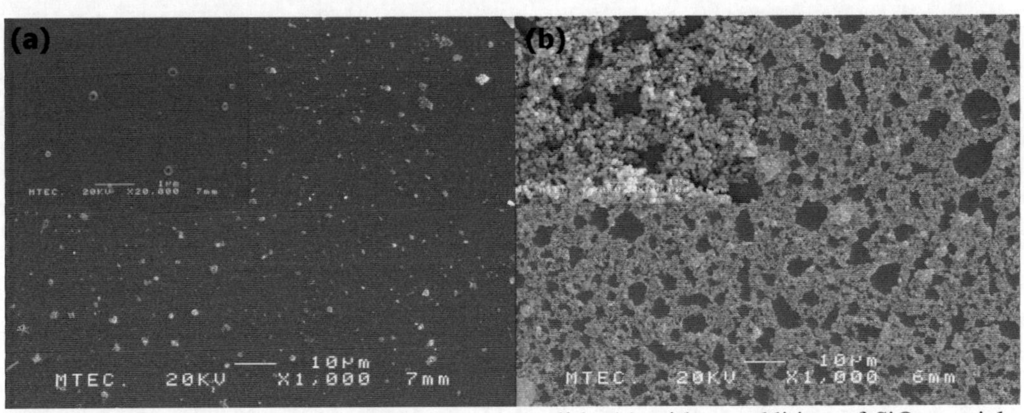

Figure 3. SEM images of the coating on glass slide (a) without addition of SiO_2 particle, and (b) with addition of SiO_2 particle.

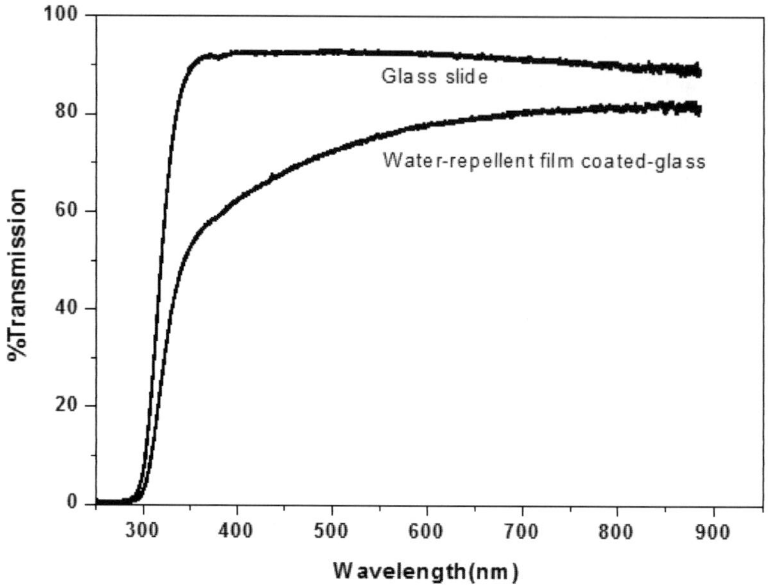

Figure 4. Transmittance spectra of bare and coated glass slides.

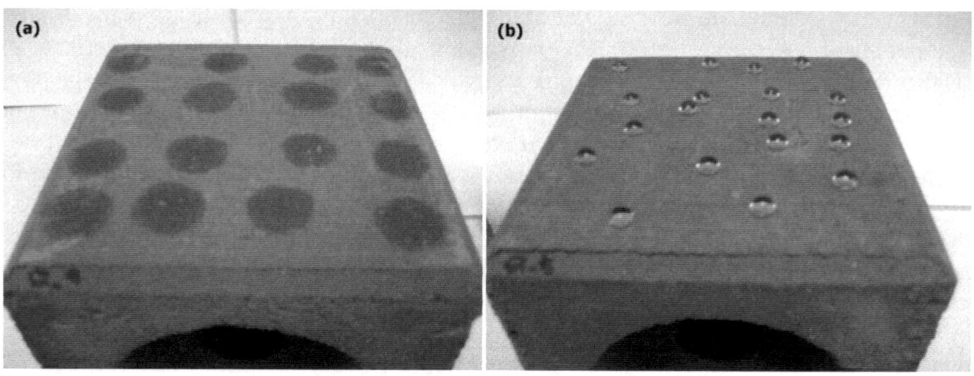

Figure 5. Photographs taken after dropping water on the uncoated and coated blocks.

3.3. Accelerated weathering and algae inhibition tests

The result of accelerated weathering test (figure 6) clearly indicated that the coating was stable as the water contact angle remained virtually unchanged after accelerated weathering exposure of 325 h which is comparable to actual weathering exposure for ~7 weeks. For the samples subjected to algae inhibition test (figure 7), it is clear that green algae can grow on the uncoated sample but cannot grow on the coated sample due to no moisture absorbed on its surface. In addition, the samples were subjected to outdoor weathering test by placing them in the garden. It is clearly observed after a rainfall that the coated blocks were in their original light brown color, while the uncoated blocks were dark brown because they absorbed water. The green algae and lichen grew on the uncoated blocks within 1-2 months turning their brown appearance to green. According to the result after 6 months of the field test which was at the end of the rainy season, only small amount of green bio-organisms was observed on the coated blocks in the lowest row which was in contact with the ground and had been occasionally flooded. After the field test, the blocks coated with the superhydrophobic coating can preserve their original appearance. It is very obvious from this field test that the coating was efficient for growth inhibition of the algae.

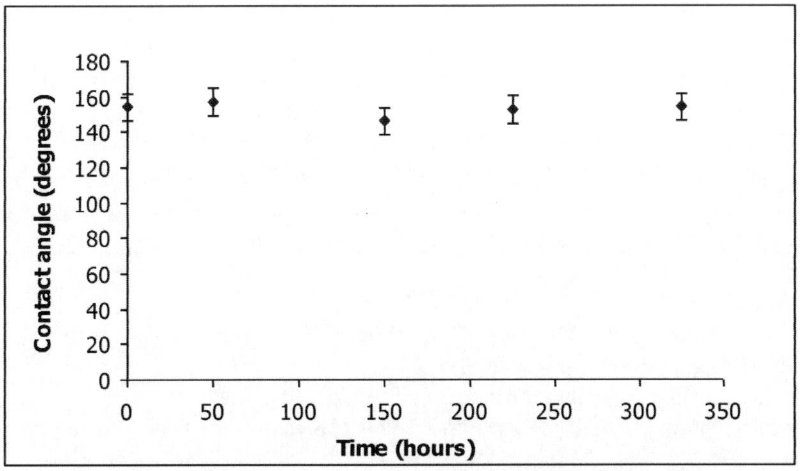

Figure 6. Accelerated weathering test of the coating on glass.

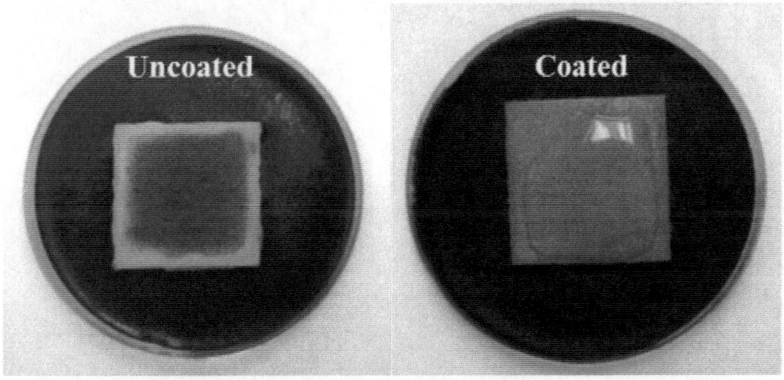

Figure 7. Photographs of uncoated sample (left) and coated sample (right) after algae growth inhibition test.

4. Conclusions

The growth of bio-organisms such as algae on the interlocking clay blocks can be mitigated by coating their surface with the superhydrophobic coating having surface property similar to that of the lotus leaf. The water absorption and porosity of the coated blocks substantially reduced while the bulk and apparent densities did not change. Thick layer of green algae can grow on the uncoated blocks but cannot grow on the blocks coated with the superhydrophobic coating. The coating was of sufficient transparency, preserving natural appearance of the blocks. The accelerated weathering test also revealed that the coating had good stability. It is feasible to apply the superhydrophobic coating on other construction materials or artefacts to obtain a similar effect.

Acknowledgments

This research was supported by the National Metal and Materials Technology Center, the National Science and Technology Development Agency, Thailand (Grant# MT-B-51-CER-07-211-I).

References

[1] Thanoon W A, Jaafar M S, Kadir M R A, Ali A A A, Trikha D N and Najm A M S 2004 *Constr. Build. Mater.* **18** 445-54

[2] Piacenti F 1994 *Sci. Total Environ.* **143** 113-20

[3] Alessandrini G, Aglietto M, Castelvetro V, Ciardelli F, Peruzzi R and Toniolo L 2000 *J. Appl.*

Polym. Sci. **76** 962-77
[4] Suzuki H, Takeishi M and Narisawa I 2000 *J. Appl. Polym. Sci.* **78** 1955-63
[5] Kuisma R, Fröberg L, Kymäläinen H-R, Pesonen-Leinonen E, Piispanen M, Melamies P, Hautala M, Sjöberg A-M and Hupa L 2007 *J. Eur. Ceram. Soc.* **27** 101-8
[6] Corcione C E, Striani R and Frigione M 2014 *Prog. Org. Coat.* **77** 803-12
[7] Cao Y, Salvini A and Camaiti M 2017 *Prog. Org. Coat.* **111** 164-74
[8] Licchelli M, Malagodia M, Weththimunia M L and Zanchi C 2013 *Prog. Org. Coat.* **76** 495-503
[9] Favaro M, Mendichi R, Ossola F, Russo U, Simon S, Tomasin P and Vigato PA 2006 *Polym. Degrad. Stab.* **91** 3083-96
[10] Barthlott W and Neinhuis C 1997 *Planta* **202** 1-8
[11] Neinhuis C and Barthlott W 1997 *Ann. Bot.* **79** 667-77
[12] Wenzel R N 1936 *Ind. Eng. Chem.* **28** 988-94
[13] Cassie A B D and Baxter S 1944 *Trans. Faraday Soc.* **40** 546
[14] Watson G S and Watson J A 2004 *App. Surf. Sci.* **235** 139-44

IOP Conf. Series: Materials Science and Engineering **283** (2017) 012014 doi:10.1088/1757-899X/283/1/012014

Development and application of free pretreatment container steel

Y Yang[1,2*], Y Liu[2], B Han[2], B Wei[2] and S Z Wang[2]

[1]The State Key Laboratory of Rolling and Automation, Northeastern University, Shenyang 110004, Liaoning, China
[2]Wuhan Branch Baosteel Central Research Institute (Technology Center of Wuhan Steel Ltd.) Wuhan 430080, Hubei, China
*E-mail: yangyi_2008@aliyun.com

Abstract. Due to economic and environmental advantages pre-treatment containers have good big development prospects, which can avoid shot blasting processes, and decrease the noise and dust pollution. By analyzing requirements of the container steel surface quality, target oxide scale structure of free pretreatment container steel has been determined. Trial process was carried out, and test results showed that the oxide scale achieved the desired objects, oxide scale with outer thin Fe_3O_4 layer and inner eutectoid α-Fe+Fe_3O_4. Salt spray test, second adhesion test, and modeling performance basically corroborated the container feasibility.

1. Preface

The container reduces the cost of transporting goods, and attains the goods transport standard. It has the advantages of being safe, rapid, especially suitable for the transportation of high added value, and easy crashing goods [1, 2]. Based on the container transportation, a logistics system of ships, ports, routes, roads, bridges, tunnels, transfer station, multimodal intermodal matching in the world is gradually established, and the world economy has changed as a result [3]. Cu, P, Cr, Ni and other alloy elements are added into the weathering steel for container to make them have good corrosion resistance in the atmosphere. Because of the bad service environment, the requirements of the container surface are very strong corrosion resistance, it requires not only the elements in the matrix of weathering steel to improve corrosion resistance, but also through the container coating system it can further improve the corrosion resistance of coating. Because of surface coating of the container, the outside water and oxygen, electrolyte is separated from steel box, which can stop and retard the corrosion of steel structure box, prolong the service life of the container. The container often uses 3 layers coating system for external and commonly used 2 layer coating system for inner. The quality of surface treatment before application of the coating is the greatest impact of coating life (i.e. antiseptic effect), moreover, it is the main factor affecting coating performance [4, 5].

The traditional coating methods of containers are shot blasting pretreatment before coating, and its main purpose is to remove impurities such as oxide scale, rust and other impurities on the strip surface, and provide the roughness that is favorable to protect performance of the coating. However, shot blasting pretreatment will lead to greater dust pollution in the process of production, and the noise is greater, seriously worsening the operating environment, also shot blasting pretreatment will increase the process cost. The development of container steel without pretreatment can be eliminated shot blasting pretreatment. It can be coated directly on the oxide scale layer. The two coating methods are shown in figure 1.

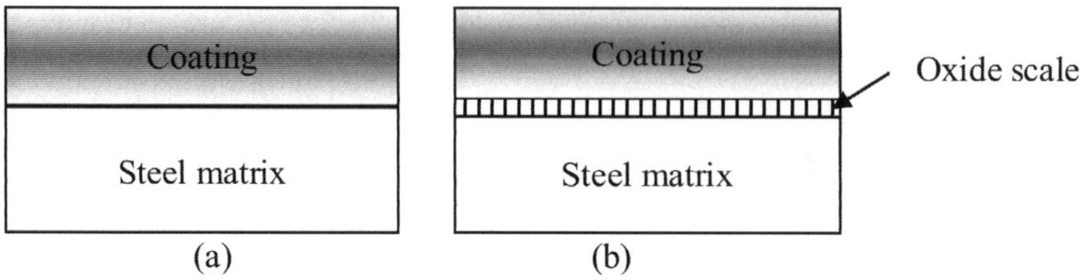

Figure 1. Comparison of traditional and free shot blasting coating methods of containers. (a) Traditional coating method and (b) Free blasting coating method.

2. Requirements on surface oxide scale of pretreatment container steel

Controlling surface oxide scale structure and the thickness of hot rolling strip, it can deformed with matrix in the deep processing process without spalling, and the pickling process is canceled, which has been successfully applied to the free pickling automobile beam steel production [6,7]. The container field mainly focuses on the coating technology and coating researching, but the adjustment and improvement of the oxide scale technology for container free blasting pretreatment is not the scope of related studies and reports.

Because of the corrosion, container coating strictly requires on the surface quality. Making the container steel directly coating without blasting pretreatment, it needs oxide scale layer which is compact and clean, the surface roughness is good to the coating, and the oxide scale is good compatibility with the coating system, so it does not affect the performance of coating. At the same time, in the production of container, the steel surface oxide scale needs to adapt to the roller, pressing, cutting, welding and other processes, there is no massive spalling that affects subsequent coating. Different oxide scale structures are of different properties, and the mechanical properties of various types of oxide scale are shown in table 1.

Table 1. Mechanical properties of different oxide scale structures at room temperature.

Oxide scale phase	Tensile strength /MPa	The integrity of oxide scales
Fe_2O_3	10	Poor
Fe_3O_4	40	Medium
FeO	0.4	Poor
α-Fe+Fe_3O_4	>40	Good

As can be seen from table 1, Fe_3O_4 and eutectoid α-Fe+Fe_3O_4 are higher tensile strength relative to Fe_2O_3 and FeO, and they can bear a certain deformation [8, 9]. In addition, it was reported that the corrosion rate and corrosion oxide scale are different with different structures [10]. The larger the eutectoid reacts, the more the eutectoid structure is, the greater the corrosion's weight is, and the faster the corrosion rate occurs. Eutectoid structure is a lamellar structure composed of α-Fe and Fe_3O_4. The inhomogeneous microstructure will lead to the formation of micro batteries, making electrochemical corrosion. The whole lamellar Fe_3O_4 or FeO single phase structure has a certain physical shielding effect, so it is relatively better corrosion resistance. Therefore, considering the oxide scale integrity and corrosion resistance, the following scale structure is better comprehensive properties, outer thin Fe_3O_4 layer, inner completely eutectoid structure, as shown in figure 2.

Steel surface after shot blasting treatment can form a wave crest. Through the micro roughness, it is better combination with the coating, ensuring corrosion resistance and adhesion. While free shot blasting pretreatment, with controlling the thickness and structure of oxide scale, the roughness of oxide scale should be controlled so as to improve the quality of the coating.

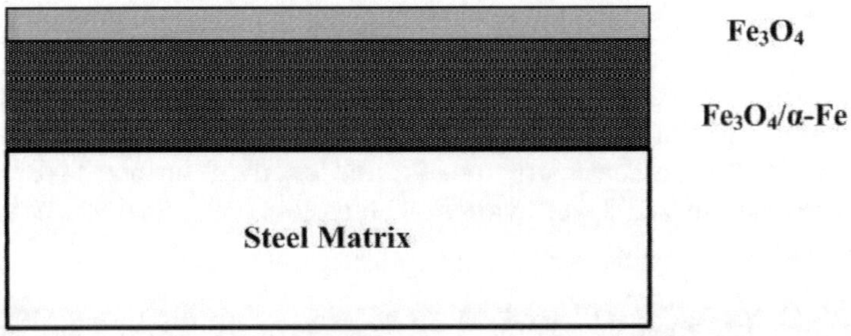

Figure 2. Target oxide scale structure.

In addition, for free shot blasting pretreatment container steel, there are still many problems to be solved in container industrial application, such as anti-corrosion in the transportation and storage process of steel, oxide scale protection in the process, etc. [11,12].

3. Production and application of free pretreatment container steel

3.1. Industrial trial manufacture and production process

The chemical composition of the steel for free pretreatment containers is shown in table 2.

Table 2. Chemical composition, wt%.

Element	C	Si	Mn	P	S	Cu	Ni	Cr
Range	0.05~0.20	0.20~0.50	0.30~0.60	0.08~0.12	≤0.015	0.20~0.50	0.05~0.40	0.30~0.50

Figure 3. Working roll after shot blasting.

The surface of the working roll of the end frame of the hot finishing mill is treatment with shot blasting as shown in figure 3. There is a certain roughness on the surface of working roll, and increasing rolling reduction rate, it will be transferred to the strip surface roughness because of the roll surface. It can lift the surface roughness of steel container, which can make up for the surface roughness of shot blasting processing canceled. The method can increase the strip surface roughness Rz from about 6 μm to 20 μm, and help improve the quality of coating.

Discharging temperature of slabs is 1180~1200℃, the final finishing rolling temperature is 860~880℃, interstand water is used in F1-F3 frame work, coiling temperature is 560~590℃, the preceding rapid cooling is used, and wind cooling + packing cooling are controlled between 570~300℃ after coiling, it can reduce the total consumption of Fe_3O_4 under high temperature with long time, and it can be a thin layer of Fe_3O_4 layer, and there is enough α-Fe+ Fe_3O_4 eutectoid structure.

3.2. Oxide scale characters

When the steel coil is cooled to room temperature, samples are carried out when the steel coil is temper rolling. Samples are prepared along the width of the sheet, 30 mm of the edge, 1/4 of the sheet width, and the center of the sheet width. The morphology, structure and thickness of the oxide scales are observed and analyzed by electron probe. The structure and thickness of oxide scale in each part are shown in figure 4. From the scanning results of the oxide scale, the thickness of the oxide scale is thinner, the thinnest is 4.4 μm, and the thickest is 8 μm. The thinner oxide sheet can guarantee better integrity and adhesion. The oxide scale structure is composed of surface layer Fe_3O_4 and bottom eutectoid Fe/Fe_3O_4, and the surface layer without eutectoid layer can reduce the electrochemical corrosion caused by eutectoid structure.

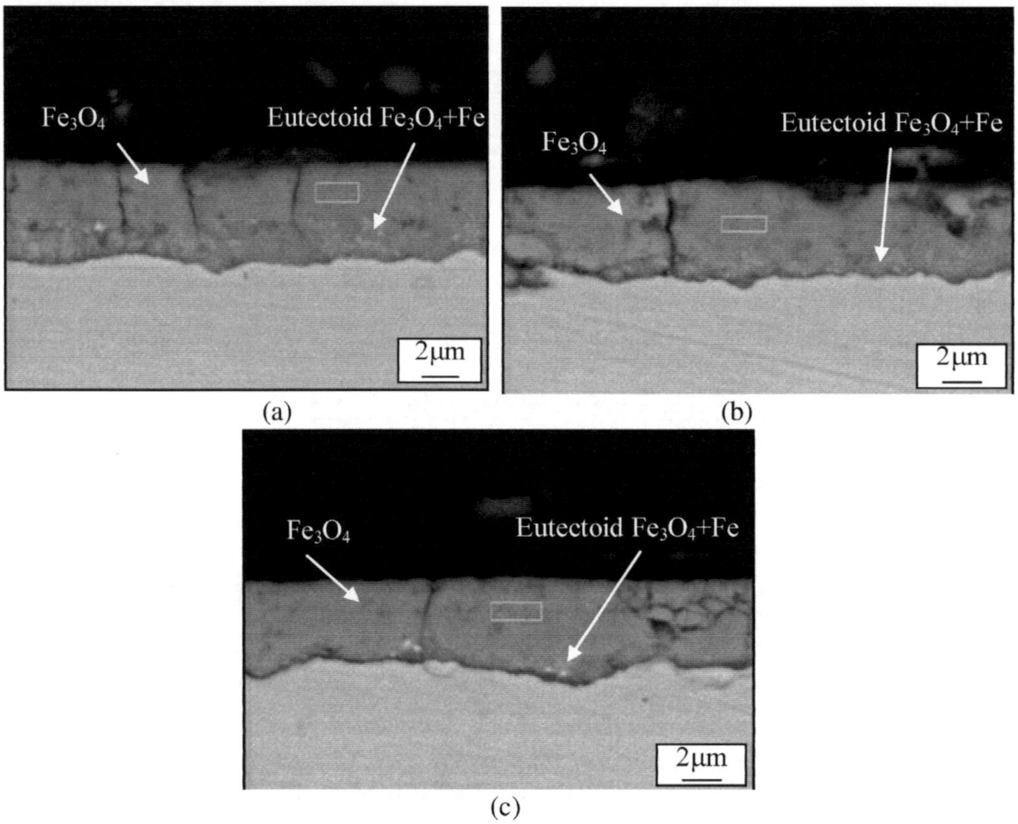

Figure 4. Oxide scale morphology and structure of free pretreatment container steel along sheet width. (a) Side of steel sheet, (b) 1/4 of steel sheet width and (c) 1/2 of steel sheet width.

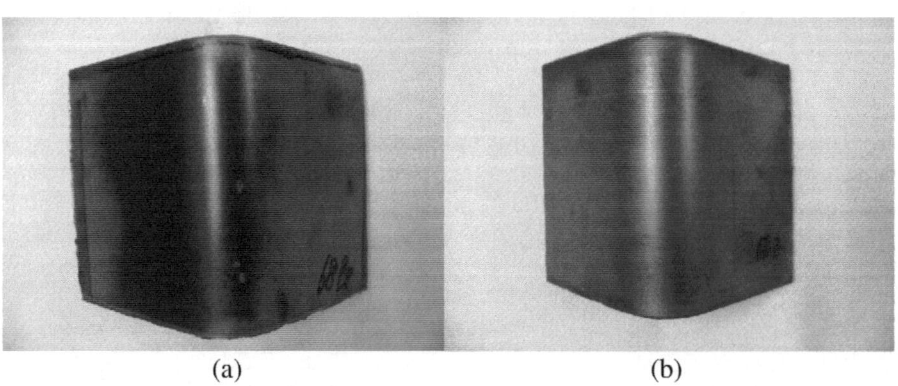

Figure 5. Cold bending surface of free pretreatment container steel. (a) Sample at broadside and (b) Sample at 1/2 width.

IOP Conf. Series: Materials Science and Engineering **283** (2017) 012014 doi:10.1088/1757-899X/283/1/012014

3.3. Examination of adhesion for oxide scale

The adhesion of oxide scale is tested by 90° cold bending test, the sample specification is 100 ×150 mm, and cold bending core diameter is d=5 mm. According to the cold bending results, there is no massive spalling of oxide scale on the sheet surface during deformation, and the integrity of oxide scale is good, as is shown in figure 5.

3.4. Performance test after coating

3.4.1. Salt spray test. According to the standard of ASTM B117-2011, salt spray test results are shown in figure 6, the test results show that after 1000 hours salt spray test, inner and external coating samples are qualified with corrosion resistance, but slight difference in second adhesion, mainly in the coating spalling after 1000 h salt spray test.

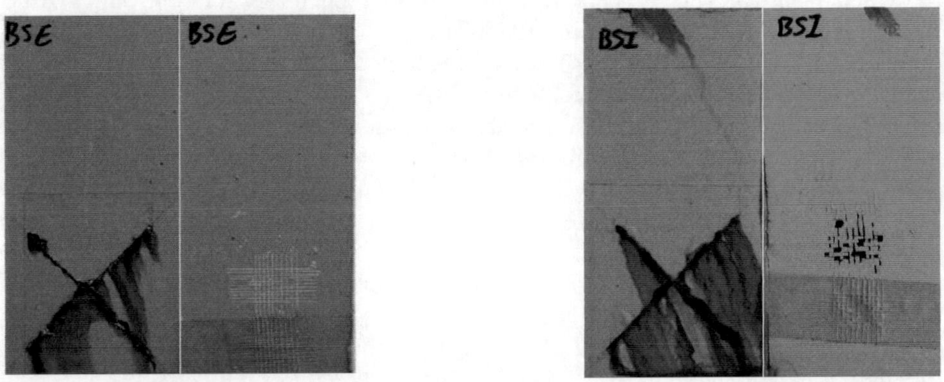

(a) (b)

Figure 6. Salt spray test. (a) External coating sample and (b) Inner coating sample.

3.4.2. Molding performance. During the stamping and roller process of the free pretreatment container steel, the free shot blasting steel sheet also exhibits good formability and adhesion to oxide scale, as shown in figure 7.

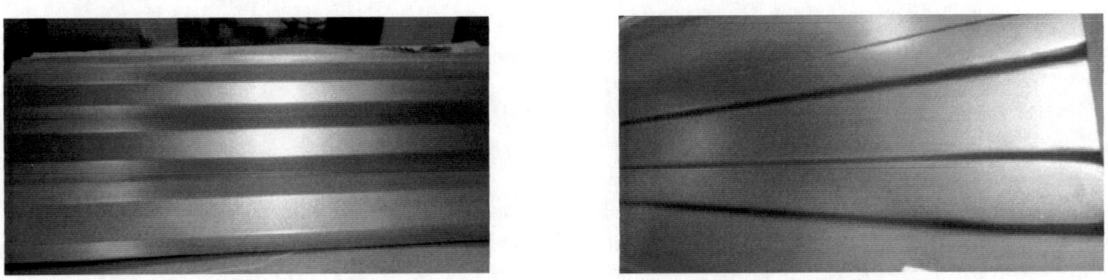

(a) (b)

Figure 7. Forming of free pretreament container steel. (a) Roller forming and (b) Stamping forming.

4. Conclusions

The free shot blasting pretreatment container steel has the advantages of traditional container steel in economic and environmental protection, and has great prospects for development. The test results show that the structure of oxide scale is composed of surface layer Fe_3O_4 and bottom eutectoid α-Fe/Fe_3O_4 layer, and the structure control of oxide scale basically achieves the expected goal. Oxide scale integrity and adhesion is good after coating, basically satisfies the container industry needs, but there is some corresponding improvement to do in the future work.

References

[1] Li M F and Zhang Q 2008 Analysis on development status of China container coatings *China Paint* **23(05)** 36-41
[2] Zhao J, Yao X M and Liu D 2010 Research and coating technology of coatings for containers, *Coat App* **40(01)** 16-21
[3] Levinson M 2016 *Container Changes the World* (Beijing: China Machine Press)
[4] Liu H C 2010 Application of new coating process in container industry *Modern* Pain *& Finish* **13(10)** 41-4
[5] Yu Z J, Wang Y S 2010 Wang X L Brief introduction of container coating process *Tianjin Sci & Tech* **37(3)** 9-12
[6] Liu Z Y, Yu Y, Guo X B, *et al* 2009 Techniques of oxide scale control during hot strip rolling *Steel Roll* **26(1)** 5-9
[7] Liu Z Y and Wang G D 2011 The R & D progress on technology for controlling oxide scale on hot rolled steel products *Angang Tech* **2011(02)** 1-5
[8] Chen R Y and Yuen W Y D 2003 Review of the high-temperature oxidation of iron and carbon steels in air or oxygen *Oxid Metals* **59(5-6)** 433-68
[9] Chen R Y and Yuen W Y D 2000 a study of the scale structure of hot-rolled steel strip by simulated coiling and cooling *Oxid Metals* **2000(5)** 539-60
[10] Gu Q D, Dong C F, Li X G, *et al* 2009 Corrosion behavior and structure of oxide scales formed on hot rolled strips after coiling with different oxygen supplies *J University Scien Tech Beijing* **31(12)** 1564-8
[11] Qi H B, Qian Y H, Ou P, *et al* 2007 The corrosion behavior and influence factors of hot rolled carbon steels *Steel World* **2007(06)** 1-5
[12] Qi H B, Qian Y H, Zhang H, *et al* 2009 Rusting behavior and preventative measures of hot-rolled steel plates for automobile applications *Baosteel Tech Res* **03(02)** 3-6

The effect of temperature characteristic variation due to thermal control design on the antenna of satellite laser communication terminals

K Mei[1], Y W Song[1], L Liang[1*], J D Du[1*], S Fu[1*] and Y J Jiang[1*]

[1]Shanghai Radio Equipment Research Institute, Shanghai 200090, China
*E-mail: 99961784@qq.com/ dujindan802@sina.com/ fusensam@126.com/ opticsjiangyijun@sina.com

Abstract. Cassegrain telescopes are widely used as optical antenna in satellite laser communication terminals, which are assembled outside the satellite when operating in orbit. The thermal environment in space will result in temperature variation and thermal deformation of the telescopes. Then tracking-pointing performance and communication performance will be affected consequently. So, thermal control design of telescope must be taken to make the temperature distribution vary in a certain range, which is permitted to assure the performance and stability of optics. For laser communication terminals in satellite of geosychronous Earth orbit, the effects on figure error of telescope due to three kinds of heating type of thermal control system are analyzed in this paper. It is shown that radial heating type brings the smallest figure error to telescope, followed by circumferentially-equal-area heating type. circumferentially-equal-radius heating type brings the biggest error of all. The result of this paper will bring certain reference to the thermal control design of telescopes in laser communication terminals on satellites in orbit.

1. Introduction

Laser communication system (LCS) is expected to be an important development direction for ultra-high-data-rate satellite communication because of many advantages: huge communication capability, high security, low weight and consumption [1-3]. The asymmetric temperature distribution and variation on telescope will bring thermal deformation to telescope in space, which will result in tracking-pointing performance and communication performance degradation. Usually, thermal control system is used to keep the temperature a certain range on telescope which is receptive to assure performance. So it is necessary to study the temperature distribution on telescopes and thermal aberration after thermal control. The thermal control method and type are also very important for the performance of telescopes.

Researchers began to study the thermal deformation of optics caused by variation of temperature field in 1970s. Applewhite researched the temperature distribution effect on some space cameras in 1992 [1]. Volkmei assessed the temperature features of the solar telescope in 2008 [2]. Myung *et al* analyzed the performance of TMT telescope using finite element method by the software of ANSYS and I-DEAS in 2009 [3]. Song *et al* studied the influence to the SiC reflectors in periscopic laser communication terminals due to temperature variation in 2010 [4]. Segato *et al* put forward a new theory to study the thermal deformation of space optical system in 2011 [5], while Ron *et al* [6] assessed the optical performance of astronomical telescope composed by lightweight reflectors at low

IOP Conf. Series: Materials Science and Engineering **283** (2017) 012015 doi:10.1088/1757-899X/283/1/012015

temperature.

Although many researchers focus their further studies on the thermal deformation of optics due to temperature field, no study on performance contrast of different heating methods to telescope in satellite laser communication was reported yet. Given this, we study three different heating methods to the telescope in LCS, which are radial heating method, circumferentially-equal-area heating method, and circumferentially-equal-radius heating method. For these three methods, the numerical simulation of optical performance of telescope is performed in this study.

2. Numerical analysis of the thermal deformation due to stable temperature field

2.1. Model construction

The CAD model of optical antenna and support structure are elaborated in this paper, as is shown in figure 1. The primary mirror is fixed in the base which gives around-support to the primary mirror. The second mirror is fixed by three brackets which are connected with the base with screws. For the simulation reason, the screws are not given in the model. The boundary conditions will be set instead of the screws. The primary mirror and the second mirror need to be well coaxial to keep a good imaging quality for telescopes. For the telescope of LCTs in space, the materials of the telescope are very important. For less temperature sensitivity and good mechanical property, the material of the primary mirror and second mirror is Zerodur, the material of the primary mirror base is Titanium alloy, and the material of the brackets is invar.

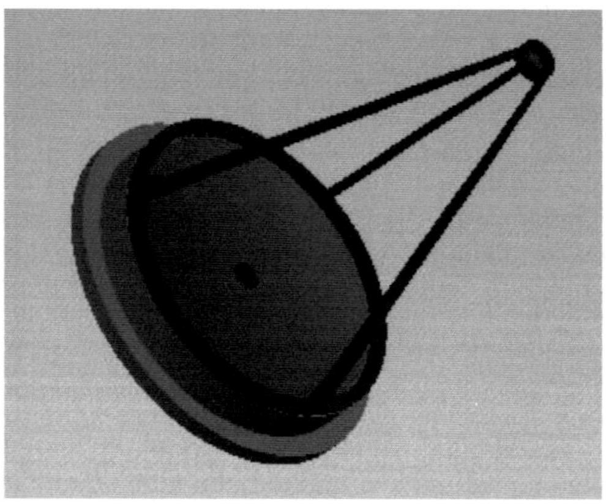

Figure 1. The CAD model of the telescope in LCT.

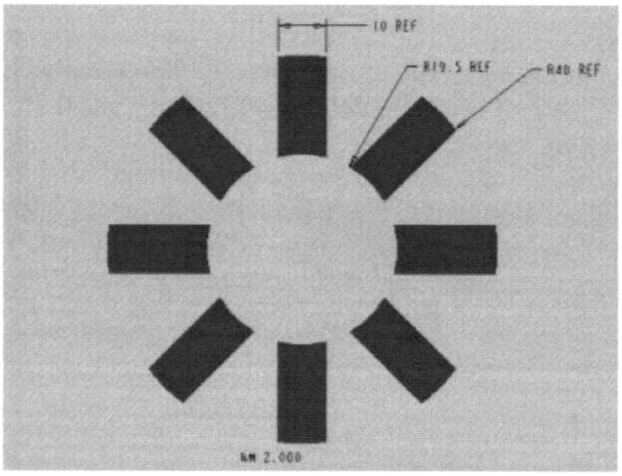

Figure 2. The model of radial heating plates.

2.2. Simulation of the figure of telescope under radial heating

The model with real size of the radial heating plates for telescope is shown in figure 2. When working in space, the environment is usually cold, although there is first order thermal control system. In this paper, the working temperature of telescope is set as 15℃ after first order thermal control for simulation. The main thermal passing progress of the telescope of LCTs in space is thermal conduction and thermal radiation. The convection is not considered in space. The solar radiation is not considered because of the cold environment to simplify the calculation. The parameter of heating plates in the simulation is set to be the temperature directly for the easy display of the temperature field. The symbol of temperature is simplified to be T in this paper. In the simulation, the environment temperature is set to be 15℃, and the value of T is between 25℃ and 29℃. The interval of temperature ΔT is set to be 0.5℃ in the analysis result. The back-uniform-heating method is used to do the thermal control for the second mirror because of its small aperture. The aimed temperature of heating is set to be 25℃.

The temperature field distribution and the reduced thermal deformation are shown in figures 3 and figure 4 under the radial heating method. It is shown from the result data, the highest temperature of the primary mirror rises from 22.6℃ to 25.8℃ as the temperature of the heating plates on the back of the primary mirror changes from 25℃ to 29℃. At the same time, the lowest temperature of the edge of the primary mirror rises from 19℃ to 20.5℃. As the temperature rises from 25℃ to 29℃, the biggest value of the deformation of surface of the primary mirror changes from 2.1 μm to 35.1 μm. Simultaneously, the smallest value of surface of the primary mirror rises from 0.71 μm to 14.1 μm.

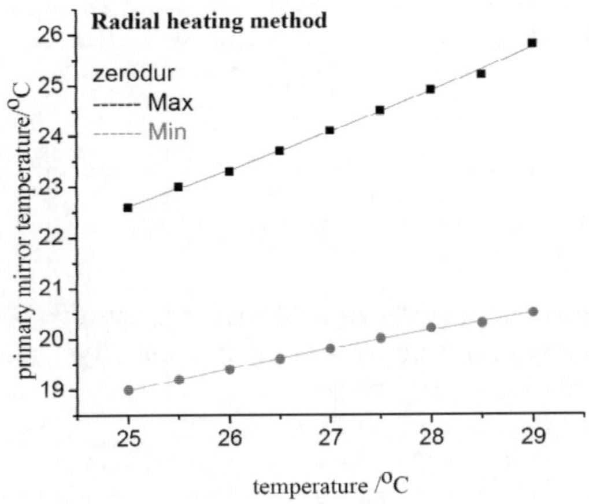

Figure 3. The temperature distribution curve of primary mirror under the radial heating method.

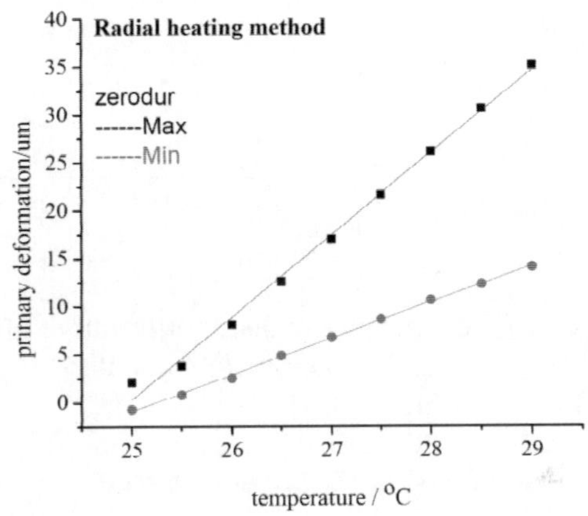

Figure 4. The thermal deformation curve of primary mirror under the radial heating method.

2.3. *Simulation of the figure of telescope under circumferentially-equal-area heating method*

The model with real size of the circumferentially-equal-area heating plates for telescope is shown in figure 5. The temperature field distribution and the reduced thermal deformation are shown in figures 6 and figure 7 under the circumferentially-equal-area heating method.

It is shown from the result data of figures 6 and figure 7, the highest temperature of the primary mirror rises from 22.3℃ to 25.2℃ as the temperature of the heating plates on the back of the primary mirror changes from 25℃ to 29℃. At the same time, the lowest temperature of the edge of the primary mirror rises from 19.2℃ to 20.8℃. As the temperature rises from 25℃ to 29℃, the biggest value of the deformation of surface of the primary mirror changes from 3.6 μm to 40.7 μm. Simultaneously, the smallest value of surface of the primary mirror rises from 0.2 μm to 16.4 μm.

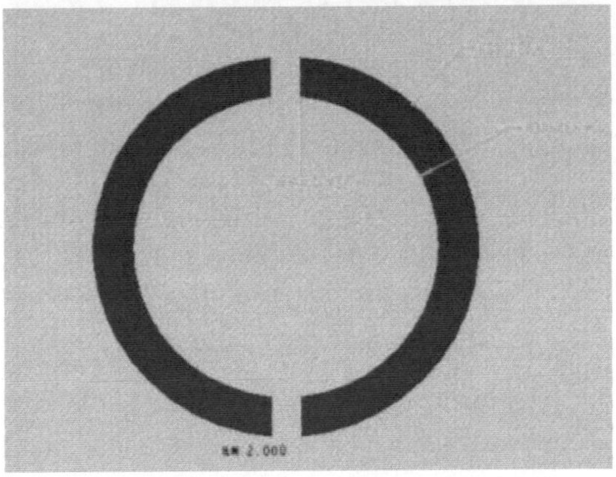

Figure 5. The model of circumferentially-equal-area heating plates.

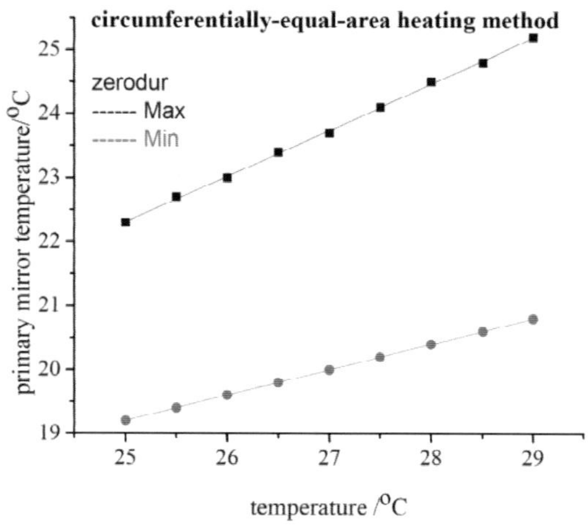

Figure 6. The temperature distribution curve of primary mirror under the circumferentially-equal-area heating method.

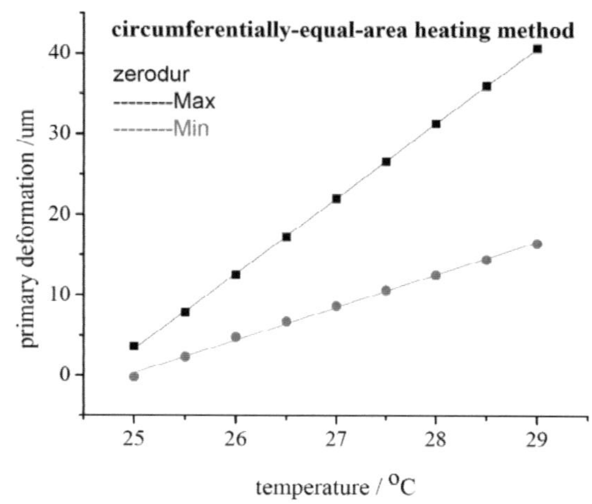

Figure 7. The thermal deformation curve of primary mirror under the circumferentially-equal-area heating method.

2.4. Simulation of the figure of telescope under circumferentially-equal-radius heating method

The model with real size of the circumferentially-equal-radius heating plates for telescope is shown in figure 8. The temperature field distribution and the reduced thermal deformation are shown in figures 9 and 10 by the circumferentially-equal-radius heating method.

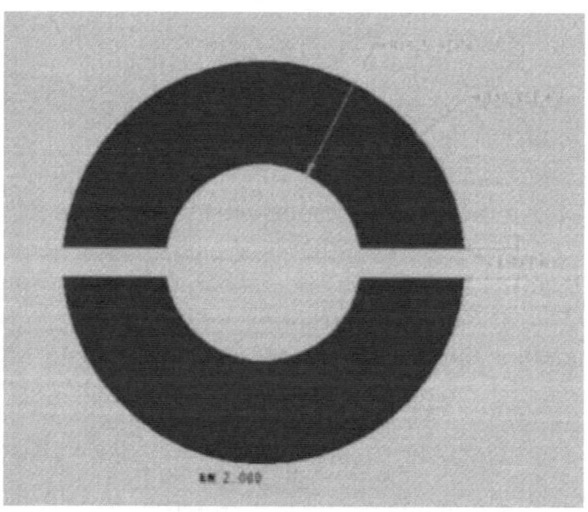

Figure 8. The model of circumferentially-equal-radius heating plates.

It is shown from the result data of figures 9 and 10, the highest temperature of the primary mirror rises from 23.2℃ to 26.5℃ as the temperature of the heating plates on the back of the primary mirror changes from 25℃ to 29℃. At the same time, the lowest temperature of the edge of the primary mirror rises to 21.5℃ from 19.7℃. As the temperature rises to 29℃ from 25℃, the biggest value of the deformation of surface of the primary mirror changes from 14.1 μm to 55.0 μm. Simultaneously, the smallest value of surface of the primary mirror rises from 5.7 μm to 23.6 μm.

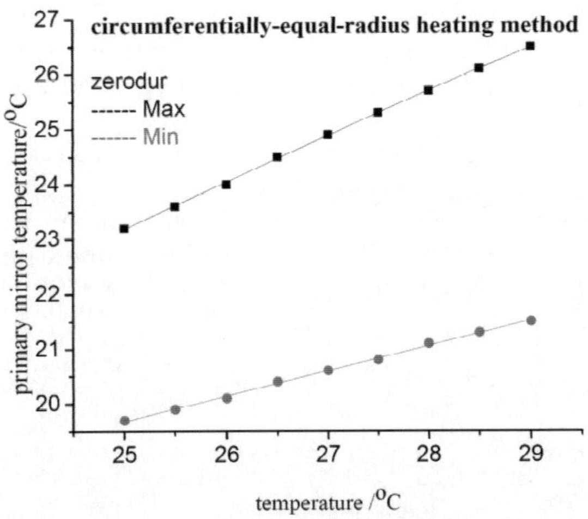

Figure 9. The temperature distribution curve of primary mirror under the circumferentially-equal-radius heating method.

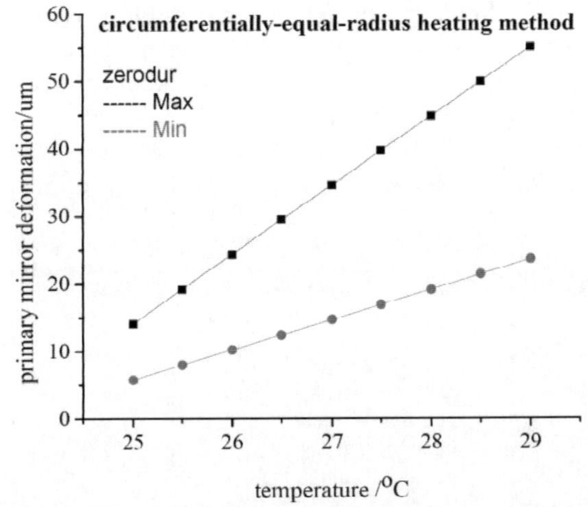

Figure 10. The thermal deformation curve of primary mirror under the circumferentially-equal-radius heating method.

3. Numerical analysis of the impact on performance of the telescope due to stable temperature field

This chapter mainly studies on the performance of the optical system with temperature variation. Because of the influence of the temperature field, the deformation of surface of the telescope occurs, resulting in the change of performance of telescope. This paper gives the wave-front aberration of telescope due to the thermal deformation under the three different heating methods above, including RMS value and P-V value. After that, the radial energy distribution and the MTF value are also analyzed.

3.1. Wave-front aberration analysis of telescope under three heating methods

Figures 11 and 12 give the RMS value and P-V value of the telescope at zero-FOV (field of view) under the three heating methods with the environment temperature being 15℃. This is because the comparability of RMS value and P-V value between different FOV values with a same wave length.

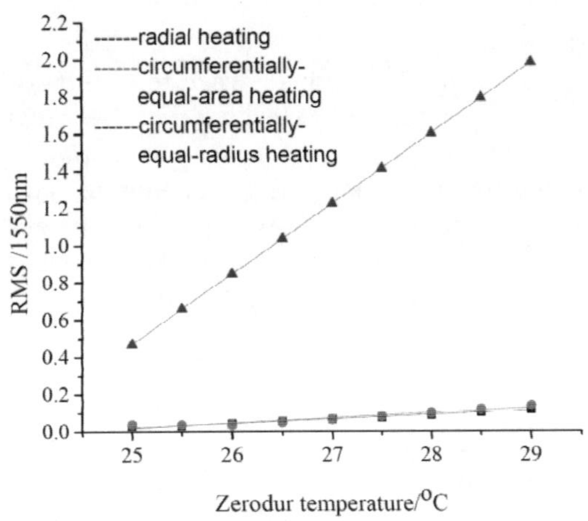

Figure 11. The RMS value variation curve at zero-FOV under the three heating method.

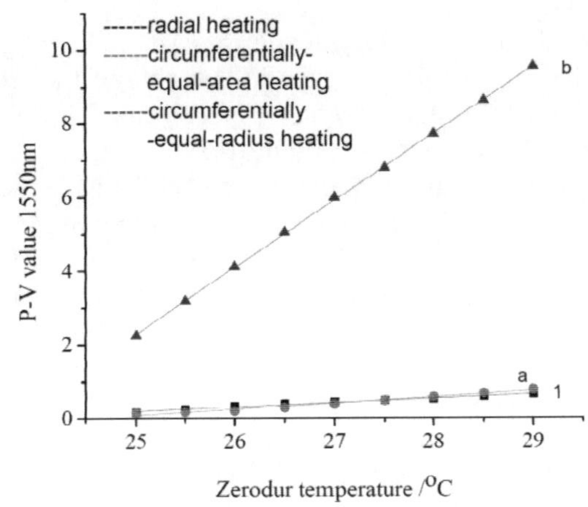

Figure 12. The P-V value variation curve at zero-FOV under the three heating method.

It is shown that the RMS value and P-V value are almost the same under radial heating method and circumferentially-equal-area heating method, and varies within a small range. However, the RMS value and P-V value under circumferentially-equal-radius heating method are much bigger and vary within a much larger range.

3.2. Radial energy distribution analysis of telescope under three heating methods

The reference to measure radial energy distribution is the diameter of the light which has nighty percent energy of the optical field, which is called 'shaped diameter' below. Figure 13 gives the shaped diameter variation with temperature under the three heating methods.

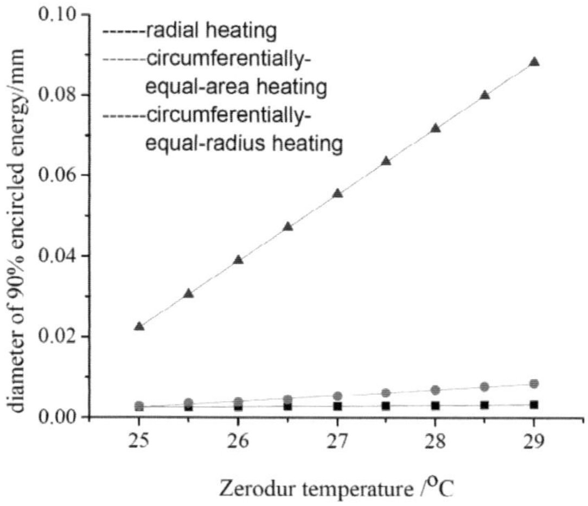

Figure 13. The shaped diameter variation curve under different heating methods.

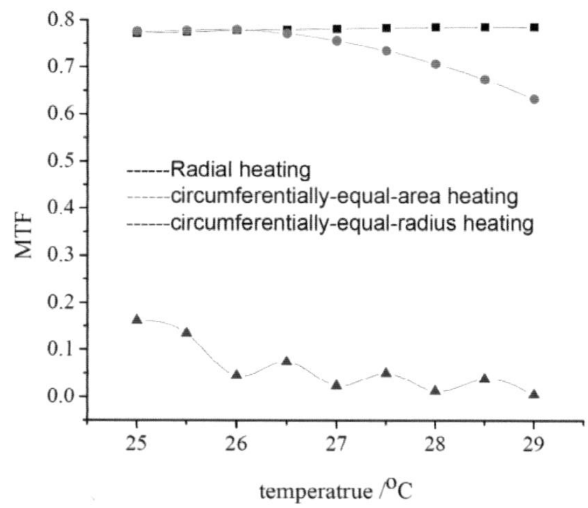

Figure 14. The MTF variation curve under different heating methods.

It is shown that the shaped diameter is almost the same under radial heating method and circumferentially-equal-area heating method when the temperature is small. As temperature grows, the difference of shaped diameter between the two heating methods becomes larger, and the circumferentially-equal-area heating method is the larger one. However, the shaped diameter of the circumferentially-equal-radius heating method is much bigger and varies within a much larger range.

3.3. MTF analysis of telescope under three heating methods

For the contrast purpose, the MTF value of the telescope is evaluated by the max space resolution of the COMS at 68lp/mm whose pixel size is 7.4 μm×7.4 μm. As the same, the MTF value of telescope at the zero-FOV is given in figure 14.

From the results, we can know that the MTF value is almost the same under radial heating method and circumferentially-equal-area heating method when the temperature is small. As temperature grows, the difference of MTF between the two heating methods becomes larger, and the circumferentially-equal-area heating method is smaller. However, the MTF value of the circumferentially-equal-radius heating method is much smaller and varies periodically.

4. Conclusions

The FEM simulation of telescope in LCTs is performed in this study using CAD and FEM software. By analyzing temperature and thermal deformation distribution, the variation result and difference of the optical performance of the telescope are given under the three heating methods. The variation regularity of optical performance is derived by different heating methods, including temperature distribution, RMS value of the surface, P-V value, shaped diameter of the light, and MTF value. It is shown that at the same temperature the radial heating method exerts the minimal effect on performance, the circumferentially-equal-area heating method is the second, and the circumferentially-equal-radius heating method has the strongest effect. We hope this work can contribute to the thermal control of telescopes in satellite laser communication terminals.

Acknowledgments
This research was supported by the National Natural Science Foundation of China (NSFC). The authors are grateful for a grant from NSFC.

References

[1] Applewhite R W 1992 The effects of thermal gradients on the mars observer camera primary mirror *J. SPIE* **1690** 376-86
[2] Volkmer R 2008 Thermal characteristics of the solar telescope Gregor *J Proc. of SPIE* **7012** 1-9
[3] Myung C, Andrew C, Konstantinos V, *et al* 2009 Thermal performance prediction of the TMT telescope structure *Proc. of SPIE* **7427** 1-15
[4] Song Y W, Yu S Y, Tan L Y, Ma J, *et al* 2010 The effect of temperature distribution in space on the figure of reflectors *J. Journal of Astronautics* **2010(3)** 868-74
[5] Segato E, Da Deppo V, Debei S, *et al* 2011 Method for studying the effects of thermal deformations on optical systems for space application *J. Applied Optics* **50(18)** 2836-45
[6] Ron E, William R and Arnold 2013 Cryogenic optical performance of a lightweighted mirror assembly for future space astronomical telescopes correlating optical test results and thermal optical model *J. Proc. of SPIE* **8837(0B)** 1-13

IOP Conf. Series: Materials Science and Engineering **283** (2017) 012016 doi:10.1088/1757-899X/283/1/012016

Heat resistance study of basalt fiber material via mechanical tests

Y Q Gao*, C Jia, L Meng and X H Li
Hebei University of Science and Technology, China
*E-mail: gao_yiqiang@163.com

Abstract. This paper focuses on the study of the relationship between the fracture strength of basalt rovings and temperature. Strong stretching performance of the rovings has been tested after the treatment at fixed temperatures but different heating time and then the fracture strength of the rovings exposed to the heating at different temperatures and cooled in different modes investigated. Finally, the fracture strength of the basalt material after the heat treatment was studied. The results showed that the room-temperature strength tends to decrease with an increase of the heat treatment time at 250°C, but it has the local maximum after 2h heating. And the basalt rovings strength increased after the heat treatment up to 200°C. It was 16.7 percent higher than the original strength. The strength depends not only on the temperature and duration of the heating, but also on the cooling mode. The value of the strength measured after cold water cooling was less by 6.3% compared with an ambient air cooling mode. The room-temperature breaking strength of the rovings heated at 200 °C and 100 °C for 2 hours each increased by about 14.6% with respect to unpretreated basalt rovings.

1. Introduction

With the development of industrial science and technology, the new fiber materials of high performance emerge, and the range of their applications expands. Among them basalt fiber materials are well known due to very high strength and excellent high temperature resistance. The studies of thermodynamic properties of basalt and glass fibers performed by Zhou 1] have shown that due to Fe_2O_3 and FeO compounds in the basalt yarn the working temperature of the basalt fiber can reach as high as 600°C, while glass fibers lose their strength at heating above 400 °C. The morphology and thermal stability of the basalt and glass fibers were analyzed and compared by Hao *et al* [2]. Using the projection microscope and scanning electron microscopy, they found similar structure of the two types of fibers, but the thermal stability of basalt fibers is higher. After study the high temperature resistance of the basalt fibers Li [3] concluded that at 200 °C they possess the best mechanical properties. Generally, the heat resistance of basalt fibers is not well investigated. Nevertheless, basalt fiber products are extensively used at high temperature in many industrial applications, and their functional temperature and heating time are different in different industrial sectors. In this study, we designed and constructed a testing system for the fracture strength measurements of the basalt fiber materials after their heat treatment at various temperatures, duration of heat exposure and different cooling modes, in order to find out the relationship between the mechanical properties and parameters of the heat treatment of basalt fiber materials. The results obtained by this method will be used for the development of basalt fiber products and their practical applications.

2. Test materials, instruments, and test conditions

2.1 Test materials

264tex basalt non-twist roving, single fiber diameter 13 μm, fracture strength of 41.51 cN/Tex, provided by Sichuan Tuxin Basalt co., Ltd. (China).

2.2 Equipment and instruments

YG-01 electrical thermostatic drying oven (temperature control range of 0-400^{0}C) was produced by the Tianjin city Test Instrument Co., Ltd. (China)

UTM5305 electronic universal power machine with WGDY - 7350L high-temperature box; Shenzhen established company

Test reference standard: GB/t25045-2010

Test conditions: temperature 20±2^{0}C, the relative humidity 64%.

3. Experimental study

3.1 Study of the relationship between the heating time and the material fracture strength

We made a series of 6 cm long samples of basalt rovings and put them in the electrical thermostatic drying oven, set the oven temperature 250^{0}C, and the processed samples in the drying box for 0.5 h, 1 h, 1.5 h, 2h, 2.5 h, 3h, 3.5 h and 4h, then placed them out. After cooling, the tensile fracture strength of each specimen was tested and presented in table 1, 10 samples per group, then we calculated the average and plotted the fracture strength of the basalt roving as a function of the heating time at T=250 °C as shown in figure 1.

Table 1. Breaking strength of basalt fiber rovings after different heating time

Time(h)	0.5	1.0	1.5	2.0	2.5	3.0	3.5	4.0
Mean breaking strength(cN/tex)	39.1	35.9	38.5	42.9	39.3	32.8	31.5	28.6
Variation(%)	9.8	9.0	5.7	7.0	5.5	10.3	7.5	9.3

As shown in figure 1, after the heating in the oven at a temperature of 250^{0}C, the fracture strength of the basalt rovings changed with the change of the heating time. There was a certain decrease after the first hour treatment, then the breaking strength increased after heating up to 2 hours when the strength reached the maximum, but then it dropped again.

According to figure 1, we suppose that a strength loss in the initial stage of heating took place due to mass agglomeration of the surface treatment agent and the decrease of the cohesive force. The fracture strength increased after 1 hour heating probably due to better consolidation of the fibers as a result of volatilization of small molecules like as water and solvents, and the fiber surface became rougher. And the fracture strength decreased almost linearly after 2 hours heating, because the surface treatment agent volatilized greatly, and the fiber structure, interatomic chemical bonds destroyed. After 4 hours heating, the strength of the rovings decreased by 31.6%.

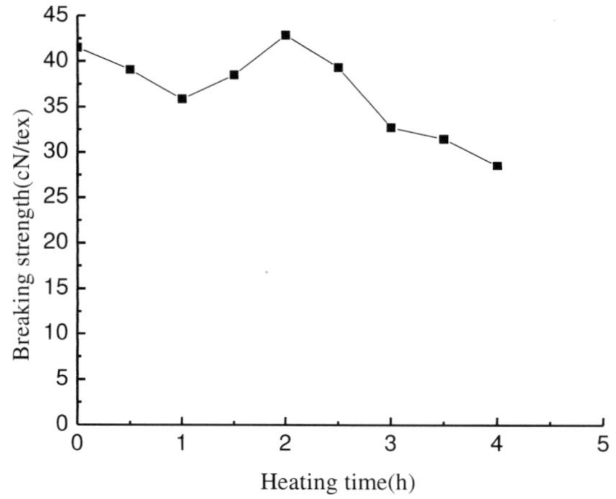

Figure 1. Room-temperature fracture strength of the basalt rovings as a function of the heat treatment time at T=250 °C

Nevertheless, after basalt rovings heating for 3 hours at 250°C temperature their strength still was equal to about 80% of the original one. So far, the basalt fibers heat resistance is quite good.

3.2 Analysis of the fracture mechanism of the basalt fiber material after the heating

The cross-section electron microscopy of the basalt fiber bundles is shown in figure 2. A photo of the short stretch break of basalt roving is shown in figure 3.

Figure 2 implies that basalt fibers differ from conventional fibers or the other organic high performance fibers. The fracture is very smooth, the texture is hard, the same as a stone material. So after the basalt stone was processed into the fibers they are still a very brittle material. It can be seen from figure 3 that the basalt roving is broken; the beam fracture appears like an explosion. At drawing the monofilament elongation is small but the energy accumulation of a tension is high, after approaching the fiber strength limit the tensile force exceeds the limit of each single filament, all of the filaments will break simultaneously and form an explosive fracture, which is very different from the traditional way of breaking a yarn.

Figure 4 shows the displacement curve of the basaltic roving at the loading. As we can see from the curve, the displacement is very small until the fracture, and no more than 3 mm after the strength reaches the maximum and it falls off as a cliff. This means that the monofilaments were fractured almost simultaneously when the breaking strength reached the maximum. The load decreased slowly at initial stage, probably due to entanglement between fibers resulting in friction.

Figure 2. Cross-sectional morphology of basalt fiber fracture

Figure 3. A photo of the moment of the crack of the basalt roving

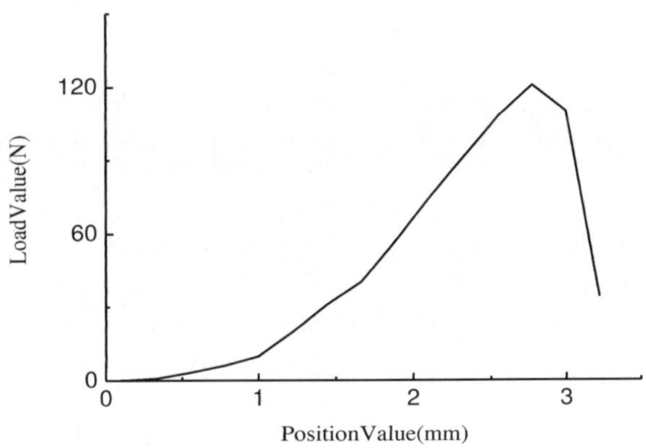

Figure 4. Tensile fracture curve of the basalt roving

3.3 Relationship between the material fracture strength and the temperature

According to the described above sample preparation method, we prepared a series of samples at -20, -10, 0, 50, 100, 150, 200, 250, and 300 °C during 2 hours.

Table 2. Breaking strength of the basalt fiber under different heating temperature

Temperature(℃)		-20	-10	0	50	100	150	200	250	300
Natural cooling	Mean breaking strength(cN/tex)	43.6	45.1	44.2	43.6	45.3	46.8	48.5	42.6	26.9
	Variation (%)	5.5	10.9	11.1	6.1	7.3	8.2	7.3	8.7	6.8
Water cooled	Mean breaking strength(cN/tex)	42.5	42.8	38.5	39.5	40.6	42.7	40.9	40.1	30.8
	Variation (%)	2.0	4.0	4.0	7.8	4.7	9.4	9.1	9.4	14.3

Then they were dealt with in two different ways: one series was cooled quickly in cold water and the other one series was cooled at ambient conditions. Then the relationship between the temperature of the treatment and the room-temperature fracture strength were tested and the results are shown in table 2 and figure 5.

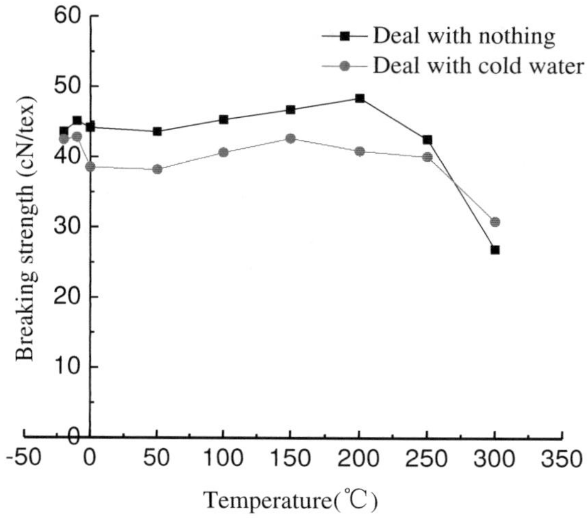

Figure 5. The room-temperature fracture strength of the basalt rovings as a function of the temperature of the treatment

As the natural cooling curve in figure 5 shows, the fracture strength of the basalt roving materials increased gradually with the increase of the processing temperature up to 200°C. The breaking strength after the treatment at 200°C reaches the maximum value, which is higher than that of untreated material by 11.1%. Then in the range between 250 - 300 °C, the strength reduced by 44.4% referring to the highest value. The reason of a sustained decline in the fracture strength of basalt rovings is a change in the internal structure of the fibers and the volatilization of the surface treatment agents at the heating. Ying *et al* [4] have shown that basalt fiber can still maintain about 70% of the initial strength at 400 °C. This may take place due to different basalt composition and the surface treatment agent.

As is seen in figure 5, the test results are different with different cooling ways after heating. The fracture strength of the basalt rovings cooled by cold water are generally less than the strength of the ones cooled at room temperature, excluding the case of the heating at 300 °C. The difference of the test values between the two different processing methods is about 16.3% for the samples heated at 200 °C.

As seen in figure 5, the strength of the basalt rovings cooled by cold water may be higher than those cooled at ambient conditions if samples were treated below -20 °C. However, the low temperature range of the experimental equipment was limited, so the low-temperature range needs further study in the future.

3.4 Study of the strength retention of materials after heat treatment

The results described above show that the breaking strength reached the maximum increase of 16.7% after the heating at 200 °C for 2 hours. In order to study the retainability of the strength gain, the following tests were carried out. Firstly, the basalt rovings were placed for 2 hours in a 200 °C environment. Then after the cooling, some of them were heated again for 2 hours at 50 °C, 100 °C, 150 °C and 200 °C. Finally, after cooling at room temperature they were tested and two curves of the breaking strength were plotted in figure 6 according to the results presented in table 3.

Table 3. Breaking strength of the basalt fiber after pretreatment and without pretreatment

Temperature(℃)		50	100	150	200
No pretreatment	Mean breaking strength(cN/tex)	43.6	45.3	46.8	48.5
	Variation (%)	6.1	7.3	8.2	7.3
Treated at 200 °C in advance	Mean breaking strength(cN/tex)	45.9	52.2	46.2	43.1
	Variation (%)	4.9	7.2	8.1	6.8

As is seen in figure 6, the fracture strength of the pretreated samples increased firstly and then decreased after reaching the maximum value at 100 °C. In this particular case the strength improved by about 14.6% with respect to unpretreated rovings. However, after the repeated treatment at 200° oC the fracture strength decreased by 6.9%, as comparied to the unpretreated basalt rovings. Nevertheless, basalt rovings heated two times at 200 °C, 2 hours each, still possess 93.1 percent the breaking strength of the original samples. This makes basalt fibers and rovings more suitable for the development of new temperature-resistive high-strength industrial materials than other raw materials.

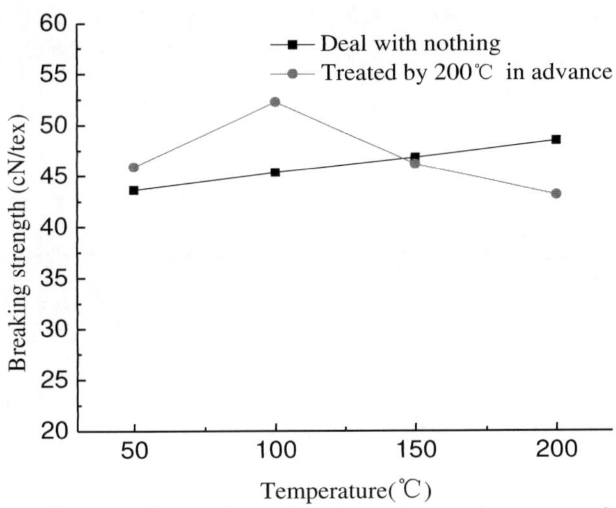

Figure 6. Comparison of the strength variation curves after two different treatment types

4. Conclusions

We made the thermal testing system on the basis of UTM5305 electronic universal testing machine and electrical drying oven and used it for the studies of heating effect on basalt rovings fracture strength. The mechanical tests have been carried out at room temperature after basalt rovings treatment at different temperatures with different heating time and different cooling modes. The following main conclusions are drawn from the research and analysis:

(1) Basalt fiber material has a very good high temperature resistance, after three hours temperature processing at 250°C, the fracture strength still maintain 78.9% of the initial value.

(2) The strength of the basalt fiber varies with the processing temperature, its fracture strength reaches the maximum after treatment at 200°C, it exceeds the initial one by 16.7%.

(3) The average fracture strength of the basalt fiber rovings cooled by cold water after the heat treatment decreased on about 6.3% comparing with unprocessed sample.

(4) The fracture strength of the basalt fiber rovings treated at 200°C temperature in advance, increased again after the next heat treatment up to 100 °C, its fracture strength increased by 14.63% comparing to the samples which were not pretreated at 200 °C temperature.

References

[1] Zhou R W, Song P F and Wang Q M, 2010 Study on properties of basalt yarn, *Journal of Functional Materials*, **51** (03): 3-5.

[2] Hao L C and Yu W D, 2009 Comparison of the morphological structure and thermal properties of basalt fiber and glass fiber, *Journal of Xi'an Polytechnic University*, **23** (2): 327-332.

[3] Li F Z and Li G C, 2015 Research on properties on high temperature resistance of basalt fiber yarn, Journal of Functional Materials, **03** (46): 3060-3063.

[4] Ying S N and Zhou X D, 2013 Chemical and Thermal Resistance of Basalt Fiber in Inclement Environments, *Journal of Wuhan University of Technology-Mater. Sci.* Ed. June: 560-565.

IOP Conf. Series: Materials Science and Engineering **283** (2017) 012017 doi:10.1088/1757-899X/283/1/012017

Experimental study of the continuous casting slab solidification microstructure by the dendrite etching method

X G Yang, Q T Xu*, C L Wu and Y S Chen
Shenyang QIGUANG Science and Technology Co. Ltd, Liaoning, Shenyang 110000, Liaoning, China
*Corresponding author: E-mail: angangxqt@163.com

Abstract. The relationship between the microstructure of the continuous casting slab (CCS) and quality defects of the steel products, as well as evolution and characteristics of the fine equiaxed, columnar, equiaxed zones and crossed dendrites of CCS were systematically investigated in this study. Different microstructures of various CCS samples were revealed. The dendrite etching method was proved to be quite efficient for the analysis of solidified morphologies, which are essential to estimate the material characteristics, especially the CCS microstructure defects.

1. Introduction

Quality of the continuous casting slab (CCS) is closely related to its solidification microstructure and it has a direct influence on defects of the steel product. By releasing the superheat, latent heat and sensible heat during the cooling process down to the room temperature, the typical solidification microstructure of CCS is usually composed of: (i) fine equiaxed grains, (ii) columnar grains, (iii) equiaxed grains, and (iv) crossed dendrite, as shown schematically in figure 1. With respect to the blooms and heavy slab, crossed dendrite as the transition zone can be observed between the columnar and equiaxed region, while crossed dendrite is rarely found in the billets and thin slab due to the fast cooling rate.

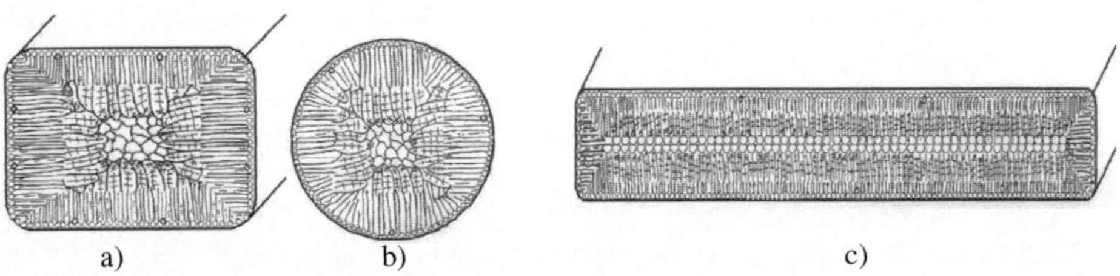

Figure 1. Schematic diagram of solidification microstructure of continuous casting: a) rectangular bloom; b) round billet; c) blank

The measurements of the deflective angles of the columnar grains and distances of the secondary dendrite are important for characterization of the solidification microstructure. In this respect, very instrumental is the dendrite etching of the solidification microstructure, which is based on the recommendations of Chinese standard GB/T 226-2015, whereas the assessment of the solidification microstructure accords to the methods given by GB/T 24178-2009.

In general, five macro-examination methods including sulfur print, hot etching, electrolytic etching, cold etching, and dendrite etching can be used to the macro-examination. Among these, the dendrite etching method is applicable for characterization of the solidification microstructure, which is further illustrated by several examples.

2. The relationship between the solidification microstructure and defects.

The solidification microstructure especial the distribution and abundance of various regions of CCS is of great concern due to their important effect on the defects of the steel products.

2.1 The relation between solidification structure and defect of continuous casting

The situation that the fine equiaxed region is too thin or inhomogeneous, especially the strong columnar grains below the surface depression, and absence of the chilled zone [1], may lead to the formation of surface longitudinal crack in the CCS and even to catastrophic failure. The strong columnar grains generally aggravate the central segregation of the CCS, thus deteriorating the properties of steel products. For example, low content of equiaxed grains in the strong columnar region of CCS may lead to an increase in the inhomogeneity and degree of deformation to effect the seamless steel tube internal buckling.

The high abundance of the equiaxed region, where the equiaxed grains are embedded into each other and firmly combined, implies a better processability to ensure isotropic mechanical properties. On the contrary, intense columnar grains with a parallel orientation to each other exhibit the intergranular segregation, banded structure, and anisotropy [2], therefore deteriorating the transversal mechanical properties and tenacity.

2.2 Influence of the solidification condition on the microstructure of CCS

It is generally believed that microstructure of CCS significantly depends on the solidification conditions. However, for the continuous casting process the latter are hard to be monitored and can be alternatively assessed and modified, according to the identified CCS microstructure patterns

3. Solidification microstructure of CCS

3.1 Fine equiaxed grains

During the solidification process, the fine equiaxed grains crystallize firstly at the wall of the mold and contribute to the surface of CCS. This region is fine, dense, and anisotropic, while its thickness is usually inhomogeneous and has a size of about 2-8 mm. In addition, the composition of the fine equiaxed region is similar to that of the liquid steel.

The fine microstructure of the fine equiaxed zone [3] is ascribed to the fact that nucleation rate much exceeds the growth rate because of very high cooling rate (100^0 C/s) caused by the strong endothermic and heat dissipation effects of the copper wall at the mold meniscus. As illustrated in figure 2, the fine equiaxed zone is located between the lower edge and the black line of CCS, above which the fine columnar zone can be observed.

The thickness of the fine equiaxed zone is mainly controlled by the liquid steel superheat: higher temperature corresponds to larger thickness and vice versa. In addition, the thickness of the fine equiaxed zone is also controlled by the cooling ability of the mold and the casting speed.

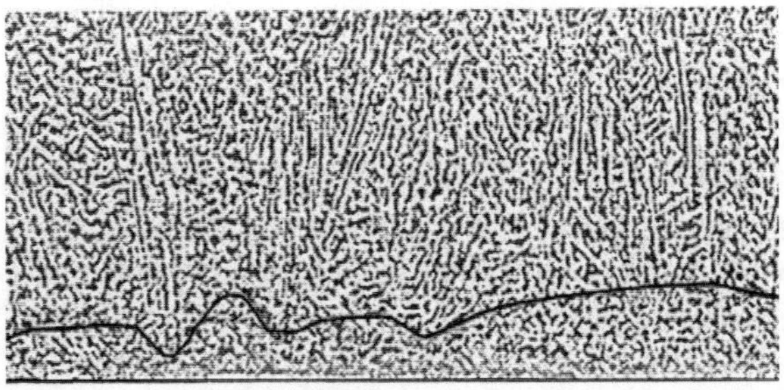

Figure 2. Micro-equiaxed and micro-columnar crystals in the transverse section of 300mm x 1650mm blank from 16Mn steel.

3.2 Columnar crystal

The columnar grains are actually composed of a primary dendrite, secondary dendrite, tertiary dendrite, and multiple dendrites. These dendrites generally grow in parallel to the same direction and have a pillar-like morphology [4]. The columnar zone distributes between the fine equiaxed zone and the equiaxed zone (central region).

Most of the columnar grains are formed in the secondary cooling zone where the intense cooling of water or vapor gives rise to a temperature gradient vertical to the surface of CCS and inhibits the grain growth parallel to the surface, and thus columnar grains vertical to the surface are achieved, as displayed in figure 3.

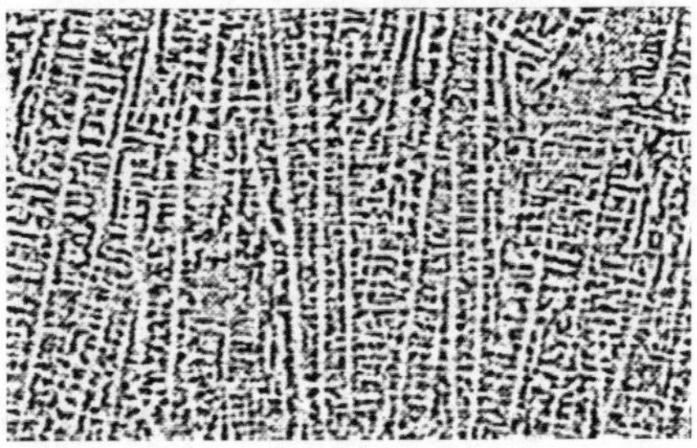

Figure 3. Micrograph of columnar crystal of the secondary cooling zone in transverse section of 380 mm×280 mm rectangular bloom of steel 20.

Fine columnar grains with the absence of branch and deflection can be observed in the region near the fine equiaxed zone (chill zone), which is attributed to the high gradient of temperature. Then, the number of columnar grains decreases, and multiple dendrites develop. Finally, the columnar grains become thicker, and the section morphology changes from simple to complicated one.

At the center and lower parts of the mold, air gaps appear due to shrinkage and separation of the billet shell from the mold. Herein, the thermal resistance increases, heat transfer slows down, and the heat flux flows along the direction perpendicular to the mold wall, and the fine columnar grains begin to grow. Therefore, the fine equiaxed grains are formed only on the mold top. To prevent distortion and breakout of CCS in the mold lower part, the billet shell thickness should reach a certain size (usually over 8-10 mm for a small billet and 15-25mm for bloom and heavy slabs).

3.3 Equiaxed grain

The equiaxed zone locating at the center of CCS presents various morphologies such as round, oval, polygon and strip-liked grains and is anisotropic. At the later stage of the columnar grains, a decrease of both the thermal transfer rate and temperature gradient caused by the increase in thickness of the shell stops the growth of the columnar grains and also the crystallization of the residual liquid steel. At the same time, some dissociative nucleus and broken dendrites are brought into the residual liquid, which decreases the temperature and becomes the new nucleus for solidification and promotes the formation of the central equiaxed zone. The central equiaxed grains are coarse for the low thermal transfer and adequate grain growth, as illustrated in Figure 4.

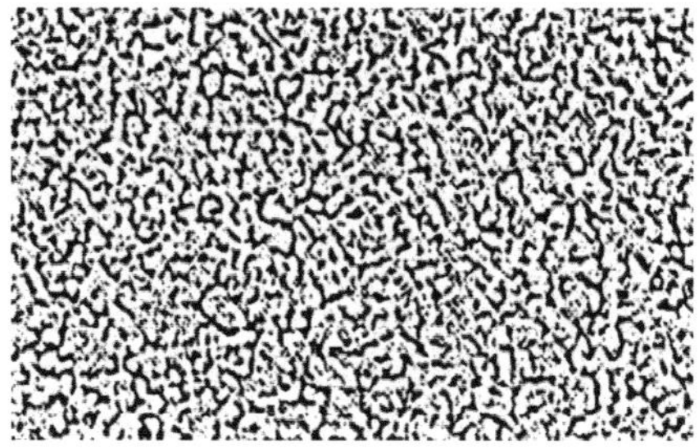

Figure 4. Equiaxed grain band: horizontal area of 300mm×1650 mm blank from medium carbon steel (0.23%W(C)).

3.4 Cross dendrite

Heat transfer and a decrease in vertical temperature gradient during the later stage of the solidification process result in the formation of the crossed dendrites, which limits the directional growth of the columnar grains. Furthermore, some broken dendrites with various growth orientations also promote the crossed dendrites. A typical morphology of the crossed dendrites can be seen in figure 5. The crossed dendrites, which are located between the columnar and central equiaxed grains, can be often observed in the bloom, heavy slab or billet with high alloying contents, but seldom in thin slabs and small billets.

The crossed dendrites with no directional distribution are beneficent for the isotropy in steel products. The crossed dendrites embed with each other and distribute uniformly thus act in a similar manner as equiaxed grains do. So according to GB/T 24178-2009, the method for the assessment of microstructure for the crossed dendrites should be identical to the one used for equiaxed grains.

In general, a fine equiaxed zone exists at the surface of CCS: the center is the equiaxed zone, and columnar grains are located between the center and the surface. The crossed dendrites are located in the middle of the columnar and central equiaxed grains or the trigonum of the blooms.

As to heavy slabs or blooms produced by the arc continuous caster, the columnar grains at the inner arc side are obviously longer than that at the outer arc side, although the cooling intensity is similar. Besides, the equiaxed and crossed dendrite zones on the inner arc side are both thinner than those on the outer arc side. The phenomenon is ascribed to the fact that growth of columnar grains is limited by some dissociative nuclei and broken dendrites brought into the outer arc side under the effect of gravity. Those broken dendrites can act as nuclei of crossed dendrites, and thus make them stronger.

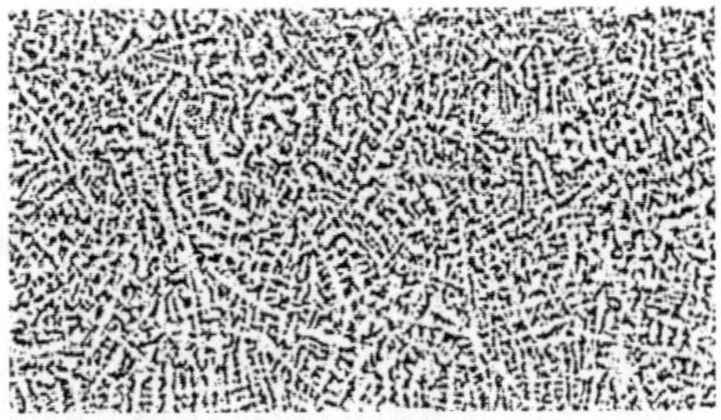

Figure 5. Cross dendrite (horizontal area) of 230mm×1650mm blank of Q235B steel.

4. Examples of different solidification microstructures

4.1 Comparison of the morphologies by different etching methods

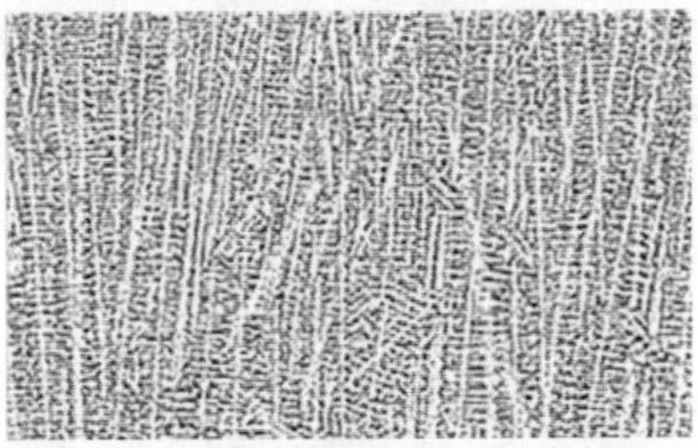

a) b)

Figure 6. Casting slab solidification structures (horizontal area) for 180 mm×1200 mm blank 510L steel: a) Dendrite etching (with the solidification microstructure and secondary dendrite details); b) Cold etching (with no solidification microstructure details).

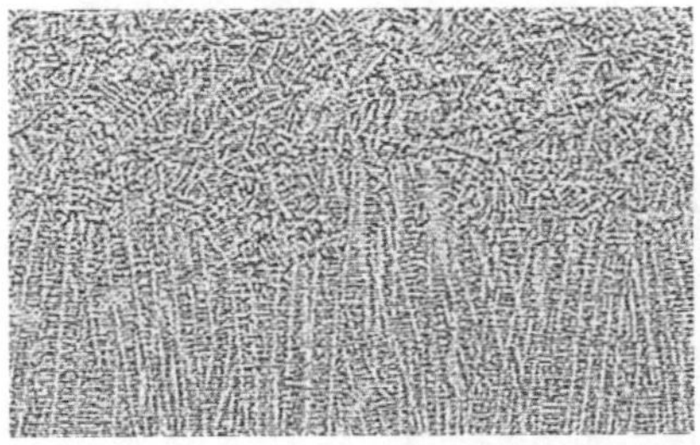

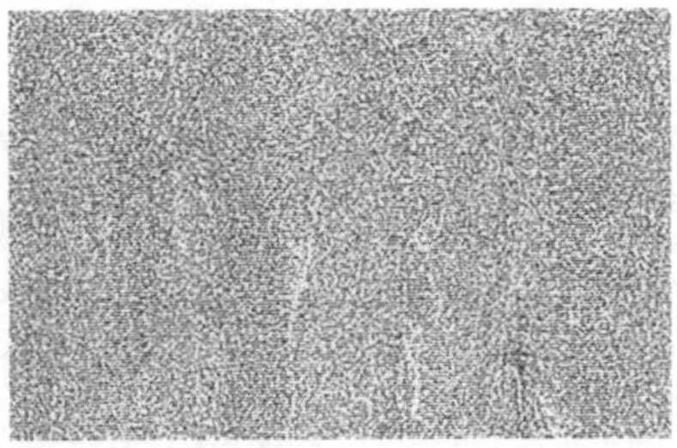

a) b)

Figure 7. Casting slab solidification structures (horizontal area) for 180 mm×1200 mm blank Q345B steel: a) Dendrite etching with detailed morphologies of the crossed dendrites (upper half) and columnar grains (lower part) and the secondary dendrite; b) Cold etching (with no solidification microstructure details).

4.2 Analyses of solidification microstructures

4.2.1. Deflection of the column crystal. The inclination/deflective angle of the columnar grains, which are mainly influenced by flow velocity of the liquid steel can be used to derive the flow condition, and modify shape of the submerged nozzle, insertion depth, size, and angle of the side hole, and thus check the CCS purity and availability of inclusions [6]. There is a deflective angle of 20° between the columnar grains and vertical direction of the CCS surface (heat flow) direction) in #1 and #2 samples, as can be seen in figure 8(b). However, no deflection is observed in the central part of #3 sample, except for the liquid flow edge.

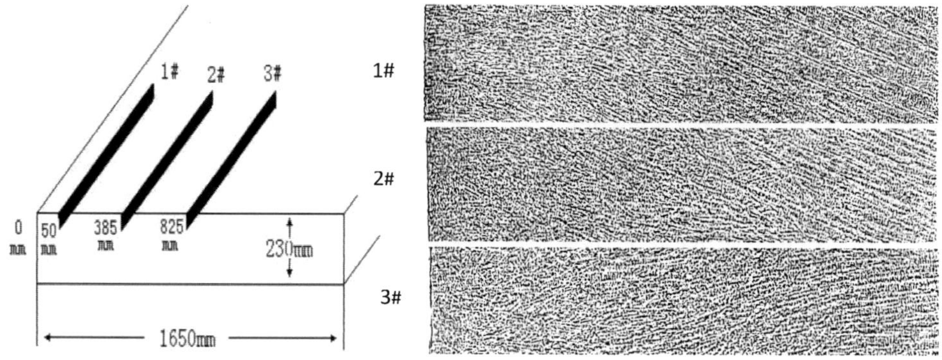

a) b)

Figure 8. Continuous casting slab column crystal's dendrite etching picture [5] for 230mmx1650mm blank of medium carbon steel ((0.15%W(C),0.38%W(Si),1.34%W(Mn), 0.013%W(P), 0.017%W(S) at casting temperature of 1542℃ and casting speed of 0.79m/min:
a) Surface of the inner arc side of SSC with positions of samples #1, #2, and #3 samples;
b) Column crystal inclination angles (vertical area) for #1, #2, and #3 samples.

4.2.2. Measurement of the secondary dendrite distances. It has been reported that the cooling rates in different locations of CCS during solidification are hard to be measured. However, the cooling rate ε (K/S) can be calculated by the following empirical equation that $\lambda_2 = a\varepsilon^{-n}$ [7], where λ_2 is the secondary dendrite distance, a = 109.2, and n = 0.44.

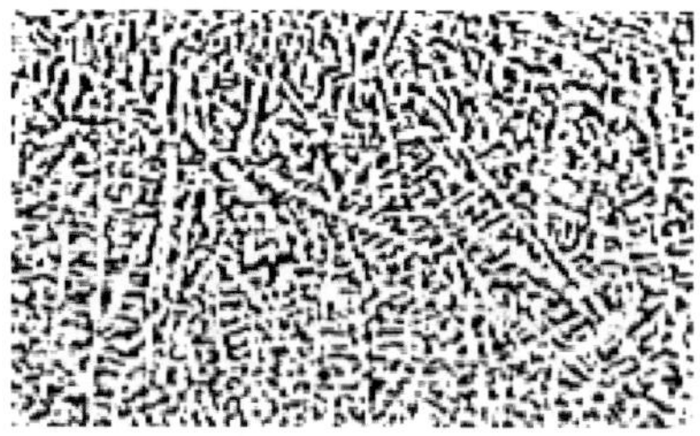

Figure 9. Micrograph of the secondary dendrite in the transverse section of medium carbon steel (0.16%W(C) in 230mmx1650mm blank.

It is also necessary to analyze the central segregation based on the secondary dendrite distance as is shown in figure 9. When the cooling rate is low, an increase in the secondary dendrite distance leads to a dramatic elevation of permeability, promotes the mother liquid enriched with the solute elements flowing into the center of CCS, and thus aggravates the central segregation. However, this does not occur at high cooling rates. This effect is more pronounced for billets of high-carbon steel (150mm×150mm), where low central segregation and fine grains can be obtained when the secondary dendrite distance is suppressed.

4.2.3. Determining the electromagnetic stirring effect (S-EMS) at secondary cooling zone. Figure 10a) depicts the phenomena of sinking equiaxed grains with a poor S-EMS, where the central segregation is assessed as grade 2, while figure 10b) shows a symmetrical distribution of equiaxed grains for a good S-EMS, and the central segregation is assumed to be of grade 1.

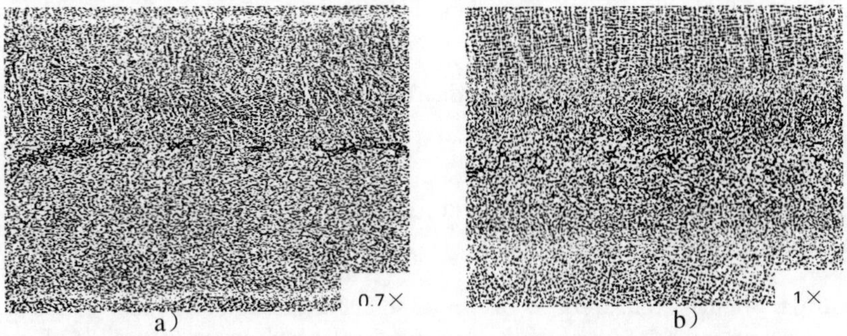

Figure 10. Assessment of the metallurgical effect of S-EMS in transverse section of the medium carbon steel: a) 300mm×1650mm blank; b) 230mm×1650mm blank.

5. Conclusions

Continuous casting is an accelerated process of steel production, for which the quality control of steel products in terms of their solidification microstructure is crucial. Abundant metallurgical data can be obtained during the continuous casting process by assessing the solidification microstructure. It is proved in this study that the dendrite etching method is instrumental to examine the solidified morphologies and estimate the material characteristics, especially the defects of the CCS microstructure, and then modify/optimize the casting technology for the particular practical purposes.

References:

[1] Zhao H, Yang Z C and Meng Y, Reason for chilled layer disappearance on surface sag of unquenched low carbon steel corner, *Physics Examination and Testing*, 2006, **24**, No.1, 59-62.

[2] Cai K K, Controlling Strand quality for steel continuous casting, Beijing, Metallurgical Industry Press 2010, 278

[3] Feng J and Shi X H, Chief Editor, Production of Steel Continuous Casting, Metallurgical Industry Press, 2005, **154**.

[4] Shi C X, Compilation, Quality of Steel Continuous Casting, Metallurgical Industry Press, 1980, 27-31.

[5] Xu Q T and Wang W Z, The Low Times Structure Inspection of the Continuous Casting Slab's Atlas, Beijing, Standards Press of China, 2009, 16-17.

[6] Guo Y G, Editor, Steel Continuous Casting, Metallurgical Industry Press, 1985, 11-12.

[7] Feng J and Chen W Q, Effect of high secondary cooling intensity on solidification structure and center carbon segregation of high carbon steel billet, *Special Steel*, 2006, **27**, No.4, 42-44.

Multi-objective optimization of swash plate forging process parameters for the die wear/service life improvement

X F Hu[1,2], L G Wang[1*], H Wu[1] and S S Liu[2]

[1]School of Materials Science and Technology, Jiangsu University, Zhenjiang 212013, China
[2]Kunshan Huizhong Machine Co. Ltd., Kunshan 215331, China
*E-mail: lgwang@mail.ujs.edu.cn

Abstract. For the forging process of the swash plate, the author designed a kind of multi-index orthogonal experiment. Based on the Archard wear model, the influences of billet temperature, die temperature, forming speed, top die hardness and friction coefficient on forming load and die wear were numerically simulated by DEFORM software. Through the analysis of experimental results, the best forging process parameters were optimized and determined, which could effectively reduce the die wear and prolong the die service life. It is significant to increase the practical production of enterprise, especially to reduce the production cost and to promote enterprise profit.

1. Introduction

Auto air-condition compressor as an important part of a car has undergone several generations of development [1, 2]. Although the latest generation is the scroll compressor, swash plate compressor still accounts for 60% of the compressor market [3]. Swash plate as the core component of this type compressor directly affects the reliability and life of the compressor. However, the swash plate is characteristic of its small size and complex shape. It is worth noting because it has high requirement of precision and quality, which makes manufacture difficult and leads to a short life of the die. Therefore, precision forging process may be a good way to solve for the company [4, 5].

To improve the performance of swash plate, many researchers have made endeavors. Wan *et al* developed an indirect squeeze casting with pressurization and local pressurization method to reduce defects such as air bubble after heat treatment, slag etc. [6]. In order to solve the seizing problem of automotive air conditioning compressor, Wu carried out the experiments of improving the lubricating property on the swash plate surface by changing the coating and substrate material [7]. Furthermore, Chen *et al* reported that the total squeeze casting pressure of 100 MPa, the delayed time of 2.8 s, and the central thick area of swash plate was pressurized locally with 180 MPa for 6 s, which could eliminate shrinkage defects in swash plate [8, 9]. Of course, die life is also a crucial factor. Behrens [10] studied the hardness evolution due to thermal softening of the tool material, and performed an accurate estimation of die wear by the proposed model. The latter proved to be applicable to wear estimation of hot forging dies over a large number of operating cycles. Based on the abrasive wear model developed by Archard, the thermo-mechanical coupled FE model of the hot forging process was built by Shi *et al* to estimate the influence of initial die temperatures, forging rate and heat treatment method on the wear depth of die [11,12]. Zheng *et al* used the DEFORM wear analysis module to analyze the die wear of typical forward cold extrusion process. The wear profile of points of

IOP Conf. Series: Materials Science and Engineering **283** (2017) 012018 doi:10.1088/1757-899X/283/1/012018

the die work surface was obtained, and the maximum die wear position was pointed out [13, 14].

In this paper, the wear failure of swash plate was analyzed, the forming numerical simulation was carried out and the orthogonal experiment was designed. Influences of the billet initial temperature, die temperature, die hardness, forming rate, and the friction coefficient between the work piece and dies on forming load and die wear were analyzed comprehensively based on the Archard wear model [15, 16]. As a result, relatively reliable optimal processing parameters were obtained, which has a certain guiding significance for the formulation of reasonable forging process and die life prediction.

2. Forging process numerical simulation

2.1. Experimental model

Taking a type of swash plate as our experiment object, processing die and billet size were designed according to the actual requirements. The swash plate and die are shown in figure 1. It is made of A390 alloy, which has good wear resistance, corrosion resistance, low coefficient of linear expansion, good thermal conductivity, especially good mechanical properties and good dimensional stability at high temperature. Meanwhile, after modification, the grain size of silicon will be refined, so that the tensile strength, yield strength and plastic properties of the metal will be greatly improved. Therefore, it is suitable for forging process and the part can obtain good mechanical properties [17]. Through the analysis of the characteristics and technical requirements of forging, closed die forging can be used. The Die forging process consisted of the following steps: blanking – heating – forging - grinding burr - T6 heat treatment - surface cleaning.

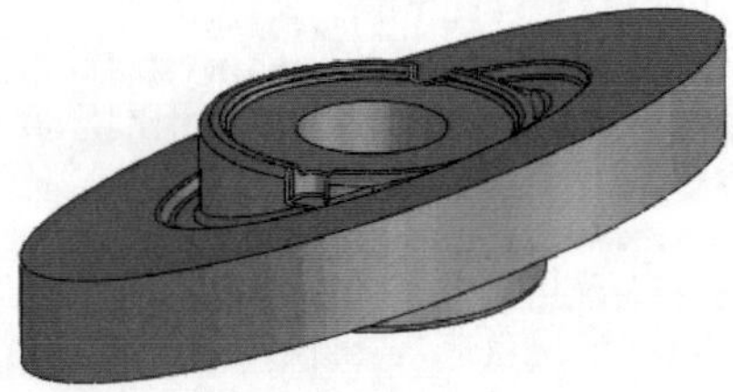

(a)

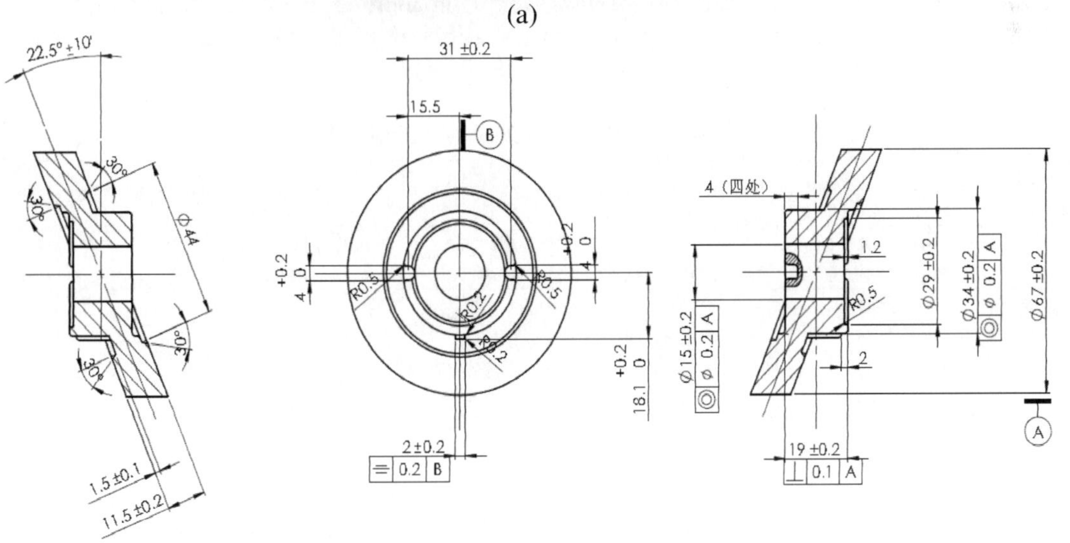

(b)

Figure 1. Forging and the geometries of the swash plate. (a) Forging of the swash plate and (b) The geometry of the swash plate.

IOP Conf. Series: Materials Science and Engineering **283** (2017) 012018 doi:10.1088/1757-899X/283/1/012018

2.2. Simulation mechanism

The Archard wear model was adopted to calculate the wear of die cavity in the deformation, and the computational formula of wear was defined as follows:

$$W = \int K \frac{p^a v^b}{H^c} dt \tag{1}$$

where W is wear depth (mm), p is positive pressure inside the mold cavity interface (N), v is relative sliding speed between die and modified material contact point (mm·s-1), H is hardness of mold material (HRC), dt is unit time of deformation, while K, a, b, c are mold material constants.

3. Design of multi-objective orthogonal experiment

In the actual production and scientific experiment, the result of experiment cannot be comprehensively evaluated by any single index, so it is particularly meaningful to design the multi-objective experiment [18]. There are two common analysis methods to analyze multi-objective orthogonal experiment, and they are comprehensive balance method and comprehensive evaluation method. In this paper, the comprehensive evaluation method was adopted to solve the problem of the multi-objective orthogonal experiment. This method analyzes the test results and gives each test a score as the total index of it. According to the total index (score), the experimental results of the single index are analyzed and the best test plan is determined. In other words, multiple indexes can be reduced to a single index so as to get the result of multi-index experiment.

3.1. Selection of the experiment factor levels and indicators

In hot forming, the basic forms of die failure are wear, deformation, fracture, fatigue, etc. Deformation and wear are the most common failure forms of the hot forging die. When deformation resistance of the billet exceeds the strength of the die material, plastic deformation will occur. When the wear of die reaches a certain degree, it leads to the die failure [19].

The selections of experiment factors and levels are very important, so main influence factors in the forging should be chosen and the selected level should be in accordance with the actual forming conditions. Combined with practical production experience and practical operability, reasonable processing parameters were selected and listed in table 1. In addition, the billet initial temperature, die temperature, forming rate, die hardness and friction coefficient in the current production process are about 430°C, 110°C, 11 mm·s^{-1}, 52 HRC and 0.4, respectively.

Table 1. Processing parameters/factors

Level number	Factors				
	A Billet initial temperature, °C	*B* Die temperature, °C	*C* Forming rate, mm·s^{-1}	*D* Die hardness, HRC	*E* Friction coefficient
1	390	80	7	45	0.3
2	420	100	10	50	0.4
3	450	120	13	55	0.6
4	480	140	15	60	0.7

3.2. Results of the orthogonal experiment

Regardless of the interaction among different affecting factors, the L_{16} (4^5) orthogonal table was selected to conduct the experiment for five factors and four levels. DEFORM software was used in the simulation analysis and the forming load and upper die wear can be obtained from it in table 2 [20,21].

First of all, it is necessary to compare and score the experiment index of every experiment number. On the basis that forming and wear indicator are as small as possible, Y_{jmin} represents 100 points, while Y_{jmax} represents 0 points. The formula for calculating Y_{ij} is as follows:

$$Y_{ij} = 100(Y_{jmax} - Y_j)/(Y_{jmax} - Y_{jmin}) \quad (2)$$

Table 2. Orthogonal experiment results.

Test number	Factors					Target		Compre-hensive weighted
	A	*B*	*C*	*D*	*E*	Forming load *F*, N*10^6	Upper die wear W, mm*10^{-5}	Y^*
1	1	1	1	1	1	2.01	0.36	43.74
2	1	2	2	2	2	2.01	0.33	48.30
3	1	3	3	3	3	2.02	0.36	42.86
4	1	4	4	4	4	1.93	0.29	61.58
5	2	1	2	3	4	2.03	0.49	16.51
6	2	2	1	4	3	2.26	0.38	22.43
7	2	3	4	1	2	1.62	0.37	67.79
8	2	4	3	2	1	1.61	0.37	68.67
9	3	1	3	4	2	1.73	0.27	78.51
10	3	2	4	3	1	1.53	0.29	89.92
11	3	3	1	2	4	1.81	0.45	39.57
12	3	4	2	1	3	1.68	0.50	39.73
13	4	1	4	2	3	1.70	0.33	68.96
14	4	2	3	1	4	1.69	0.47	44.55
15	4	3	2	4	1	1.61	0.23	95.21
16	4	4	1	3	2	1.72	0.34	65.88
K1	196.48	207.72	171.62	195.81	297.53	The sum of quality targets in each relative level of various factors		
K2	175.40	205.21	199.75	225.50	260.48			
K3	247.72	245.43	234.59	215.19	173.99			
K4	274.61	235.85	288.25	257.72	162.21			
k1	49.12	51.93	42.91	48.95	74.38	The average of quality targets in each relative level of various factors		
k2	43.85	51.30	49.94	56.37	65.12			
k3	61.93	61.36	58.65	53.80	43.50			
k4	68.65	58.96	72.06	64.43	40.55			
R	24.80	10.05	29.16	15.48	33.83	Extreme value		

Thus, the score Y_1 and Y_2 of forming load and die wear can be calculated shown in table 3. Weighted composite score values Y^* can be obtained by the formula:

$$Y_i^* = \alpha_1 Y_{i1} + \alpha_2 Y_{i2} + \cdots + \alpha_j Y_{ij} \quad (3)$$

From this formula, the weight factor coefficient of the test index is expressed by α_j, which represents the weight that each index should take in the comprehensive weighted score. The sum of all index weight factor coefficients is equal to 1, and their determination should be based on analysis of the quality of the part, production efficiency and manufacturing cost, the weight factor coefficient of important test indices should be increased. Based on the actual requirements, the forming load factor coefficient α_1 is 0.5 and the wear weight factor coefficient α_2 is 0.5. At this point, the dual indices have been transformed into a single target index and then Y_i^* can be calculated. In addition, In order to get the influence law of each factor at different levels, we also calculate the sum of quality targets in each relative level of various factors (K_{ij}) and the average of quality targets in each relative level of various factors (k_{ij}) from Y_i^*. Meanwhile, the extreme value (R_i) also can be obtained from k_{ij}.

IOP Conf. Series: Materials Science and Engineering **283** (2017) 012018 doi:10.1088/1757-899X/283/1/012018

Table 3. Scores of forming load and wear values.

Test number	Y_1	Y_2
1	34.25	53.23
2	34.25	62.36
3	32.88	52.85
4	45.21	77.95
5	31.51	1.52
6	0	44.87
7	87.67	47.91
8	89.04	48.29
9	72.60	84.41
10	100	79.85
11	61.64	17.49
12	79.45	0
13	76.71	61.22
14	78.08	11.03
15	90.41	100.00
16	73.97	57.79

The performed analysis of tables 2 and 3 provided no obvious evidence of any correlation between the forming load and die wear parameters. Besides, the combinations of five factors at different levels have different influences on load and die wear. However, simply from the experimental results, we cannot directly obtain the curve of the influences of the five factors at different levels of load and die wear. Thus, the experimental results should be further analyzed in more detail.

3.3. Analysis of the orthogonal experiment results

To further analyze the above data, the influence of experimental comprehensive index on the forming load and die wear was drawn as a trend chart [22] (shown in figure 2) on the basis of data of each factor levels in table 2. In table 2, extreme values of each factor were ordered as follows: $R_E > R_C > R_A > R_D > R_B$, which determined the importance of the influence of the various factors on swash plate forming load and die wear [23,24]. The impact degree is ranked as follows in the decreasing order: the coefficient of friction between the work piece and its dies, forming speed, billet initial temperature, die hardness, and die temperature.

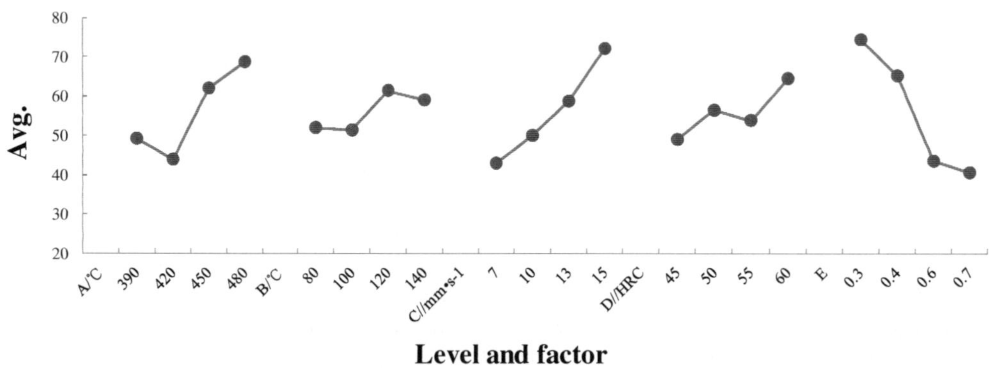

Figure 2. Influence of experiment comprehensive index on the forming load and die wear.

Through the analysis of the results on test No.16 in table 2, the forming load *F* and the die wear *W* in test No. 15 were the smallest, the parameter combination of which was A4B3C2D4E1. Comparing the average of comprehensive indices in each relative level of various factors in table 2, the highest were k14, k23, k34, k44, k51, while the corresponding parameters' combination was A4B3C4D4E1. Thus, DEFORM was used again, yielding the forming load of $1.60*10^6$ N and upper die wear of $2.33*10^4$ mm, as compared to $1.61*10^6$ N and $2.34*10^4$ mm, respectively. It is clear that A4B3C4D4E1 is the best combination of parameters. Therefore. the processing parameters combination was finally selected as billet initial temperature 480°C, die temperature 120°C, forming speed 15 mm·s-1, die hardness 60 HRC and friction coefficient between the work piece and its dies 0.3. The curve of forming load and upper die wear obtained by DEFORM can be seen in figure 3, we can see from that wear easily occurs in the middle circular region. The area of φ29-φ34 can be seen in figure 1 [25-27].

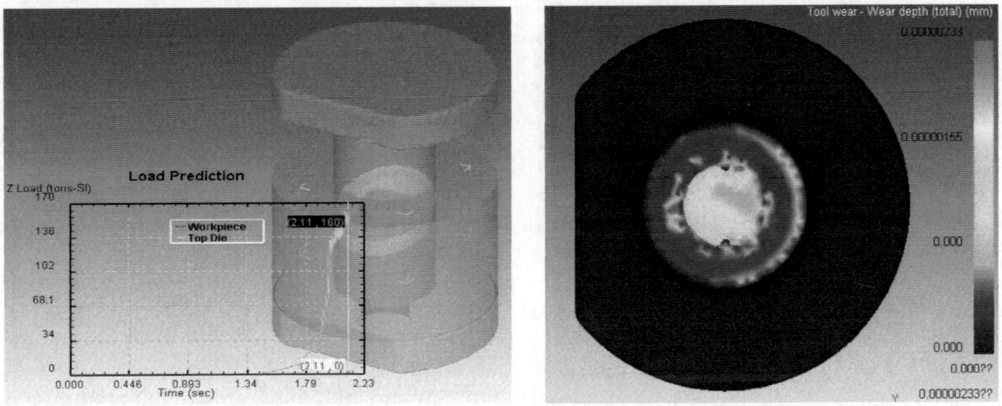

(a) (b)

Figure 3. Simulation results of optimum parameters by DEFORM. (a) The curve of forming load and (b) Upper die wear.

4. Practical application of the optimized parameters

In order to check the optimizing effect of the simulated parameters on the die life, we put the five optimal parameters into practical production and compare the die life of the optimized die with the non-optimized one.

4.1. One test of die wear

Figure 4. A hydraulic press with a die.

Referring to previous die swash plate forgings average life of 8500 pieces, for the first test, we made two same dies in operation in the same machine to produce swash plate forgings which used optimized processing parameters and un-optimized parameters separately. Then the surface states of the two dies were observed when they both finished producing 8000 pieces of forgings. The machine used in this test is hydraulic press, the type of which is LYF-500SA. In the test, the upper and bottom die were put into the hydraulic press simultaneously, as shown in figure 4.

Figure 5 shows the surface states of the two dies after the production of 8000 pieces of forgings. Figure 5(a) is the upper die with un-optimized parameters while figure 5(b) depicts the upper die with the optimized parameters. It is obvious that both dies were worn after the production. By comparison, the die wear of the former is more serious than the latter, indicating that the optimized process can significantly decrease the die wear and prolong the die service lifetime.

(a) (b)

Figure 5. The surface state of the die working for non-optimized processing parameters (a) and optimized ones (b).

4.2. Comparison of die service life

Table 4. The number of actual forgings before the die failure.

Test number	Hydraulic machine	Die label	Number of forgings before die failure
1	LYF-500SA	03001	10523
2	LYF-500SA	03001	9536
3	LYF-500SA	03001	11324
4	LYF-500SA	03001	11230
5	LYF-500SA	03001	9876

However, results of a single test are not reliable enough to draw the conclusions since the two dies do not fail. On the basis of the experience of production and previous die failures, the verticality of the plane based on the datum plane A is 0.1 mm, which is perpendicular to another plane with the length of 19±0.2 mm (see figure 1). The verticality is very critical because it may affect the stability and assembly of the swash plate. When the wear reaches 0.05 mm, the surface of the die has obvious pits and deformation, it may cause further damage if it continues to use. So just to be on the safe side, the failure of the die is defined when the die wear amount reaches 0.05 mm [28]. Thus, we used the optimal parameters to produce the swash plate forgings for 5 times. Five same dies labeled of 03001 were used in this experiment and detected the dimension of the part online, it is treated as a failure when the round area (φ29-φ34) exceeds 0.05mm from the original ones. Then the number of actual forgings until the dies failed was recorded and tabulated in table 4. By calculation, the average forgings number with a set of the die can produce when the die wear reaches 0.05 mm is 10498, which increased by 23.5%, as compared to the non-optimized average die life of 8500. The above experiments and analysis demonstrate the benefit of the optimized processing parameters and provide a good reference for the real production of swash plates forgings.

5. Conclusions

- Based on the Archard wear model, the curve relationship between the billet initial temperature, die temperature, die hardness, forming rate, friction coefficient and forming load, die wear were obtained through simulation. In addition, the influence of various processing parameters on forming load and wear of die was also investigated, providing certain guidance for the design of die process parameters.
- Through orthogonal experiment, the influence of multi-targets on the swash plate forming force and die wear was comprehensively considered and the optimal processing parameters were obtained as follows: billet initial temperature 480°C, die temperature 120°C, die hardness 60 HRC and the workpiece/die friction coefficient of 0.3.
- Applying the optimized parameters into practical production, the die wear of the non-optimized one is more serious than the optimized one when the two same dies produce 8000 pieces of swash plates forgings using the same operation machine. Besides, taking five same dies as an experimental object and calculating their die life under the optimized processing parameters, the average die life is 10498 pieces, increased by 23.5%, compared with the un-optimized average die life of 8500 pieces. These results demonstrate that through the multi-objective optimization of the orthogonal test, the die wear can be reduce, greatly improving the die service life.

6. Acknowledgments

The authors would like to thank the financial support provided by the National Natural Science Foundation of China (Grant no. 51755249).

7. References

[1] Jinhua Q, Enyuan G J and Liang Q 2015 Analysis report of China refrigeration industry development in 2014 *Refrigeration Technology* **A01** 11-5

[2] Liu L Z, Luo J and Liang Q 2003 Present situation and prospect of automotive air conditioning compressor *J. Refrigeration* **22(1)** 31-4

[3] Xia X Y 2017 Present situation and development of scroll refrigeration compressor *Process and Equipment* **2** 129-30

[4] Wang Z L and Zhao G Q 2009 Research status and development trend of precision forging *Precision Forming Engineering* **1** 32-8

[5] Lin Y C and Chen M S 2009 Numerical simulation and experimental verification of microstructure evolution in a three-dimensional hoot upsetting process *Journal of Materials Processing Technology* **209** 4578-83

[6] Li W, Luo J and Liang Q 2008 Production of aluminum alloy swashplate for compressor of air conditioner for the car by squeezing casting *Special Casting & Nonferrous Alloys*

[7] Bu-Xian W U 2010 Research of lubrication on the swash plate surface of automotive air conditioning compressor *Mechanical Engineer*

[8] Chen M, Chen Y and Wu Y, *et al* 2014 Squeeze casting A390 alloy swash plate for automotive air-conditioner compressor *Special Casting & Nonferrous Alloys* **34(5)** 504-3

[9] Xinyong, Liu D L and Wu X Y, *et al* 2010 Experiment of specular injection molding process control *Polymer Materials Science and Engineering* **26(6)** 114-8

[10] Behrens B-A 2008 Finite element analysis of die wear in hot forging processes *CIRP Annals-Manufacturing Technology* **57(1)** 305-8

[11] Shi Y J, *et al* 2016 Experimental analysis and numerical study on wear failure of hot forging die for automobile flange *Tribology* **36(2)** 215-25

[12] Kim D H and Lee H C, *et al* 2005 Estimation of die service life against plastic deformation and wear during hot forging process *Journal of Materials Processing Technology* **166(3)** 372-80
[13] Zheng W G, *et al* 2013 Simulation analysis and research on the wear of the extrusion die based on Archard *Forging & Stamping Technology* **38(5)** 148-51
[14] Kang J H, Park I W, Jae J S, *et al* 1999 A study on a die wear model considering thermal softening (I): construction of the wear model *Journal of Materials Processing Technology* **96** 53-8
[15] Xu W, *et al* 2014 Influence of hot forging process parameters of connecting rod on die wear *Hot Working Technology* **43(9)** 99-105
[16] Wang L G, Huang Y, *et al* 2006 Wear analysis of extrusion die based on archard theory *Lubrication Engineering* **3** 10-3
[17] Kong X and Yang B C 2015 Effect of reheating process on microstructure and mechanical property of A390 aluminum alloy *Materials Science Forum* **2015** 173-9
[18] Chenyan, Cai W L, Cai Y L and *et al* 2010 Optimization of process parameters of the multi-targets injection based on CAE and orthogonal experiment *Engineering Plastics Application* **38(4)** 36-8
[19] Wu J L and Tao J C 2002 Research on optimizing drilling fluid formulation with the comprehensive weighted evaluation *Nuclear Science and Techniques* **18(2)** 45-8
[20] Kiefer B V and Shah K N 1990 Three-dimensional simulation of open-die press forging *Journal of Engineering Materials and Technology* **112(4)** 477-85
[21] Hartley P and Pillinger I 2006 Numerical simulation of the forging process *Computer Methods in Applied Mechanics and Engineering* **195(48)** 6676-90
[22] Lee R S and Jou J L 2003 Application of numerical simulation for wear analysis of warm forging die *Journal of Materials Processing Technology* **2003** 43-8
[23] Behrens B-A 2008 Finite element analysis of die wear in hot forging processes *CIRP Annals - Manufacturing Technology* **57** 305-8
[24] Siamak A, Metin A, Mustafa, *et al* 2010 Wear analysis of hot forging dies *Tribology International* **43(8)** 467-73
[25] Li J G and Tang W C 2008 3D FEM simulation of precision forging automobile clutch damping spindle sleeve *CADDM* **2008** 73-6
[26] Ou H 2004 An FE simulation and optimisation approach for the forging of aero-engine components *Journal of Materials Processing Technology* **2004** 208-16
[27] Li X F 2014 3D finite element modeling and analysis of radial forging processes *Journal of Manufacturing Processes* **2014** 329-34
[28] Lin G Y and *et al* 2009 Analysis of influence of extrusion times on die wear based on Archard theory *Journal of Central South University (Science and Technology)* **40(5)** 1245-50

IOP Conf. Series: Materials Science and Engineering **283** (2017) 012019 doi:10.1088/1757-899X/283/1/012019

Study on the coal mixing ratio optimization for a power plant

Y A Jin[1,2*], J W Cheng[1,2], Q Bai[1,2] and W B Li[1,2]

[1]State Key Laboratory of Automotive Simulation and Control, Jilin University, Changchun 130022, China
[2]College of Automotive Engineering, Jilin University, Changchun 130022, China
*E-mail: 1055236456@qq.com

Abstract. For coal-fired power plants, the application of blended coal combustion has been a great issue due to the shortage and rising prices of high-rank coal. This paper describes the optimization of blending methods between Xing'an lignite coal, Shaltala lignite coal, Ura lignite coal, and Inner Mongolia bituminous coal. The multi-objective decision-making method based on fuzzy mathematics was used to determine the optimal blending ratio to improve the power plant coal-fired economy.

1. Introduction

In recent years, blending different types of coal at pulverized coal-fired power plants are becoming increasingly common in China because of the uneven distribution of resources and transportation difficulties [1]. In power plants, different types of coals can be blended out-furnace or in-furnace [2,3]. This paper describes the mathematical analysis of the blending strategy of different coal types to find the optimized coal blend ratio.

2. Determination of the coal mixing ratio

Main coals used in a 350 MW supercritical combustion power plant are shown in table 1. Xing'an coal, Ura coal, and Shaltala coal are lignite coals, while Inner Mongolia coal is bituminous one. Design coal of the 350 MW power plant is lignite coal.

Table 1. Main coals used in a 350 MW power plant.

Coal No.	Coal Origin/Grade	Mt (wt%)	Mad (wt%)	Aar (wt%)	Vdaf (wt%)	Qnet, ar MJ/kg
1	Inner Mongolia	21.8	13.9	17.88	37.54	20.05
2	Xing'an	13.1	4.92	37.07	48.39	14.22
3	Ura	26.8	16.86	6.14	47.83	17.54
4	Shaltala	30.2	9.64	18.25	48.34	13.42
5	Design coal	27.8	9.27	12.06	45.09	17.34

Properties of coals used in 350 MW supercritical combustion power plant are shown in table 2.

Xing'an coal, Ura coal and Shaltala coal are lignite coal. They are characterized by high volatile content, easy to catch fire, low calorific value and softening temperature are lower than 1192℃, easy to get slag. While Inner Mongolia bituminous coal has lower volatile content, not easy to catch fire, and its low calorific value is higher than the design coal. The softening temperature of Inner Mongolia bituminous coal is greater than 1500℃, so it is not easy to get slag. The volatile content of Ura lignite is 52.24%, the volatile content of design coal is 45.09%, and the volatile content of Inner Mongolia bituminous coal is about 37%. The low calorific value of boiler design coal is 17.34 MJ/kg, the low calorific value of the main burning Xing'an coal, Ura coal and Shaltala coal are less than the designed coal. Therefore, a certain amount of Inner Mongolia bituminous coal should be mixed into lignite coal to meet the boiler operation requirements. The power plant has six coal mills, 5 for running and 1 for backup. Two different coal types were ground in different mills because they have different grinding coefficients. In the in-furnace blending method, different coal types were fed into the furnace from separate burners and burned together in the furnace. Three kinds of lignite and bituminous coal blending schemes were used to find the optimal blending ratio by using the multi-objective decision-making method based on the fuzzy mathematics. Blending schemes for Ura lignite coal and Inner Mongolia bituminous coal are shown in table 3.

Table 2. Properties of coals used in a 350 MW power plant.

Coal No.	Coal Origin/Grade	Cdaf (wt%)	Hdaf (wt%)	Odaf (wt%)	Ndaf (wt%)	DT(℃)	ST(℃)	FT(℃)
1	Inner Mongolia	78.86	4.95	14.43	1.76	1340	>1500	>1500
2	Xing'an	75.75	5.13	16.7	2.42	1120	1130	1140
3	Ura	74.65	3.62	19.51	2.22	1180	1190	1200
4	Shaltala	75.32	3.82	18.6	2.26	1160	1180	1190
5	Design coal	78.18	5.81	14.69	1.32	1100	1110	1130

Table 3. Ura and Inner Mongolia bituminous coal blending schemes.

Option	Ura lignite	Inner Mongolia bituminous coal
1	1	4
2	2	3
3	3	2
4	4	1

Thus, the set of evaluation schemes is:

$$U = \{ \text{I, II, III, IV} \}$$

It can be seen from the previous text that the factors affecting the mixed coal mainly include the ignition characteristics of coal, the low calorific value of coal, the slagging characteristics of coal and the price factor of coal. Therefore, the evaluation index is:

- volatile coal content;
- low coal calorific value;
- comprehensive slagging index of mixed coal;
- mixed coal prices;

According to the industrial analysis of coal, ash comprehensive index R and coal prices, according to the four coal blending program to calculate the coal price in accordance with the ratio of linear addition to calculate the indicators as shown in table 4:

Table 4. Indicators of mixed coal.

Factors		Unit	Scheme 1	Scheme 2	Scheme 3	Scheme 4
Volatile content	f1	%	40.48	43.42	46.36	49.3
Low calorific value	f2	KJ/kg	18760	17470	16180	14890
Slagging index	f3	1	1.02	1.09	1.12	1.23
Price of coal	f4	yuan	534	489	442	396

The set of four evaluation factors is:

$$V = \{f_1, f_2, f_3, f_4\}$$

The four factors of each factor index matrix F are:

$$\Delta = \begin{pmatrix} 40.48 & 43.42 & 46.36 & 49.3 \\ 18.76 & 17.47 & 16.18 & 14.89 \\ 1.02 & 1.09 & 1.12 & 1.23 \\ 534 & 489 & 442 & 396 \end{pmatrix}$$

The weight of the four factors can be obtained according to the actual operating experience of the power plant. Because of the relatively low heat generation and slagging characteristics of the coal plant in the actual operation process, the weight of the power plant is relatively large, and its weight is relatively large in price and volatile The weight set A is:

$$A = (a_{1,} a_2, a_3, a_4) = (0.2, 0.3, 0.3, 0.2)$$

Next, we apply the weighted relative deviation distance to the minimum method for decision making. According to the factors of each scheme, the optimal index of each factor of each scheme is taken as the standard vector of matrix F, so the standard vector of matrix F is:

$$f^0 = \{49.3, 18760, 1.02, 396\}$$

Make

$$\delta_{ij} = \frac{|f_i^0 - f_{ij}|}{f_{i\max} - f_{i\min}}$$

The fuzzy matrix of the relative deviation of each factor is derived as follows:

$$\Delta = \begin{pmatrix} 0 & 0.24 & 0.48 & 1.0 \\ 1.0 & 0.54 & 0.38 & 0 \\ 1.0 & 0.53 & 0.18 & 0 \\ 0 & 0.25 & 0.50 & 1.0 \end{pmatrix}$$

According to the weighted relative deviation distance formula,

$$d_j = d_j(u_j, f^0) = \frac{1}{a}\sqrt{\sum_{i=1}^{n}(a_i \cdot \delta_{ij})^2}$$

Inferred

$$d_j = \{0.67, 1.85, 0.86, 1.06\} \quad i = 1, 2, 3$$

The minimum method of weighted relative deviation distance is based on the smallest value as the optimal solution of the matrix. Therefore, the program is the optimal solution, that is to say when Ura coal and Inner Mongolia bituminous coal ratio is 4: 1, it is the best ratio program. In the same way, the optimal ratio of Xing'an coal, Tara coal and Inner Mongolia bituminous coal is studied by using multi-objective decision-making method in fuzzy mathematics.

Different proportions of coal are obtained by fuzzy mathematics in tables 4 and 5.

Table 5. Xing'an coal and bituminous coal blending program.

	Xing'an coal: Inner Mongolia bituminous coal				Best option
	1:4	2:3	3:2	4:1	
d_j	1.32	1.02	0.86	1.26	3:2

Table 6. Tara coal and bituminous coal blending scheme.

	Tara: Inner Mongolia bituminous coal				Best option
	1:4	2:3	3:2	4:1	
d_j	1.03	0.79	0.98	1.12	2:3

3. Numerical simulation analysis of blended coal
Finally, the analysis of coal blending is performed, and the results are shown in table 7.

Table 7. Analysis of coal blending.

Industry Analysis					
		Unit	Option I	Option II	Option III
Moisture	Mt	%	20.42	13.42	24.22
Ash	Aar	%	19.76	30.55	8.7
Volatile	Var	%	24.69	24.43	30.85
Fixed carbon	FCar	%	35.13	31.6	36.23
Low calorific value		kJ/kg	17398	16552	18042
Elemental analysis					
		Unit	Option I	Option II	Option III
Carbon	Cdaf	%	77.44	76.99	75.49
Hydrogen	Hdaf	%	4.5	5.06	3.89
Oxygen	Odaf	%	16.1	15.79	18.49
Nitrogen	Ndaf	%	1.96	2.16	2.13

4. Conclusion
Based on the multi-objective decision-making and multi-level evaluation model in fuzzy mathematics, the volatiles, low calorific value, coking index, water, and coal price standard were determined. When the Ura coal and Inner Mongolia bituminous coal mixed combustion, their best blending ratio is 4: 1. When the Shaltara coal and Inner Mongolia bituminous coal mixed combustion, their best blending ratio is 2: 3. When the Xing'an coal and Inner Mongolia bituminous coal mixed combustion, their best blending ratio is 3: 2. So coal power plant can be blended in accordance with the proportion of this research in the actual combustion process, to achieve the best coal-fired economy.

Acknowledgments
This project "Improving the Combustion Mode and Security Running Control Strategy to Decrease the Pollution in Northeast of China" (UK-CIAPP\201) was supported by the Royal Academy of Engineering under the UK-China Industry-Academia Partnership Programme scheme. The authors are grateful for the support of the Royal Academy of Engineering and the UK Government's Newton Fund. This project was supported by the China Scholarship Council (No. 201706175069).

References

[1] Spalding D B 1976 Mathematical models of turbulent flames: A review *Combustion Science and Technology* **13(1-6)** 3-25

[2] H Q Lu 2015 Boiler coal blending ratio optimization *Shandong Electric Power Technology* **42** 76-80

[3] Abbes A S and Lockwood F C 1986 Prediction of a comer-fired utility boiler twenty-first *Symposium (Int.) on Combustion*

[4] Long tacit study based on the combustion characteristics of blended coal blending different ways of Changsha University of Technology 2009

Simulation of one-dimensional heat transfer system based on the blended coal combustion

Y G Jin[1,2*], W B Li[1,2], Z S Cheng[1,2], J W Cheng[1,2] and Y liu[1,2]

[1]State Key Laboratory of Automotive Simulation and Control, Jilin University, Changchun 130022, China
[2]College of Automotive Engineering, Jilin University, Changchun, 130022, China
*E-mail: 694833871@qq.com

Abstract. In this paper, the supercritical boiler thermodynamic calculation model is studied. Three types of heat exchangers are proposed, namely furnace (total radiation type), semi-radiation and convection, and discussed. Two cases were simulated - mixing of two bituminous coals and mixing of a bituminous coal and lignite- order to analyze the performance on the flue gas side. The study shows that the influence of flue air leakage and gas distribution coefficient on the system.

1. Introduction

Domestic coal production has declined since 2013, while the consumption of coal resources has risen slightly. Coal accounts for more than 60% of the total energy consumption in China, and power generation plants account for about 50% of the total domestic coal consumption [1]. However, while bituminous coal and anthracite - extraction did not grow, lignite and peat production increased [2]. The earliest report about coal blending was in the 1960s, and the early coal blending research was mainly used in coking industry [3]. Chen and Lin used the mass unit matrix analytical method, regarding the actual power plant thermodynamic system as a series of steam and water mixed heat systems. Each part of the power plant is decomposed into condenser parts and heating parts, which establishes an overall equation which has to be solved. Based on this, a mass cell method for heat exchanger based on the distribution equation of heating circulating steam/water is established [4].

Auray *et al* used a combination of linear and nonlinear constraints to determine the optimum coal blending ratio [5]. Shi applied the Cybersim simulation platform (North China Electric Power University) to design and calculate the simulation model of a 660 MW supercritical boiler [6].

Gan *et al* took coal species from specific regions as the research target, using mixed coal element analysis, industry analysis and the calorific value of the raw material [7].

2. Thermal system boundary condition

We direct our analysis to the boiler flue gas flow simulation based on the 660 MW supercritical boiler, which is a one-dimensional model. The process includes furnace, low-temperature superheater, final superheater, screen superheater, platen superheater, high-temperature superheater, high-temperature reheater, and low-temperature reheater. Model inputs include different kinds of coal. The boiler flue is arranged in a way of PI (∏), which is the most commonly used power plant boiler arrangement. The first vertical section is the furnace, which produces vertical upward smoke. Superheater and reheater are arranged in the horizontal component. The second vertical section is the flue, which includes

IOP Conf. Series: Materials Science and Engineering **283** (2017) 012020 doi:10.1088/1757-899X/283/1/012020

economizer and air preheater.

In table 1 the main information of the boiler is presented, while in table 2 the ultimate analysis of the design coal and lignite are shown. The coal blend was obtained from 80% design coal and 20% lignite. Table 3 shows the ultimate analysis of blended coal, which was used as the numerical one-dimensional heat transfer system input.

Table 1. Boiler information.

Furnace volume 18705 m^3	Horizontal section of the furnace 342 m^2
Low-temperature superheater 29 MPa	Total heating area of the furnace 7944 m^2
Final superheater 28 MPa	Screen superheater 28.7 MPa
Platen superheater 3374 m^2	High-temperature superheater 2989 m^2
High-temperature reheater 4175 m^2	Low-temperature reheater 114.3 mm
High-temperature reheater 228.6 mm	BMCR main steam flow: 2150 t/h

Table 2. Fuel properties (ultimate analysis).

Element	Design coal	Lignite	Element	Design coal	Lignite
C_{ar}	58.85	50.38	A_{ar}	19.66	5.67
H_{ar}	3.76	2.45	M_t	7.7	26.8
O_{ar}	8.43	13.17	V_{daf}	39.69	48.03
N_{ar}	0.56	1.12	$Q_{net,ar}$	22.83	17.54
S_{ar}	1.03	0.38			

Table 3. Ultimate analysis of blended coal.

Element	C_{ar}	H_{ar}	O_{ar}	N_{ar}	S_{ar}	A_{ar}	M_t	V_{daf}	$Q_{net,ar}$
Data	57.15	3.499	9.379	0.6733	0.9001	16.86	11.52	41.358	21.77

The design coal of the boiler and lignite were mixed. For the boiler stable operation, lignite ratio should not be too large, so we put the proportion of lignite 20%.

3. Mathematical model and thermodynamic system model

For flue gas flows to tube transversely, it is forced convection and it can be calculated by press type:

$$K = \frac{\psi}{\frac{1}{\alpha_1} + \frac{1}{\alpha_2}} = \frac{\psi \alpha_1}{1 + \alpha_1 / \alpha_2} \qquad (1)$$

In the formula, ψ is the coefficient of thermal effectiveness, α_1 is flue gas side heat transfer coefficient, $\mathrm{W/(m^2 \cdot {}^0C)}$; α_2 is steam side heat transfer coefficient, $\mathrm{W/(m^2 \cdot {}^0C)}$.

$$\psi = 1 - \frac{\varepsilon \alpha_1}{1 + \left(\varepsilon + \frac{1}{\alpha_2} \right) \alpha_1} \qquad (2)$$

According to heat transfer formula

$$q = K(T_1 - T_2) \qquad (3)$$

In the case of known heat flow q, the coefficient of heat transfer K, inlet temperature of heat exchanger T_1, we can calculate the heat exchanger outlet temperature T_2, thus the heat transfer model of the thermodynamic system is established [8].

According to the smoke flow, the block diagram of thermodynamic calculation is divided as shown in figure 1, as a one-dimensional model of the thermal system based on the MATLAB SIMULINK module. The flue gas flow is not connected in series. The flue gas of the two processes is finally aggregated and flows through two air preheating units, to complete the whole flue gas flow. The whole model is too huge to fit the paper, so I select part of the whole. There are many parameters for the inlet and outlet, so we select the signal transmission module as [A] > From and [A] Goto to simplify the model. The red line is connected to the Memory module to achieve the purpose of the early assignment.

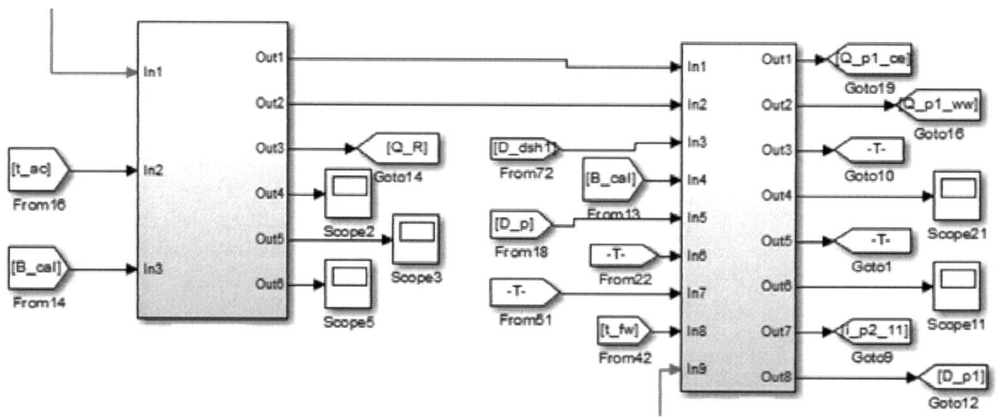

Figure 1. Thermodynamic system model (part of whole).

4. Simulation results and discussion

4.1. Influence of furnace air leakage on system

Since negative pressure boiler flue air leakage is known to reduce the boiler efficiency, an appropriate selection of the boiler leakage coefficient is important. With an increase in the flue gas temperature, leakage of the furnace and flue gas flow rate, the heat transfer coefficient of the flue gas will increase, which will lead to a rise in the steam temperature.

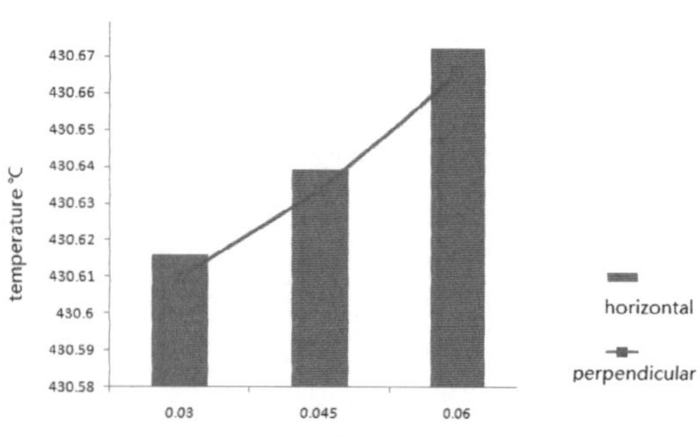

Figure 2. Temperature changes of the low-temperature superheater.

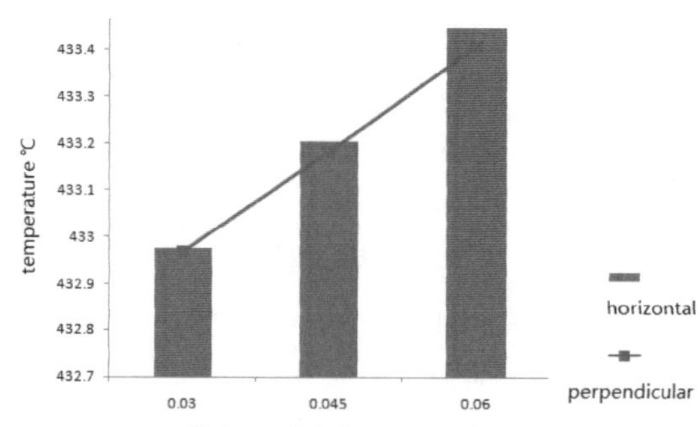

Figure 3. Temperature changes of low-temperature reheater.

As seen from figures 2 and 3, the steam temperature in both upstream and downstream heat exchanger units is increased. The steam temperature in the upstream heat exchanger outlet is higher than in the downstream heat exchanger. That is, the upstream heat exchanger is more sensitive to leakage of the furnace.

The average radiation heat load of the furnace decreases with the leakage of air increase, the radiation heat absorption of water wall decreases, from the law of conservation of energy, the enthalpy of the flue gas at the outlet of the furnace can be increased, so the flue gas temperature increases.

When the leakage coefficient of the furnace is changed from 0.03 to 0.06, the radiant heat load of the furnace varies from 108.9478629 kW/m^2 to 108.7899275 kW/m^2. The changing trend is shown in figure 4. From figure 5, it can be seen that the thermal efficiency of the boiler decreases with the leakage coefficient.

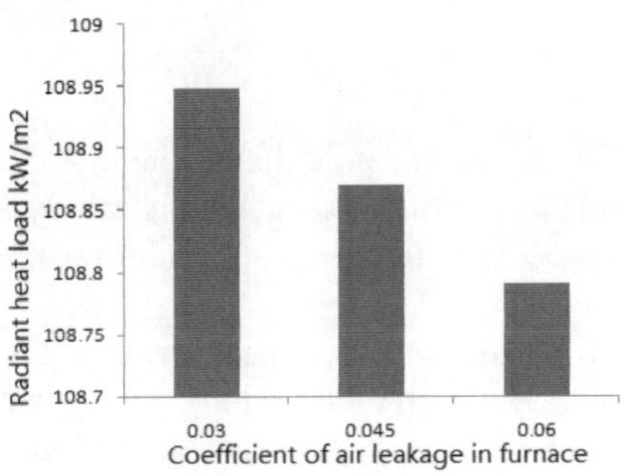

Figure 4. Radiant heat load change in the furnace.

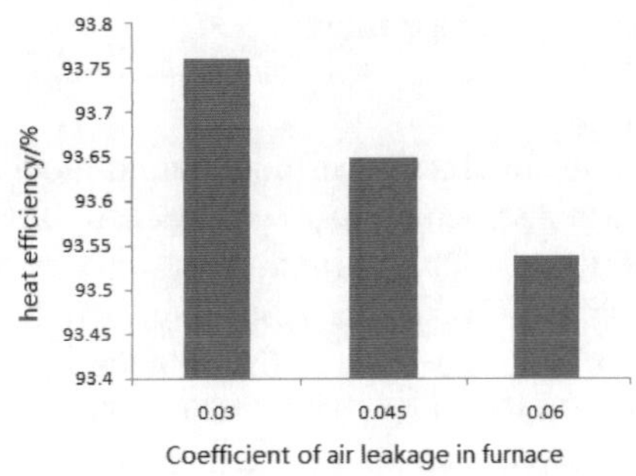

Figure 5. Heat efficiency change of boiler.

4.2 The influence of flue air leakage on the system

The influence of the flue gas pipe on the radiative heat transfer coefficient of flue gas is relatively clear, the radiation heat transfer coefficient of flue gas is mainly affected by the flue gas temperature. The change of convection heat transfer coefficient and of radiation heat transfer coefficient can be seen in figures 6 and 7.

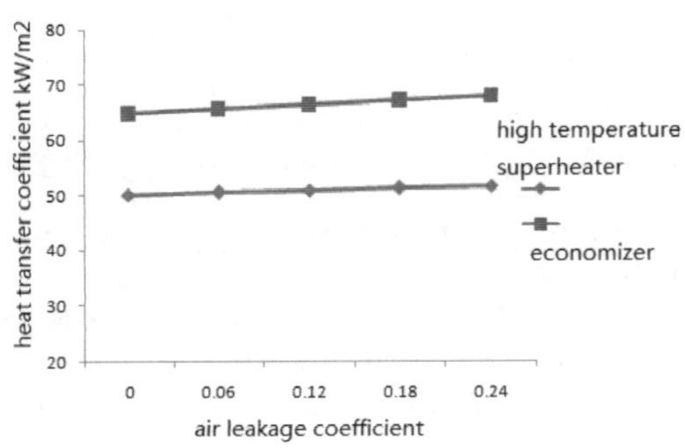

Figure 6. Change of convection heat transfer coefficient.

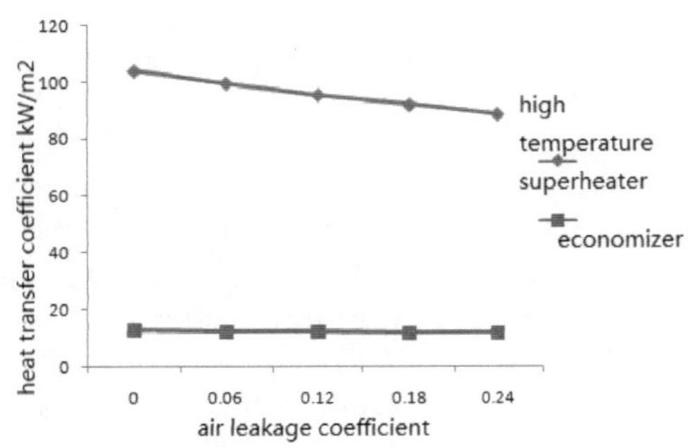

Figure 7. Change of radiation heat transfer coefficient.

4.3 Influence of flue gas distribution coefficient in tail flue gas pipe on system

With an increase in flue gas distribution coefficient, the change of smoke temperature in the front smoke pipe is bigger and the smoke flow through the front flue increases. These two aspects cause the boiler exhaust gas temperature to rise and boiler efficiency to decrease, as shown in figures 8 and 9.

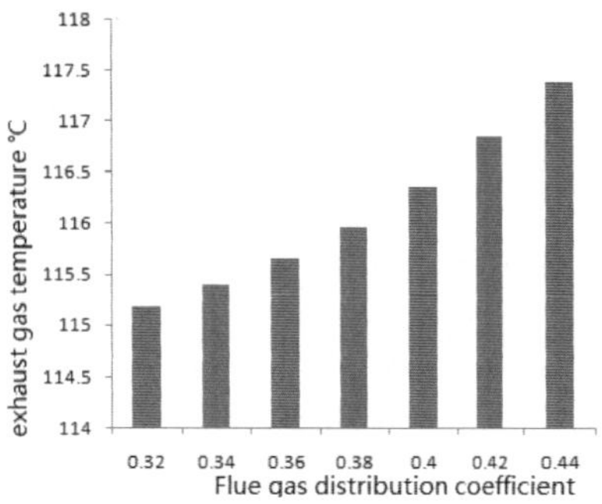

Figure 8. Change of exhaust gas temperature.

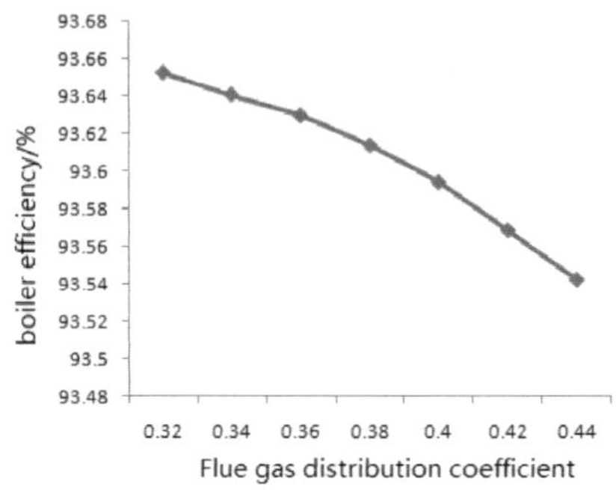

Figure 9. Change of boiler efficiency.

5. Conclusions

In this work, we simulate a thermodynamic model under BMCR condition, calculating the influence of furnace air leakage, leakage of smoke and flue gas distribution coefficient on the thermodynamic system. BMCR is boiler maximum continuous rating, maximum output in case of the steam parameter and furnace safety. Calculation results show that:

- When the air leakage in the furnace is changed from 0.03 to 0.06, the radiant heat load of the furnace varies from 108.9478629 kW/m^2 to 108.7899275 kW/m^2. For a long time operation and a large area of heat exchange boiler, the heat loss will be very large, so a reasonable choice of air leakage coefficient has a great significance.
- The proportion of bituminous coal mixed from 32% to 44%, boiler efficiency decreased from 93.65 to 93.54. However, the flue gas temperature has risen 2.25℃, which makes the flue gas discharged into a higher atmosphere and partial pollution is reduced, so a reasonable choice of fuel mixture ratio has a great significance.

Acknowledgments

The program "Improving the Combustion Mode and Security Running Control Strategy to Decrease the Pollution in Northeast of China" (UK-CIAPP\201) was supported by the Royal Academy of Engineering under the UK-China Industry-Academia Partnership Program scheme. The authors are grateful for the support of the Royal Academy of Engineering and the UK government's Newton Fund. The project was supported by the Chinese Scholarship Council (No. 201706175069).

References

[1] The Medium-Term Coal Market Report 2016, The International Energy Agency (IEA) https://www.iea.org/textbase/npsum/MTCMR2016SUM.pdf
[2] The 13th five-year plan for national economic and social development of the People's Republic of China (2016–2020), Central Compilation & Translation Press, Beijing, China
[3] Zeng C 2006 Application of power coal blending principle in thermal power plant, Beijing, North China Electric Power University.
[4] Lin A F and Chen H P 2009 Analysis method for thermodynamic system of heating supplying unit based on mass unit *Northeast Electric Power Technology,* No.3, 33-4
[5] Auray S, de Wolf D and Smeers Y 2015 Using column generation to solve a coal blending problem *Rairo-Oper. Res.* 49(1) 15-37
[6] Shi Y T 2014 Simulation study on dynamic characteristics of 660 MW supercritical DC unit, Beijing: North China Electric Power University.
[7] Gan P X 2013 Study on combustion characteristics of blended coals which formed by lignite, bituminous coal, anthracite, Kunming: Kunming University of Science and Technology, China.
[8] Feng J K 2003 *Boiler Principle and Calculation (the Third Edition)* , Beijing, Science Press.

IOP Conf. Series: Materials Science and Engineering **283** (2017) 012021 doi:10.1088/1757-899X/283/1/012021

Improved simulation method of automotive spot weld failure with an account of the mechanical properties of spot welds

H Wu, X M Meng*, R Fang, Y F Huang and S Zhang

China Automotive Technology & Research Center, No.68, East Xianfeng Road, Dongli District, Tianjin, China

*E-mail: mengxianming@catarc.ac.cn

Abstract: In this paper, the microstructure and mechanical properties of spot weld were studied, the hardness of nugget and heat affected zone (HAZ) were also tested by metallographic microscope and microhardness tester. The strength of the spot weld with the different parts' area has been characterized. According to the experiments result, CAE model of spot weld with HAZ structure was established, and simulation results of different lap-shear CAE models were analyzed. The results show that the spot weld model which contained the HAZ has good performance and more suitable for engineering application in spot weld simulation.

1. Introduction

With the development of body connection technology, various connection methods have been applied in car bodies design, including resistance spot welding, laser welding, riveting, adhesive bonding and so on. Resistance spot welding is still the main method to assemble sheet metal part/panels for automotive bodies. The modern frameless body is the combination of metal/composite and joints, a car body contains thousands of spot welds, which occupies more than 75% of the car body assembly work [1].

According to crash experiment result, failure of spot welds has the critical influence on crash analysis [2] and is the main concern during safety design of automobiles. To improve the accuracy, reliability, and efficiency of joint failure simulation, spot welds are simplified to beam or solid elements as shown in figure 1, and failure criteria are applied in the elements according to different load conditions. Traditionally, layers of shell elements are welded by beam element through a single node, which is not able to transfer torque [3] and lead to the deviation between simulation and experimental result. While the solid weld elements connect layers through multiple modes, they are able to simulate the stress and torque at the same time. Studies of simulation methods with solid weld element were carried out by relevant researchers. Scie *at al* [4] analyzed a cluster of solid hexahedral elements from 1-Hex to a 16-Hex, respectively, and found that beam and single hex element spot welds should not be treated as mesh-independent in terms of internal forces and displacements. Wu [5] found that solid elements with complete integral shell elements had a better performance in the crash simulations.

However, a simple spot weld model failed to considered different mechanical properties of different spot weld parts, which led to inaccurate failure simulation results. Moreover, it failed to meet the stringent requirements on the accuracy for crash simulation.

In this paper, micromechanical and macro-mechanical properties of spot weld between B340-590DP and B340LA steel were tested. Based on the test results, the CAE model of spot weld was applied, and the comparison was made between testing and simulation result. The results show that the new developing spot weld modeling method was more accurate and suitable for engineering application.

IOP Conf. Series: Materials Science and Engineering **283** (2017) 012021 doi:10.1088/1757-899X/283/1/012021

(a) (b)

Figure 1. Simulation of spot weld using beam element (a) and solid elements (b)

2. Experimental equipment and methods

In the experiment, dual-phase steel B340-590DP and low alloy steel B340LA steel sheets (thickness of 2mm) were produced by Baosteel. The lap-shear spot weld specimens were made by resistance welding machine as shown in figure 2. Lap-shear specimens were tested with WDW100 universal material testing machine with the tensile speed of 2mm/min. The metallographic structure was observed by Leica optical metallurgical microscopy, micro-hardness of the spot weld was also measured by Vickers Indenter under the load of 100gf. The CAE model based on the experimental results was realized via LS-DYNA software package.

a) b)

Figure 2. Resistance welding machine (a) and lap-shear spot weld specimen (b)

3. Analysis of the mechanical properties

3.1. Metallographic structure

Figure 3 shows the cross-section of the B340-590DP/B340LA spot-weld. According to the microstructure character of the spot weld, it can be subdivided into three different regions: nugget, heat affected zone (HAZ), and base metal (BM). There is also a significant boundary between BM and HAZ. As it is shown in metallographic (figure 3), BM of B340LA is made of ferrite and pearlite predominantly. It is due to the high temperature during the welding process, The HAZ of B340LA transforms into ferrite and bainite. For B340-590DP type, the BM consists both ferrite and martensite. The contents of martensite increase significantly near the HAZ part. Nugget is the metallurgical combination of two layers' metal sheet, and its microstructure is also martensitic.

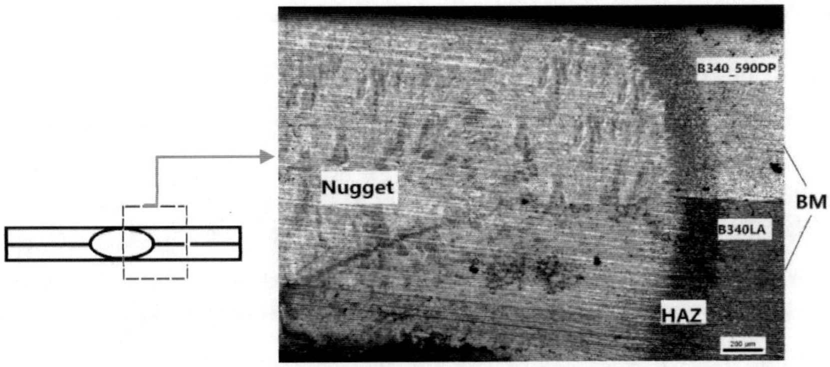

Figure 3. Microstructure of the B340-590DP/B340LA spot weld

3.2. Microhardness analysis
The microstructure of spot weld is different, so it leads to different mechanical and failure performance among BM, HAZ, and nugget. Microhardness of different part weld joints is shown in figure 4. The nugget diameter is 7 mm and the distance between neighboring measure points is 200um approximately. Microhardness of B340-590DP and B340LA base metal is 225Hv and 180Hv respectively, the average value in nugget is 425Hv. The results show that the various spot weld parts have different mechanical properties due to varying metallographic structure characteristics.

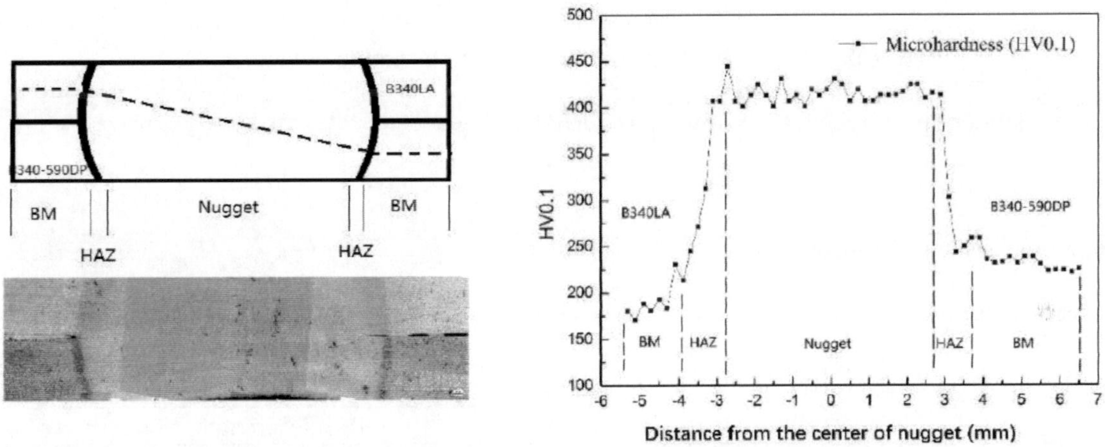

Figure 4. Structure (a) and microhardness variation (b) of B340-590DPand B340LA spot weld.

3.3. Transformation of tensile strength and microhardness
In order to analyze the difference between the material microstructure of weld area, the microhardness was converted to the tensile strength value, which could be used in calculations via the CAE model.

The relationship between tensile strength and hardness of steel is described by the international standard ISO 18265:2003 and ASTM E140-12be1 [6, 7], which can be seen from the research result of Shen [8]. The tensile strength has a linear relationship with hardness value for mild steel and alloy steel. According to the above analysis, the mechanical properties of B340-590DP and B340LA steels are shown in table 1.

The tensile strength of nugget and HAZ, which cannot be measured directly, may be also calculated by the above method. The relationship between the tensile strength and hardness of B340-590DPand B340LA is shown in figure 5.

Table 1. Mechanical properties of B340-590DP and B340LA steels

	Tensile strength, MPa	Young's modulus, GPa	Microhardness (HV0.1)
B340-590DP	598	203	225
B340LA	421	206	181

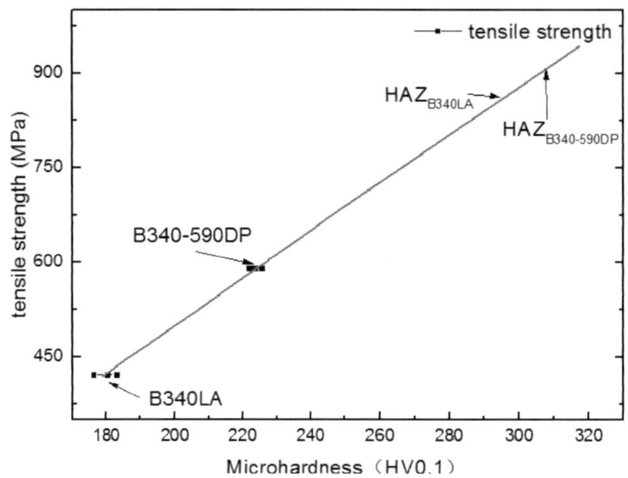

Figure 5. Relationship between tensile strength and microhardness

The relationship between tensile strength and microhardness can be described as follows:

$$\sigma = 4.02x - 307.1 \quad (1)$$

where σ is tensile strength, and x is microhardness of steel.

The tensile strength of HAZ and nugget of the spot weld is calculated by equation (1) and shown in table 2 below.

Table 2. Tensile strengths of HAZ and nugget

	Microhardness (HV0.1)	Tensile strength, MPa
$HAZ_{B340-590DP}$	309	935.1
HAZ_{B340LA}	295	878.8
nugget	425	1401.4

4. Lap-shear test and FEM analysis

4.1. Modeling approach of spot weld

According to the previous research, the solid elements have better performance than beam elements in spot weld simulation, so the simple solid element was established in a lap-shear model (Figure.6b). In order to obtain more accurate results, the HAZ part was included in the spot weld model, which was established in the lap-shear model (Figure 6c). The thickness of shell elements was defined as 2mm, the bottom of the specimen was constrained, while the top of the specimen was subjected to tensile load with the displacement rate of 2mm/min. The size of nugget and HAZ were both consistent with the results of measurement. The HAZ part was defined as a layer of the washer.

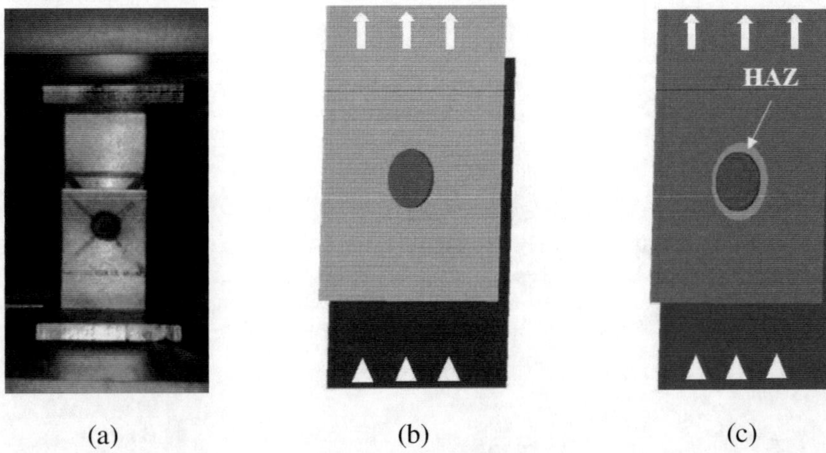

Figure 6. Lap-shear test (a) and CAE models with simple solid elements (b) and solid elements with HAZ (c)

Material parameters of HAZ and nugget were derived from the tensile strength of base metals in table 2, while true stress-strain curves for the input into the material map "MAT_24 piecewise_linear_plasticity" are depicted in Figure 7.

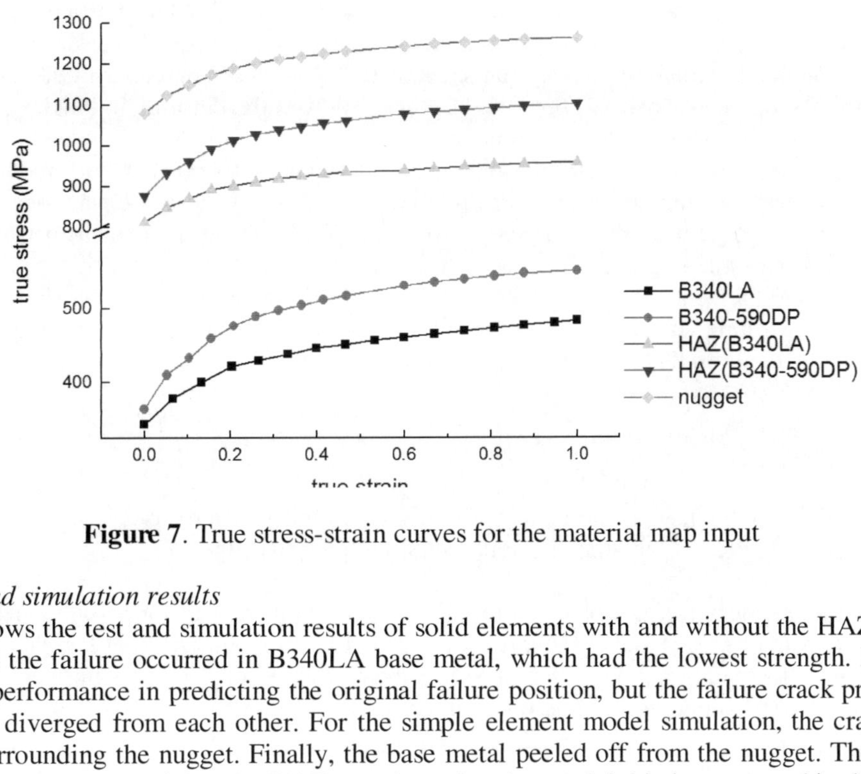

Figure 7. True stress-strain curves for the material map input

4.2. Test and simulation results

Figure 8 shows the test and simulation results of solid elements with and without the HAZ part. It can be seen that the failure occurred in B340LA base metal, which had the lowest strength. Both models have good performance in predicting the original failure position, but the failure crack propagation of two models diverged from each other. For the simple element model simulation, the crack extended path was surrounding the nugget. Finally, the base metal peeled off from the nugget. The calculation results via the model containing the HAZ part show that the crack initiation occurred in the base metal near the HAZ part, while crack propagation took place in the perpendicular direction to the tensile axial load, as it was observed in the test.

IOP Conf. Series: Materials Science and Engineering **283** (2017) 012021 doi:10.1088/1757-899X/283/1/012021

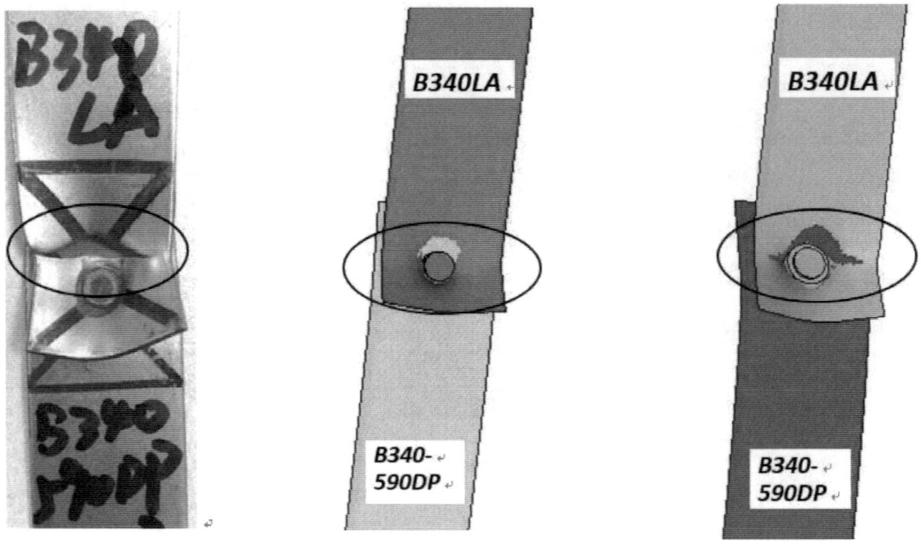

(a) (b) (c)

Figure 8. Failure modes in the test (a) and via CAE models with simple solid elements (b) and solid elements with HAZ (c)

Figure 9 shows the load-displacement curves of lap-shear test and simulation. It could be seen that at the beginning of the curve, tensile force was increase linearly due to the elastic deformation of the base metal. During the deformation process, the sheet of B340LA base metal around the nugget started to warp, and the flat curve is shown accordingly. Cracks followed the deformation and load decreased significantly until the specimen was failed completely.

By comparing the results of two simulation methods, it can be found that both models matched well with test curve in the stage of elastic deformation, regardless of the HAZ part, but the result of simple solid element model had errors in peak load value and failure displacement because of quick deleting of grids around the nuggets, correspond to the previous failure process of simple model. As for the result of solid element model with HAZ, the growth of the failure crack continued for a long period, which led to a long-term loading, so the solid element spot weld simulation model with HAZ part yielded the identical results to the experimental data.

5. Conclusions

- The hardness in the nugget and HAZ is significantly higher than the base metals of B340-590DP and B340LA due to the different metallographic structure.

- Based on the microstructure of spot weld, this paper established a simulation model which solid element of nugget and HAZ part is included. The test comparison of lap-shear test and two kinds of simulation models indicated that the model with HAZ has better performance in failure analysis, and more suitable for engineering applications.

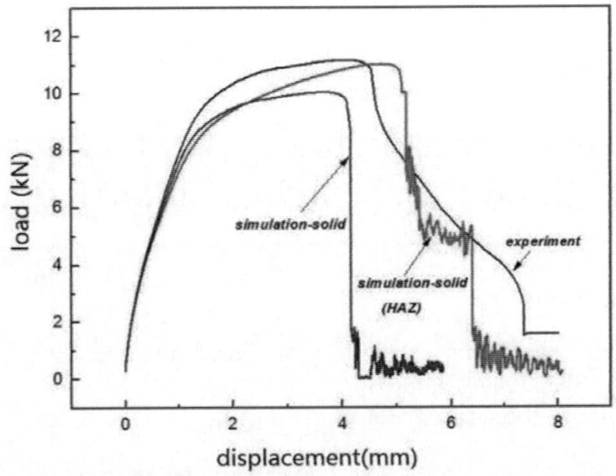

Figure 9. Load-displacement curves of test and simulation

References

[1] Deng S Z and Fan H S 1997 Automobile body manufacturing technology: Beijing Institute of Technology Press.

[2] Inoue T and Nakanishi E. 2000 A crash simulation analysis which considers the spotweld rupture. Proceedings of Annual Congress JSAE, No. 88-00, pp. 5–7.

[3] LS-DYNA Keyword User´s Manual, Version 971 2010 Livermore Software Technology Corporation (LSTC), 2010.

[4] Skye M and Nutwell E 2007 Spotweld failure prediction using solid element assemblies. 6th European LS-Dyna users' conference. Gothenburg, Sweden.

[5] Wu J Z 2009 Beam-type versus Solid-type Spot Weld in LS-DYNA, SAE Technical Paper 2009-01-0354, https://doi.org/10.4271/2009-01-0354.

[6] ISO 18265:2013 Metallic materials - Conversion of hardness values, 2013, https://www.iso.org.

[7] ASTM E140-12be1, 2012, Standard Hardness Conversion Tables for Metals Relationship Among Brinell Hardness, Vickers Hardness, Rockwell Hardness, Superficial Hardness, Knoop Hardness, Scleroscope Hardness, and Leeb Hardness, ASTM International, West Conshohocken, PA, https://www.astm.org.

[8] Shen B L, Li L and Yue C L 2012 Summarization of relationship between tensile strength and hardness of Iron-Steel Materials *Modern Cast Iron*, **32** (1):93-96.

IOP Conf. Series: Materials Science and Engineering **283** (2017) 012022 doi:10.1088/1757-899X/283/1/012022

Integrated hydraulic booster/tool string technology for unfreezing of stuck downhole strings in horizontal wells

Q Z Tian[1*]

[1]Petroleum Engineering Technology Research Institute, Sinopec Shengli Oilfield Company, Shandong, China
*E-mail: tianqizhongcs@163.com

Abstract. It is common to use a jarring tool to unfreeze stuck downhole string. However, in a horizontal well, influenced by the friction caused by the deviated section, jarring effect is poor; on the other hand, the forcing point can be located in the horizontal section by a hydraulic booster and the friction can be reduced, but it is time-consuming and easy to break downhole string using a large-tonnage and constant pull force. A hydraulic booster - jar tool string has been developed for unfreezing operation in horizontal wells. The technical solution involves three elements: a two-stage parallel spring cylinder structure for increasing the energy storage capacity of spring accelerators; multiple groups of spring accelerators connected in series to increase the working stroke; a hydraulic booster intensifying jarring force. The integrated unfreezing tool string based on these three elements can effectively overcome the friction caused by a deviated borehole, and thus unfreeze a stuck string with the interaction of the hydraulic booster and the mechanical jar which form an alternatively dynamic load. Experimental results show that the jarring performance parameters of the hydraulic booster-jar unfreezing tool string for the horizontal wells are in accordance with original design requirements. Then field technical parameters were developed based on numerical simulation and experimental data. Field application shows that the hydraulic booster-jar unfreezing tool string is effective to free stuck downhole tools in a horizontal well, and it reduces hook load by 80% and lessens the requirement of workover equipment. This provides a new technology to unfreeze stuck downhole string in a horizontal well.

1. Introduction

At present, available unfreezing methods commonly used in horizontal wells include the following components.

(1) Surface lifting equipment + underground jar tool string [1]. The surface lifting equipment quickly lifts the drill string to generate elastic energy which is stored by the delay mechanism connected to the jar at the bottom of the drill string, and then the instantaneous release of the elastic energy creates a shock to unfreeze the stuck string. However, the elastic energy may be counteracted by the friction in the horizontal wellbore, thus the jarring effect is poor.

(2) Hydraulic anchor + unfreezing tool string. The hydraulic anchor sets the unfreezing tool string against the wall of the casing, then a large-tonnage pull force is applied to the wall between the hydraulic anchor and the fish by the hydraulic booster to free the stuck string. The advantage of the technology is that the pull force applied to the wall of the casing through the anchor protects the upper tubing string from damage, so it is widely used in oil fields. The disadvantage is that the large-tonnage pull force easily forces the stuck string to break or the fish top to crack or deform [2], making more difficult the further fishing operation.

(3) Jar + back-off tool string. First, use the jar to loosen the thread on the fish, and then use the left-handed string to buckle the back-off, thus to unfreeze the stuck string. However, influenced by wellbore frictions, it is difficult to transmit the torque from the surface equipment to the downhole back-off tool [3], and accordingly, the unfreezing result is poor. In order to solve the problems above, hydraulic booster and spring accelerator have been developed. Using these tools, the pull force energy to downhole drill string from ground equipment can be converted into the elastic energy stored in spring accelerators to counteract the horizontal section friction; and under the interaction of hydraulic booster and jar tools, successful unfreezing operations are performed in horizontal wells.

2. Integrated unfreezing tool string

2.1. *The tool string structure*

The integrated hydraulic booster -jar unfreezing tool string is made up of (from bottom to top) fishing tool + jar + spring accelerator + hydraulic booster + hydraulic anchor + tubing string (figure 1).

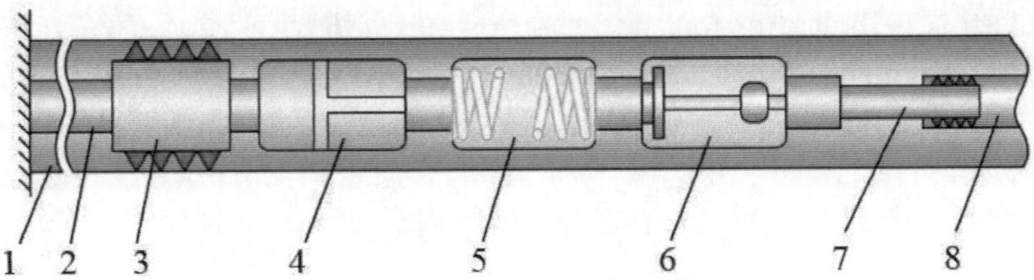

Figure 1. The schematic integrated hydraulic booster -jar unfreezing tool string.
1-Casing; 2-Tubing; 3- Hydraulic anchor; 4- Hydraulic booster; 5- Spring accelerator; 6-Jar; 7-Fishing tool; 8-Fish

2.1.1. Spring accelerator. In the case that a stuck string is in a horizontal section, in order to overcome the friction caused by the deviated borehole, first mechanical energy is converted into elastic potential energy of the spring accelerator by the hydraulic booster, and then the elastic potential energy is converted to kinetic energy of the jar piston, thus to unfreeze the stuck string. The schematic structure of the spring accelerator is shown in figure 2.

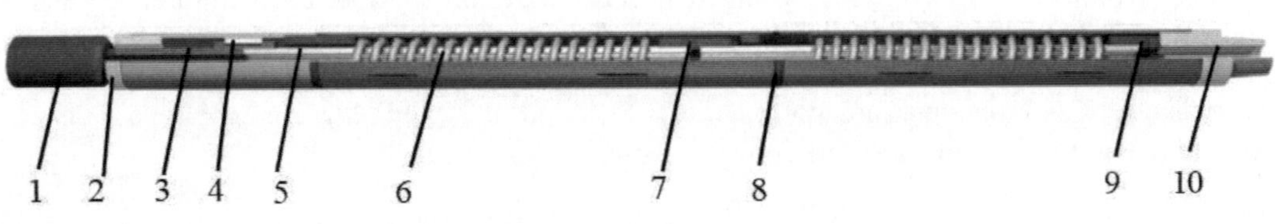

Figure 2. Schematic structure of the spring accelerator.
1 - Top joint; 2- Sleeve; 3-Mandrel; 4-Flat key; 5-Central tube; 6-Spring; 7 - Inner joint; 8-Outer joint; 9-Base; 10-Bottom joint.

Operating principle: The spring accelerator adopts two-stage spring cylinder structure which ensures two groups of springs in a parallel pattern to store elastic energy and thus doubles elastic energy. In addition, to increase the stroke of the spring accelerator and get a long jarring stroke, small single spring accelerators can be connected in series. During an unfreezing operation, the bottom of the spring accelerator is connected to the jar, and the top connected to the booster; after the fishing tool connected to the jar grasps a fish, a large hydraulic pressure is created by ground pumps, and hydraulic anchor on the booster opens and sets against the casing wall; at the same time, the cylinder of the hydraulic booster moves up under the hydraulic force, the booster outer sleeve drives the spring accelerator mandrel up with the bottom joint, and in turn the mandrel drives the base and the inner joint to compress the spring to create elastic energy through the central tube; at this point, the jar enters the delay phase, the elastic potential energy is released instantaneously as the delay mechanism is activated, and the resulting kinetic energy on the jar mandrel is applied to the stuck point.

Table 1 shows the technical parameters of the energy storage spring calculated according to the structure and performance requirements of the spring accelerator.

Table 1. Technical parameters of the energy storage spring.

Material	OD, mm	ID, mm	Free length, mm	Stiffness, $N{\cdot}mm^{-1}$	Number of coils
55CrSi	99	55	631	153	23

Preset conditions: the single-stage stroke of the spring accelerator is 110 mm, and the total strokes required to complete a shock is 360 mm; to get the length, four groups of spring accelerators are connected in series, with a single spring providing a free length of 631 mm; thus, the total free length of the four groups is 631 mm × 8 = 5048 mm; the elastic coefficient of a spring accelerator is $K_2=2K_1$, accordingly the elastic coefficient of four accelerator series is $K_3=K_2/4=K_1/2$; after installing, the compressed length is Δx_1 = (631-620) × 4 = 44 mm, the total working stroke Δx_2 = 110 × 4 = 440 mm; as a jarring stroke is 360 mm, then the total spring variation $\Delta x_3=\Delta x_1+\Delta x_2-360=124$ mm.

The initial elastic potential energy is:

$$E_j' = \frac{1}{2}k(\Delta x_1 + \Delta x_2)^2 \tag{1}$$

Substitution of K_3, Δx_1 and Δx_2 into equation (1), yields $E_j'=8960.292$ J

The final elastic potential energy is:

$$E_j'' = \frac{1}{2}k\Delta x_3{}^2 \tag{2}$$

Substitution of K_3 and Δx_3 into equation (2), yields $E_j''=588.132$ J

The converted elastic potential energy is:

$$E_j = E_j' - E_j'' \tag{3}$$

Thus, we get E_j=8372.16J

Since the hydraulic booster-jar tool string is in the horizontal section, the variation of its gravity potential energy is not considered. According to the law of conservation of energy, the elastic potential energy is converted to the kinetic energy of the jar mandrel and spring accelerator sleeve and drill pipe, and then creates a shock on the struck point.

The total mass of the jarring acceleration system is:

$$m = m_1 + m_2 + m_3 + 2m_4 \tag{4}$$

where m_1 is the mass of the jar mandrel, kg; m_2 is the mass of the leader of the jar, kg; m_3 is the total mass of the accelerator sleeve, kg; m_4 is the mass of a single drill pipe, kg.

Taking these values (m_1=62 kg, m_2=14.3 kg, m_3=230.8 kg, m_4=138 kg) into equation (4), we get m=583.1 kg.

According to the law of energy conservation, the instantaneous jarring speed is

$$\frac{1}{2}mv^2 = E_j \tag{5}$$

So E_j=5.36 m/s

According to the momentum theorem,

$$Ft = mv_2 - mv_1 \tag{6}$$

After the shock (which is very short – from 10 to 15 ms) is over, v_2=0, and equation (6) yields:

$$F = -\frac{mv_1}{t} = -312541.6 \text{ N}$$

where the negative symbol indicates that the jarring force is opposite to the direction of the jarring stroke. The instantaneous jarring force to a fish is about 313 kN.

2.1.2. Hydraulic booster. The hydraulic booster is one of the key tools in the integrated unfreezing tool string. Its role is to change the way that a stuck string is shocked by a large-tonnage pull force between the stuck point and the booster by pumping fluid through tubing string [4]. The hydraulic booster can both increase the pull force and stretch the spring accelerator to store more elastic potential energy. The jar has a delay mechanism, which can release the energy instantaneously and apply a shock on the stuck point. Figure 3 shows the schematic structure of the hydraulic booster.

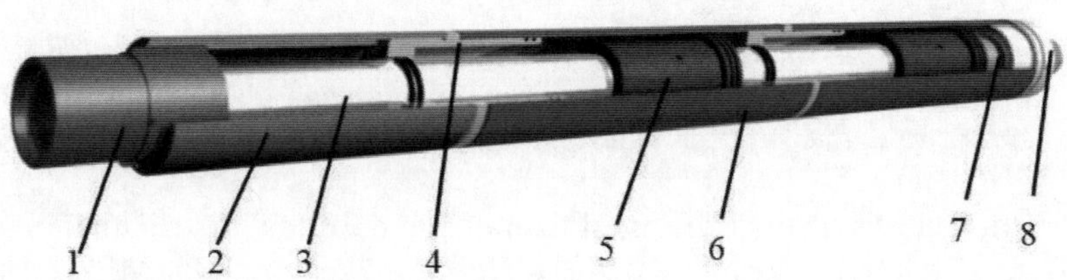

Figure 3. Schematic structure of the hydraulic booster.
1 – Top joint; 2-Top sleeve; 3-Central tube; 4-Connector; 5-Piston; 6-Cylinder liner; 7-Valve seat; 8-Bottom joint

Operating principle: The hydraulic booster consists mainly of a top joint, a central tube, a piston, a cylinder liner, a valve seat, a bottom joint, and other components. When pumping fluid into the tubing, the fluid enters the annulus between the piston and the cylinder liner; because the piston connects to the hydraulic anchor through the central tube and the top joint, and it is away at a relatively fixed distance from the casing, the connector creates an upward pull force under the action of hydraulic pressure [5], which is applied directly to the spring accelerator through the cylinder liner and the bottom joint. The booster cylinder adopts a multi-stage parallel structure, and the maximum can be up to six stages according to required pull force and speed [6].

Technical features: The hydraulic booster changes the way that the stuck string is shocked. The pull force from the ground equipment on the downhole string is converted to the hydraulic pressure in the tubing, and then converted into pull force by the booster and directly applied to the spring accelerator and jar [7]. During unfreezing operation, no friction will be produced between the tubing and the casing wall, as well as no lateral force, so the casing won't be damaged [8]. This is very suitable for unfreezing stuck string by jar or booster in horizontal and highly deviated wells.

According to the structure and performance requirements of the hydraulic booster, the strength of its parts was calculated and verified. The strength verification formula for the parts is referred to the "Mechanical Design Manual" and other relevant information.

As shown in figure 4, the OD-to-ID ratio of the cylinder is $K = \frac{D}{d} = 1.15 \leq 1.2$ (D is the OD and d is the ID), the strength verification is calculated using the thin-walled cylinder theory which assumes that the cylinder is encountered stresses in two directions, and the stress on any section is even: the radial stress $\sigma_r = 0$, and the circumferential stress is $\sigma_\theta = \frac{P(d+\delta)}{4\delta}$, $\sigma_z = \frac{P(d+\delta)}{2\delta}$. According to the first Strength theory, the theoretical strength formula for the cylinder wall thickness is:

$$P = \frac{2\times\sigma_s\times\delta}{n(d+\delta)} \tag{7}$$

where P is the body compressive strength, MPa; σ_s is the material yield strength, MPa; δ is the wall thickness, mm; d is the ID, mm; n is the safety factor, dimensionless.

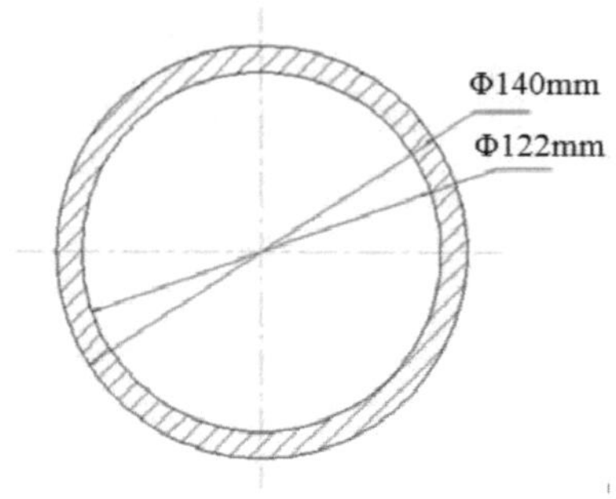

Figure 4. Cross section of the cylinder liner of the hydraulic booster.

All parts are made of 35CrMo steel, which yield strength is 835 MPa and tensile strength is 980 MPa, and safety factor is n=1.25. The internal pressure strength calculated by formula (7) is 91.8 MPa, which is much larger than the highest field pumping pressure, therefore, it meets the strength requirements.

The finite element analysis method was used to simulate the stress distribution on a single cylinder liner of the hydraulic booster under the highest working pressure (25 MPa). The ANSYS 14.0 software was used. The solid 187 unit was selected and the 35CrMo properties include the Young modulus of 2.13 × 105 MPa, Poisson's ratio of 0.286, the mass density of 7870kg/m^3. After meshing (figure 5), a 25 MPa hydraulic load was applied, and the stress was simulated (figure 6).

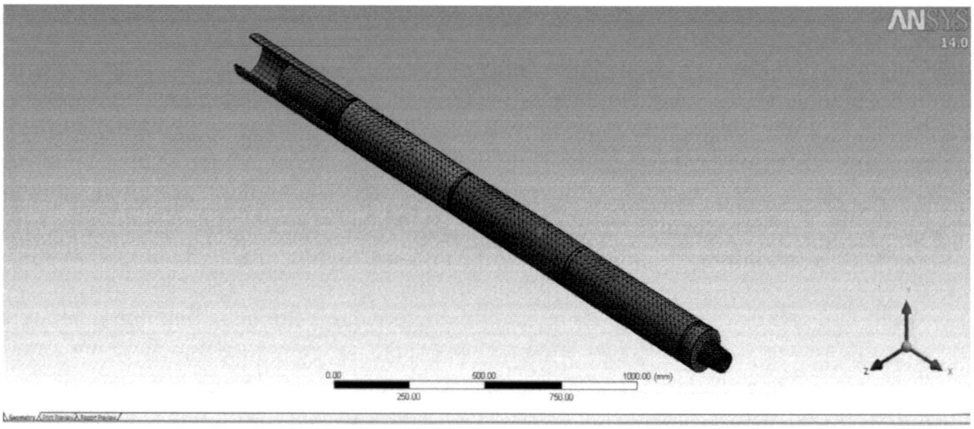

Figure 5. Schematic diagram of finite element meshing of the hydraulic booster.

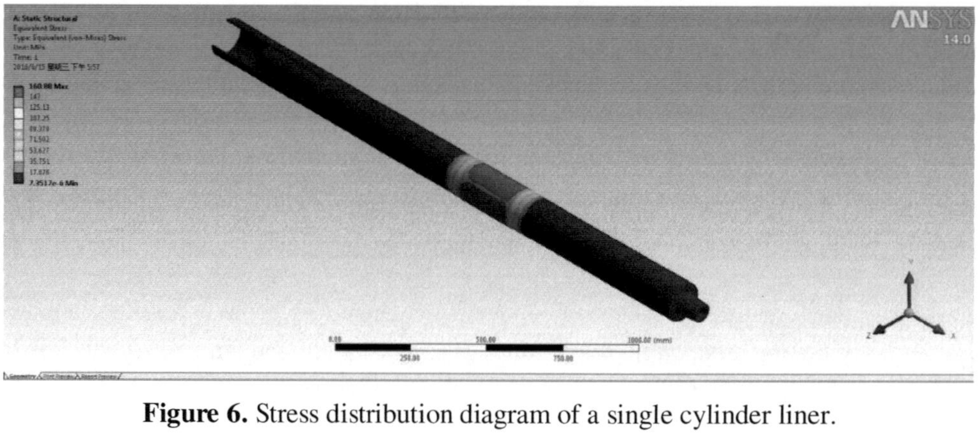

Figure 6. Stress distribution diagram of a single cylinder liner.

The stress distribution diagram shows that the maximum stress in the cylinder liner is 160.88 MPa, which is much smaller than the yield strength of the material. Therefore, the designed strength satisfies the performance requirement.

Considering field operation, the specifications and technical parameters of the hydraulic booster were determined and listed in table 2.

Table 2. Technical parameters of the hydraulic booster.

Max. OD, *D*, mm	Casing ID, *d*, mm	Rated pressure, P, MPa	Free length, *L*, mm	Rated pull force, *F*, kN
115	121-132	20	900	490
140	150-162	20	900	1000

2.2. Unfreezing principles

First catch the fish with fishing tools, then pump fluid into the tubing to open the hydraulic anchor clutches while the cylinder of the hydraulic booster moves up quickly and drives up the central tube of the spring accelerator. Because of connecting to the delay mechanism of the jar, the outer working sleeve of the spring accelerator is relatively slow, while the central tube compresses the springs in the working sleeve and stores the elastic energy. When the delay mechanism reaches the release chamber, the elastic potential energy is instantly released and converted into kinetic energy of the jar mandrel, which will shock the jarring sleeve and then drive the fishing tools to shock the stuck point. Repeat the procedures until the stuck string is released successfully.

3. Indoor tests

After the completion of the development of the single tools, the performance parameters of the tools were tested on the ground and the results meet the designed requirements. Then downhole simulation test on the entire tool string was carried out. The simulation device and parameters are shown in figure 7 and table 3.

Figure 7. Simulation test device for the horizontal well jarring tool string.
1 –Hydraulic lifting system; 2-Torsion system; 3-T-pipe; 4-Water hose; 5- Tool string

Table 3. Simulation test parameters.

Pre-tightening force, kN	Pumping rate, $m^3 \cdot min^{-1}$	Boosting time, s	Pump pressure, MPa	Delay time, s	Instantaneous jarring force, kN
50	1.2	4	5	23	287

The downhole simulation test has proved that the technical solution of the hydraulic booster + jar unfreezing string is feasible in horizontal wells, indicating that performance parameters meet the design and can be used under the field test conditions. The field technical parameters of the hydraulic booster + jar unfreezing string are determined based on the numerical simulation results and experimental data (table 4).

Table 4. Field technical parameters of the hydraulic booster + jar unfreezing string.

Casing ID, mm	Pumping rate, $m^3 \cdot min^{-1}$	Pre-tightening force, kN	Downward load, kN	Jarring force. kN
121-159	≥1.0	40-60	0-5	280-300

4. Field test

4.1. Operating procedure

- Run into the unfreezing string which is made up of (from bottom to top): releasing spear + rigid centralizer + drill pipe × 1 + jar + drill pipe × 1 + 4-stage spring accelerator + hydraulic booster + 2-stage hydraulic anchor + tubing string;
- Check the depth of the fish while slowing down the string; pressurize to 40 kN, and insert the spear into the fish cavity, and then slowly lift the string; once the jar is fully open, increase the lifting force to 60 kN. If the stuck string is not released at this moment, connect the mud line, and then test pressure at 25 MPa.
- Boost unfreezing force. Enhance pumping rate in the order from 5 MPa to 10 MPa to 15 MPa and until 20 MPa, and keep the pressure stable for 5 min at each point. Once a sudden pressure drop occurs, there is fluid flowing out from the annulus, and the hook load reduces to the original suspending weight, they indicate that the stuck string has been free [9]. Then stop the pump to release pressure, and after 5 min, unlock the hydraulic anchor and lift the string. During lifting, no increase of the load means the string has been free, then run out of the string and retrieve the fish. If the pump pressure does not drop and the hook load does not change, this means the string is still stuck; then it is necessary to restore the string to original suspending state and reset the jar, be ready to unfreeze the string by hydraulic jarring.
- Hydraulic jarring operation. Lift the string to make the original suspending weight increase to 60 kN, enhance the pump pressure to 10 MPa at the highest rate so that the hydraulic anchor can set against the casing wall [10] while the booster cylinder drives up the center tube of the spring accelerator under the hydraulic action. The delay mechanism in the jar slows down the mandrel and the working cylinder of the spring accelerator connected to the mandrel so that the center tube of the accelerator compresses the spring in the cylinder to store elastic energy. When the hydraulic delay mechanism works, the elastic energy stored in the accelerator will be instantaneously released and converted into kinetic energy of the jar mandrel, and the jarring sleeve will be shocked once and in turn the lower fishing tool to unfreeze the stuck string once.
- Repeat the steps above several times until the stuck string is free; otherwise, repeat step (3) by boosting unfreezing force, or further repeat step (4) by hydraulic jarring until the fish is retrieved and the string is free.

4.2. Application cases

The Zhong14-Ping X well is a horizontal well in the First Field of Gudao Oil Production Plant in Shengli Oilfield, China In order to improve the sand control, the well was completed with a screen liner, and gravel packing was done outside the screen liner. After packing, the packing tool string was stuck. After lifting up by 30-ton load using the workover rig, the tool string was still stuck. Then the string above the safety joint was run out of the wellbore. The fish was ready to be retrieved by the unfreezing process with hydraulic booster and jar. The string parts left in the wellbore and their data are shown in table 5.

Table 5. The string parts left in Zhong 14-Ping-X well.

No.	Tool	Length, m	Depth, m
1	Bottom packing device	2.10	1497.69
2	Tubing sub	1.01	1495.59
3	Centralizer	0.33	1494.58
4	Tubing tub	2.03	1494.25
5	Centralizer	0.33	1492.22
6	Safety joint	0.40	1491.89
7	Setting tool	0.58	1491.49
8	Tubing	54.34	1490.91
9	Packing tool	1.38	1436.57

The unfreezing string is made up of (from bottom to top): Φ76 mm releasing spear + Φ150 mm centralizer + Φ73 mm drill pipe × 1 + jar + Φ73 mm drill pipe × 1 + 4-stage spring accelerator + hydraulic booster + 2-stage hydraulic anchor + Φ88.9 mm tubing string to the wellhead.

After running the string to the stuck point, an attempt was made to free the fish by boosting hydraulic pressure up to 50 ton, but it didn't remove the sand plug. Then the jarring method was used but the plug was still present after trying for 20 times and, finally, the pull force was increased up to 50 ton. After repeating the jarring operation for 40 more times, the string was observed to move up at a small distance at the wellhead, indicating that the plug had been loosened. After increasing the force to 30 ton, the suspending weight was reduced significantly. After repeating the boosting and jarring operations, the sand plug was eventually removed.

5. Conclusions

- The integrated unfreezing technology supported with spring accelerator and the hydraulic booster is effective to unfreeze stuck string in a horizontal well by reducing the friction caused by the deviated wellbore.
- The hydraulic booster-jar unfreezing tool string can unfreeze stuck string through the jarring operation and hydraulically boost the pull force. These two mechanisms worked together to obtain the successful unfreezing result.
- During field operation, to activate the booster to drive the spring accelerator for accumulating elastic energy in a short time, the pump displacement must be higher. However, a hydraulic pressure increase can only lead to a shock and the operation is complicated. Future study will focus on the development of a jarring tool string which can apply a continuous jarring force through ground pumping, thus to improve the jarring effect.

Acknowledgments

The author would like to thank the management of Petroleum Engineering Technology Research Institute for their permission to publish this paper. The author would also like to acknowledge the contribution of the oil well intervention/workover team.

References

[1] Sun R P and Jin D S 1996 New super fishing jar *China Petroleum Machinery* **24(7)** 38-41
[2] Tian M and Wu W H 2008 Improvement and application of hydraulic jar for high-pressure ultra-deep well *China Petroleum Machinery* **36(8)** 81-3
[3] Zhou J Q, Zhang G T, Li D Y, *et al* 2009 Development and application of the fully hydraulic while-drilling bumper sub China Petroleum Machinery **37(9)** 89-91

IOP Conf. Series: Materials Science and Engineering **283** (2017) 012022 doi:10.1088/1757-899X/283/1/012022

[4] Zhang G T and Long J Q 2009 Application and improvement of a hydraulic downhole pulling tool Oil Field Equipment **38(2)** 75-9
[5] Guo J G and Zhang H Q 2004 A new hydraulic downhole fishing tool with augmenter for special wells China Petroleum Machinery **32(8)** 37-9
[6] Wang L L, Jiang Z S, Zhang J Z, *et al* 2012 Study on unfreezing and fishing technology for horizontal wells Oil Field Equipment **41(9)** 64-8
[7] Kong L W 2013 Reinforced stick-freeing and fishing techniques of the string pipe of the horizontal well Petroleum Geology & Oilfield Development in Daqing **32(3)** 93-6
[8] Wang B and Kong C L 2013 Unfreezing and fishing techniques for horizontal wells Petroleum Geology & Oilfield Development in Daqing **32(4)** 94-7
[9] Li J Y, Li M Z, Du B G, *et al* 2011 Stress analysis and strength check for hydraulic force increasing stuck-releasing strings Oil Drilling & Production Technology **33(6)** 49-52
[10] Fu J H, Luo J and Li C Y 2008 Workover operation technology for horizontal wells in China Drilling & Production Technology **31(6)** 91-3

IOP Conf. Series: Materials Science and Engineering **283** (2017) 012023 doi:10.1088/1757-899X/283/1/012023

Investigation on solubility of hydroxy dibasic acids in alkanolamine solutions

M Du[1*]

[1]School of Shipping and Naval Architecture, Chongqing Jiaotong University, Chongqing 400074, China
*E-mail: dm-god@126.com

Abstract. Solubilities of three hydroxy dibasic (adipic, suberic, and sebacic) acids in alkanolamine solutions were measured within the 30-90℃ temperature range. It is found that solubility of these acids sharply grows with temperature and concentration of alkanolamine solvent. In addition, the study substantiates the adjustment of pH to optimize the CO_2 absorption and desorption processes. The precipitation of added acids from alkanolamine solvents by cooling is found to be quite problematic, which makes the recovery of residual acids from lean alkanolamine solvents non-feasible and requires the application of alternative methods.

1. Introduction

The work is a continuation of the author's previous study [1], where the feasibility of adding hydroxy dibasic acids for reducing the energy consumption of CO_2 desorption from alkanolamine solutions was verified. The preliminary results obtained in [1] strongly suggested that addition of some hydroxy dibasic acids could decrease the energy consumption of CO_2 desorption from alkanolamine solvents, but residual acids may cause degradation of alkanolamine solvents. To check whether adipic acid could be recovered by the cooling method, the experimental study of the solubility of hydroxy dibasic acid/alkanolamine solution is required.

In this study, an attempt is made to reduce the solution pH by adding some acids to desorption solvents, to reduce the energy consumption, with the further recovery of these acids by cooling. This procedure is based on the steep solubility curves of some hydroxy dibasic acids in water. However, when the aqueous solutions are replaced by alkanolamine solvents, the situation may drastically change. Therefore, the determination of solubility in this study is instrumental to solving this problem.

2. Experimental

2.1. Experimental devices and reagents

The experimental device is shown in figure 1. The bottom is the heating agitator for heating and stirring the solution in a beaker. There are three holes on the top of the beaker. One is for measuring the pH and temperature of the solution, and the other two are for the solute addition. The intermediate interface is for placement of the condenser. An oil bath for heating the solvents is on the outside of the container. The oil heated to a certain preset temperature controlled by the oil bath flows from the upper inlet into the lower one.

Measuring devices are as follows: weight meter (accuracy ±0.001 g), pH meter (accuracy ±0.001), thermometer (accuracy ± 0.1℃), measuring cylinder (accuracy ±0.5 mL).

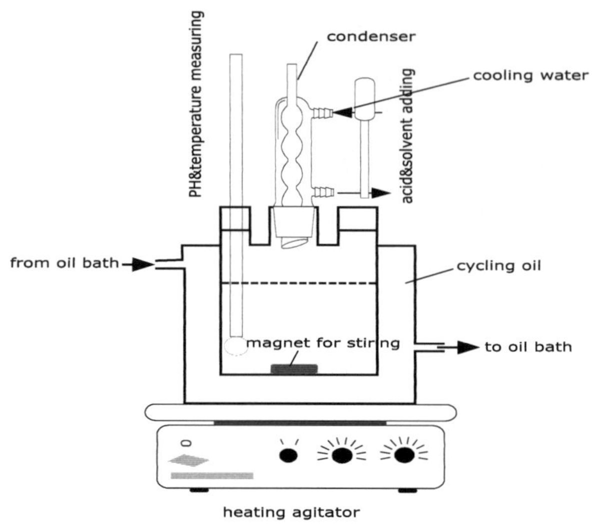

Figure 1. Schematic flow diagram of the solubility experimental apparatus.

Three hydroxy dibasic acids are used in experiments, including adipic ($(CH_2)_4(COOH)_2$, Fluka Chemie GmbH, ≥99.5%), suberic ($C_6H_{12}(COOH)_2$, Sigma-Aldrich (US); ≥99%) and sebacic acids ($(HOOC)(CH_2)_8(COOH)$, Sigma-Aldrich (US); ≥99%).

The solvents tested are Monoethanolamine (MEA, Sigma-Aldrich (US), ≥99%), Diethanolamine (DEA, Sigma-Aldrich (US), ≥98%) and Methyldiethanolamine (MDEA, Sigma-Aldrich (US), ≥99%).

2.2. Experimental process

Firstly, 20 mL water was measured and poured into a beaker with a volume of 50 mL, then the water was heated to 30℃ with real-time monitoring of temperature. Alkanolamine solvents were prepared as 50 mL sample with concentrations of 0.2, 0.4 and 3 mol/L. Then, acids were added to the solution until a trace of crystal precipitation was observed, which indicated that the solution was saturated. The mass of the acids added to the solution was recorded as well. Besides, pH of the solution was recorded in real time. Another criterion for solution saturation is that the pH does not change anymore. The above steps were repeated to obtain the data of solubility for three acids in the alkanolamine solvents at 30-90°C, then solubility curves were plotted. The uncertainty of this test results directly from the minimal portion of a solute being added to the vessel and temperature control. The experimental contents are shown in table 1.

Table 1. Experimental contents.

Acids	Solvents	Concentration of solvents (mol/L)	Temperature (℃)
Adipic acid	MEA MDEA DEA	0, 0.2, 0.4, 3	30-90
Suberic acid	MEA		
Sebacic acid	MEA		

3. Results and discussion

3.1. The solubility of adipic acid in three alkanolamine solvents

The solubility curves of adipic acid in MEA, DEA, and MDEA solvents are plotted in figures 2, 3, and 4, respectively, and the corresponding data are tabulated in tables 1-3.

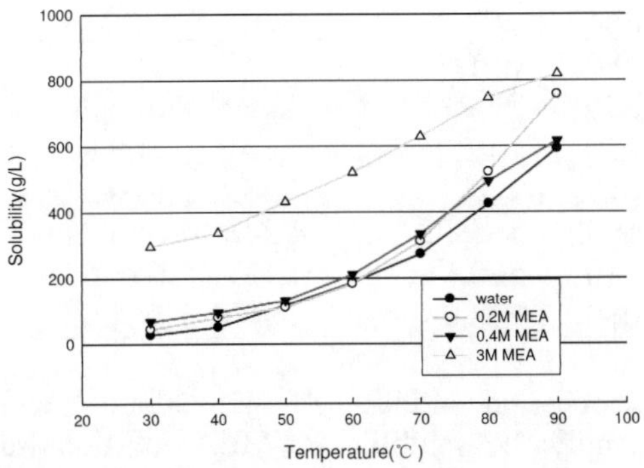

Figure 2. Solubility curves of adipic acid in water and MEA(0.2 , 0.4 , and 3 mol/L).

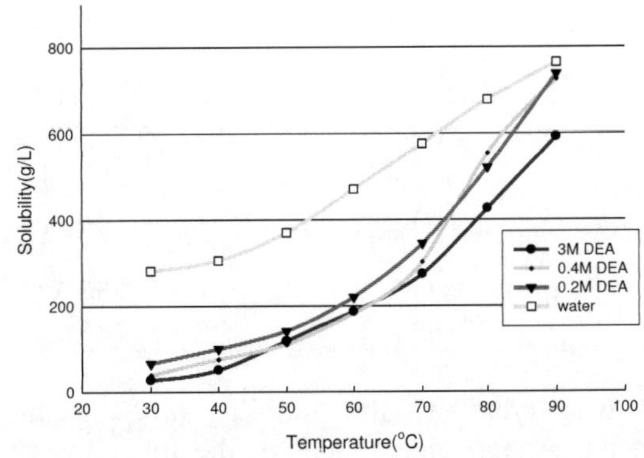

Figure 3. Solubility curves of adipic acid in water and DEA(0.2 , 0.4 , and 3 mol/L).

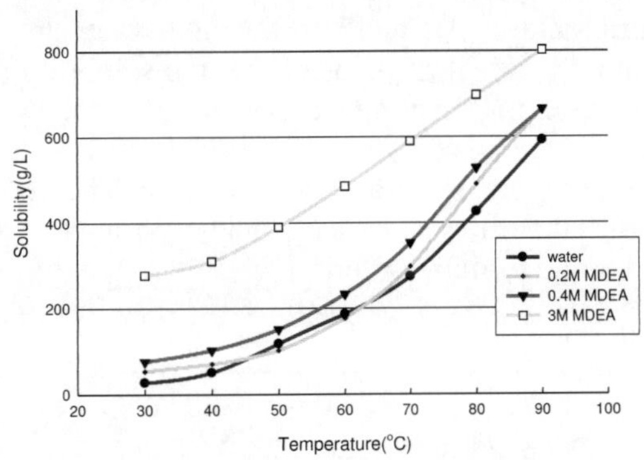

Figure 4. Solubility curves of adipic acid in water and MDEA (0.2 , 0.4 , and 3 mol/L).

Table 2. Solubility and pH of adipic acid in water and MEA (0.2 , 0.4 , and 3 mol/L).

Temperature (℃)	Water	0.2 mol/L solution		0.4 mol/L solution		3 mol/L solution	
	Solubility (g/L)	pH value	Solubility (g/L)	pH value	solubility (g/L)	pH value	Solubility (g/L)
30	27.23	4.408	44.12	4.613	67.66	5.360	296.15
40	50.65	4.153	79.52	4.436	95.20	5.139	337.31
50	117.48	3.980	110.96	4.232	132.11	4.872	431.88
60	186.41	3.730	183.54	3.969	210.70	4.641	519.87
70	273.38	3.390	311.96	3.62	334.08	4.418	628.49
80	425.37	3.120	521.09	3.327	493.73	4.197	747.54
90	592.07	2.941	757.36	3.125	615.82	4.092	819.24

The above results imply that the solubility of adipic acid in alkanolamine solvents is higher than that in aqueous solutions, and that solubility sharply increases with the alkanolamine concentration. The solubility curve of adipic acid in water is similar to that in alkanolamine solvents, i.e., the solubility increases steeply with temperature [2,3]. The pH of the saturated solution decreases with temperature, and the pH value is lower than 7.

IOP Conf. Series: Materials Science and Engineering **283** (2017) 012023 doi:10.1088/1757-899X/283/1/012023

Table 3. Solubility and pH of adipic acid in water and DEA (0.2 , 0.4 , and 3 mol/L).

Temperature (℃)	Water	0.2mol/L solution		0.4mol/L solution		3mol/L solution	
	Solubility (g/L)	pH value	Solubility (g/L)	pH value	Solubility (g/L)	pH value	Solubility (g/L)
30	27.23	4.440	38.91	4.582	65.66	5.192	281.53
40	50.65	4.040	74.97	4.336	99.09	5.059	305.34
50	117.48	3.845	109.01	4.140	140.64	4.865	369.92
60	186.41	3.560	181.90	3.876	220.20	4.589	470.58
70	273.38	3.262	301.82	3.565	343.72	4.348	576.34
80	425.37	3.000	553.32	3.280	520.63	4.137	678.55
90	592.07	2.780	725.95	3.030	738.03	4.960	764.92

It was found that the solubility of each acid in the same concentration of three alkanolamine solvents is nearly identical. For the following two acids, only the solubility in MEA solution was measured.

3.2. The solubility of suberic acid in MEA solvents

The solubility curves of suberic acid in 0.2 mol/L MEA and water are shown in figure 5, and the corresponding data are presented in table 4. It is found that the solubility of suberic acid in 0.2 mol/L MEA solvent is quite different from that in water, exceeds it at 30℃ by almost 20 times. When the temperature rises, the solubility difference is reduced. The solubility curve of suberic acid in 0.2 mol/L MEA solution also exhibits a sharp increase with the temperature. The pH values of the solutions after saturation exceed 7, which means that the solutions are acidic. When the solution temperature rises to 70℃, suberic acid starts to evaporate from the solution.

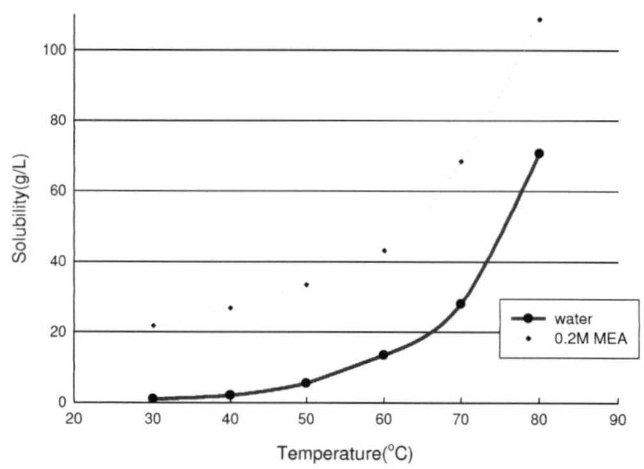

Figure 5. Solubility curves of suberic acid in water and 0.2 mol/L MEA.

Table 4. Solubility and pH of adipic acid in water and MDEA (0.2 , 0.4 , and 3 mol/L).

Temperature (℃)	Water	0.2 mol/L solution		0.4 mol/L solution		3 mol/L solution	
	Solubility (g/L)	pH value	Solubility (g/L)	pH value	solubility (g/L)	pH value	Solubility (g/L)
30	27.23	4.278	53.84	4.558	76.31	5.217	277.07
40	50.65	4.079	70.75	4.361	102.55	5.104	309.97
50	117.48	3.879	101.94	4.140	151.10	4.846	390.07
60	186.41	3.584	176.72	3.865	232.90	4.595	485.37
70	273.38	3.287	297.04	3.615	353.15	4.350	590.21
80	425.37	2.981	489.55	3.373	527.32	4.122	697.27
90	592.07	2.790	665.30	3.080	664.72	3.968	802.86

3.3. The solubility of sebacic acid in MEA solvents

The solubility curves of sebacic acid in 0.2 mol/L MEA and water are shown in figure 6 and the corresponding data are given in table 5. It is found that the solubility of sebacic acid in MEA solvents is much higher than that in water, and increases steeply with temperature, which is similar to the first two acids. As shown in table 6, in contrast to the previous two acids, the solubility of sebacic acid in alkanolamine solvents is much lower at approaches zero at 0℃. Its dissolution and crystallization processes are much slower as well. It was also revealed that the pH value of saturated solutions exceeded 7 at 30℃, and dropped below 7 when the temperature rose to 40℃.

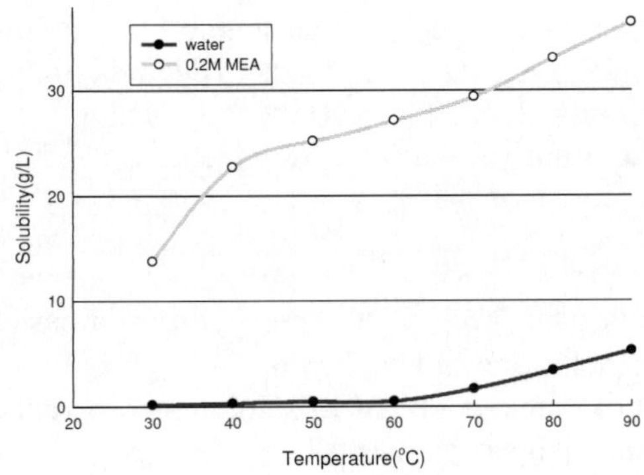

Figure 6. Solubility curves of sebacic acid in water and 0.2 mol/L MEA.

Table 5. Solubility and pH of suberic acid in water and 0.2 mol/L MEA.

Temperature (℃)	Water	0.2 mol/L solution	
	Solubility (g/L)	pH value	Solubility (g/L)
30	0.98	5.494	21.75
40	2.04	5.177	26.79
50	5.45	4.960	33.45
60	13.42	4.772	43.17
70	27.94	4.488	68.43
80	70.64	3.967	108.73

Table 6. Solubility and pH of sebacic acid in water and 0.2 mol/L MEA.

Temperature (℃)	Water	0.2 mol/L solution	
	Solubility (g/L)	pH value	Solubility (g/L)
30	0.16	9.144	13.74
40	0.29	5.779	22.71
50	0.45	5.565	25.19
60	0.49	5.413	27.14
70	1.66	5.263	29.37
80	3.37	5.140	33.03
90	5.25	4.977	36.45

3.4. Further elaboration of the findings

The results of solubilities of adipic acid in water are very close to the data from literature [4-6], which indicates that the devices and processes of this experiment are reliable. The purpose of this work is to determine the difference between the solubility of three acids in alkanolamine solvents and water, and whether the added acids can be precipitated from the desorption lean-solvents by cooling. This study allows one to optimize CO_2 absorption and desorption processes by the method of pH swing [1]. The results show that the solubility curves of the three acids in alkanolamine solvents follow the trend of "increasing steeply with the rising temperature". The solubility observed in alkanolamine solvents was higher than in water. As the concentration of alkanolamine increases, the difference is further enlarged. These are mainly due to the following factors. First, the pH of alkanolamine solvents is alkaline, and the solubility of three acids in alkaline solutions should exceed that in water; Secondly, when hydroxy dibasic acids and alkanolamine coexist in aqueous solution, some reactions may occur, and additional products can be formed [7]. For MEA and adipic acid, the following reaction may occur [8,9]:

$$HOOC\text{-}(CH_2)_4\text{-}COOH + RNH_2 \leftrightarrow RHNOC\text{-}(CH_2)_4\text{-}CONHR + H_2O$$

The occurrence of this reaction breaks the original adipic acid hydrolysis equilibrium, allowing more adipic acid to dissolve into the solution, which leads to degradation of alkanolamine solvents and reduction of CO_2 absorption capacity. Thus, the concept of recycling some weak organic acids from alkanolamine solvents by cooling is not yet feasible. New methods need to be further explored to deal with the residual acids in the alkanolamine solvents.

4. Conclusions

Through the experimental results of solubility of three hydroxy dibasic acids in MEA, DEA and MDEA solvents, the following conclusions can be drawn:

- The solubility curves of three acids in alkanolamine solvents still follow the trend of increasing with increasing temperature, as the case in aqueous solution.
- The solubility of three acids in alkanolamine solvents increases steeply with the concentration of alkanolamine solutions.
- The added acids cannot be mostly precipitated from alkanolamine solvents by cooling, which means that it is unreasonable to recovery the residual acids from lean alkanolamine solvents.

Acknowledgments

This research was financially supported by Scientific and Technological Research Program of Chongqing Municipal Education Commission (Grant No. KJ30418) and the open project of Key Laboratory of Low-Grade Energy Utilization Technologies and Systems (Chongqing University), Ministry of Education of China (Grant No. LLEUTS201401).

References

[1] Feng B, Du M and James T 2010 Reduction of energy requirement of CO_2 desorption by adding acid into CO_2-loaded solvent *Energy & Fuels* **24(1)** 213-9

[2] Fan L H 2007 Measurement and correlation for solubility of adipic acid in several solvents *Chinese Journal of Chemical Engineering* **15(1)** 110-4

[3] Gaivoronskii A N and Granzhan V A 2005 Solubility of adipic acid in organic solvents and water *Russian Journal of Applied Chemistry* **78(3)** 404-8

[4] Suren S, Sunsandee N, Stolcova M, *et al* 2013 Measurement on the solubility of adipic acid in various solvents at high temperature and its thermodynamics parameters *Fluid Phase Equilibria* **360(1)** 332-7

[5] Yu Q S, Ma Xi X and Gao W Y 2012 Determination of the solubility, dissolution enthalpy and entropy of suberic acid in different solvents *Fluid Phase Equilibria* **330(25)** 44-7

[6] Xia Q, Zhang F-B, Zhang G L *et al* 2008 Solubility of sebacic acid in binary water + ethanol solvent mixtures Journal of Chemical & Engineering Data **53(3)** 112-7

[7] Saotome Y and Miyazawa T 2006 Novel enzymatically degradable polymers comprising alpha-amino-acid, 1,2-ethanediol, and adipic acid *Chemistry Letters* **1(1)** 21-4

[8] Adams E Q 2002 Relations between the constants of dibasic acids and of amphoteric electrolytes *Journal of The American Chemical Society* **2002(8)** 1503-10

[9] Yu V M and Zapevalova N P 1977 Methods of synthesis of peptides *Russian Chemical Reviews* **46(5)** 449-51

IOP Conf. Series: Materials Science and Engineering **283** (2017) 012024 doi:10.1088/1757-899X/283/1/012024

Austenite – ferrite transformation temperature regression equations for low carbon steels with cooling rate account

P. Záhumenský, I. Kohútek* and J. Semeňák
U. S. Steel Košice, 044 54 Košice, Slovak Republic
*Corresponding author: ikohutek@sk.uss.com

Abstract. The austenite-ferrite transformation temperatures evaluated by dilatometry using thermo-mechanical simulator Gleeble 1500D are investigated in this paper. The effect of cooling rates 1, 5, 10 and 15°C/s on the upper and lower critical transformation temperatures was evaluated for 30 specimens of six material groups. Considering the cooling rate from dilatometry tests and chemical composition (C ≤ 0.2%, Mn ≤ 2%, Si ≤ 0.26%) of particular specimens, the regression equations for both transformation temperatures were derived. These relations have to be satisfied to avoid the crack formation during continuous casting, as well as to provide the hot rolling control. The proposed regression equations are compared with 32 similar ones adopted from 1961 to 2017 and exhibit a good conformity and accuracy.

1. Introduction

For the improvement of the surface quality of continuously cast steels, it is important to study the hot ductility behavior of steels considering their solidifying thermal history along with crack formation during casting. It has been reported [9] that there exist three temperature regions where typical embrittlement is observed in the carbon steels, i.e., drop from the melting temperature (T_m) to 1200°C (I), decrease from 1200 to 900°C (II), and cooling from 900 to 600°C (III). The cause of embrittlement in region (I) is the existence of residual liquid film along the dendritic interfaces. The ductility was found to be independent of the strain rate. In region (II), the precipitation of finely distributed oxy-sulfides at the austenite grain boundary weakens the boundary strength, and therefore such treatments as slow cooling, holding for certain time, or low-rate straining should be used to improve the ductility. On the other hand, the embrittlement in region (III) is manifested by a lower rate of straining. Controlling factors of this embrittlement are grain boundary sliding and the localization of strain to the pro-eutectoid ferrite film along the austenite grain boundary, which is invoked by the $\gamma \rightarrow \alpha$ (austenite to ferrite)transformation [3], 5, 10, 11].

The present work provides a tool to calculate the austenite-ferrite transformation temperatures with the account of the effect of chemical composition of low carbon steels.

2. Methods

Using a Gleeble 1500D thermal deformation simulator of physical processes, the temperature range of $\gamma \rightarrow \alpha$ transformation was defined for the steels and four different cooling rates corresponding approximately to those of slabs, transfer bars (i.e., semi-products during hot rolling between roughing mill and finishing mill), and hot rolled sheets. Standard size tensile specimens with dimensions of ∅10×110 mm taken from transfer bar were used in the experiment. The specimens were heated with heating rates of 25°C·s^{-1} and 10°C·s^{-1} to 1200°C and 1250°C, respectively. The austenitization temperature of 1250°C with holding time of 30 s was applied before cooling to room temperature by cooling rates of 1, 5, 10, and 15°C·s^{-1} , as shown in figure 1. The temperature range of $\gamma \rightarrow \alpha$

transformation during cooling was identified using dilatometry for the four different cooling rates. Although two methods (intersectional and tangential) were used for estimation of the upper (Ar_1) and lower (Ar_3) critical transformation temperatures in °C, only the results obtained by the intersectional one were used for the regression curve construction and estimation of transformation temperatures, which are plotted in figure 2. Due to a problem with accurate estimation of transformation temperatures for grades with high Si contact (e.g., dynamo, non-oriented electrical steels and/or some HSLA specimens), specimens with Si > 0.3% were excluded from the analysis. Moreover, Ar_1 values for steels with carbon content below 0.05% were also excluded, since the eutectoid transformation was not considered to occur for such low carbon content.

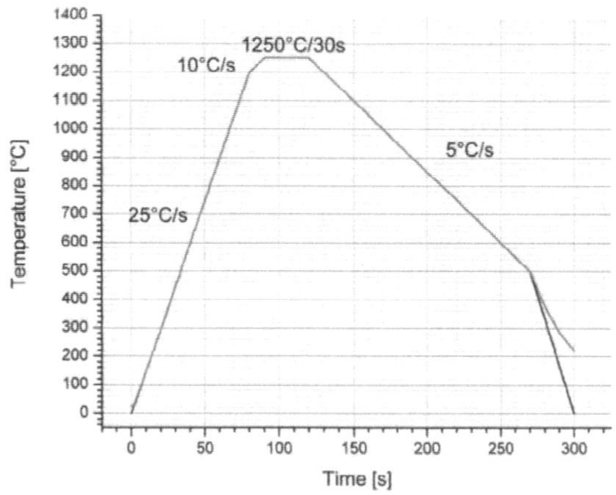

Figure 1. Heating and cooling curves of test specimens by Gleeble equipment

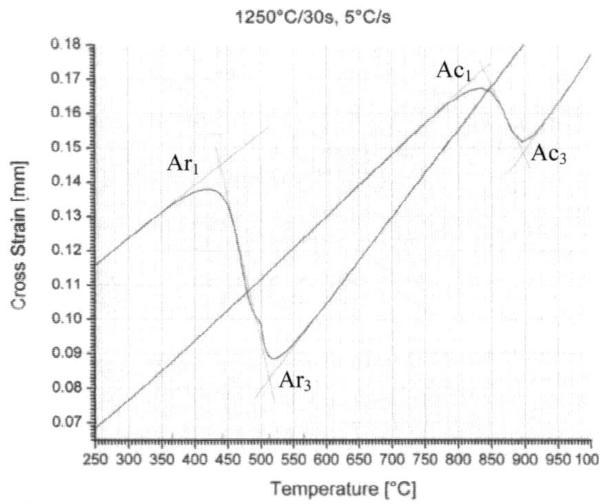

Figure 2. Dilatation curves used for the transformation temperature characterization in the region with a typical embrittlement pattern

The chemical composition of the steels has been determined using a LECO analyzer by Optical Emission Spectral Analysis. Full spectra of 22 chemical elements were measured in total.

3. Material

In total, 30 specimens of 6 material groups were tested, as shown in Table 1.

Table 1. Number of specimens in six evaluated material groups.

Group No	Material group	Short description	Number of specimens
1	C-Mn	C-Mn	9
2	Drawing	Draw	9
3	High-Strength Low-Alloy	HSLA	5
4	Interstitial Free	IF	3
5	Dual-Phase	DP	3
6	Transformation-Induced Plasticity	TRIP	1
	Total		30

The contents of C, Mn, and Si (in weight %) for the evaluated materials are depicted in figures 3, 4, and 5, respectively, while the histogram of cooling rates is plotted in figure 6.

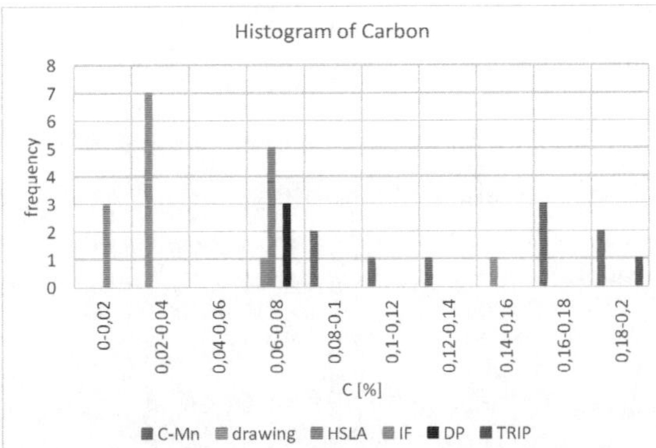

Figure 3. Carbon content in 6 material groups

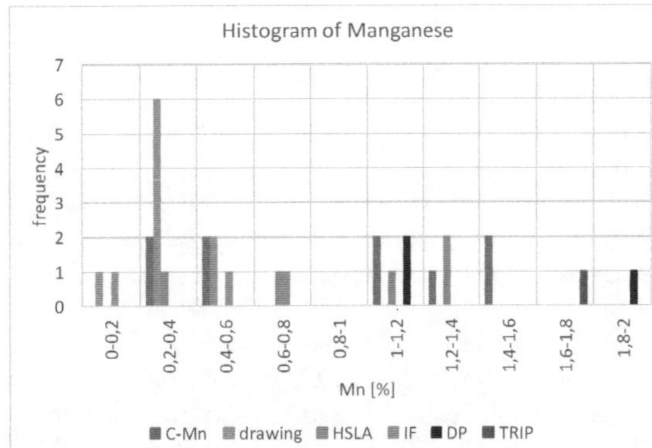

Figure 4. Mn content in 6 material groups

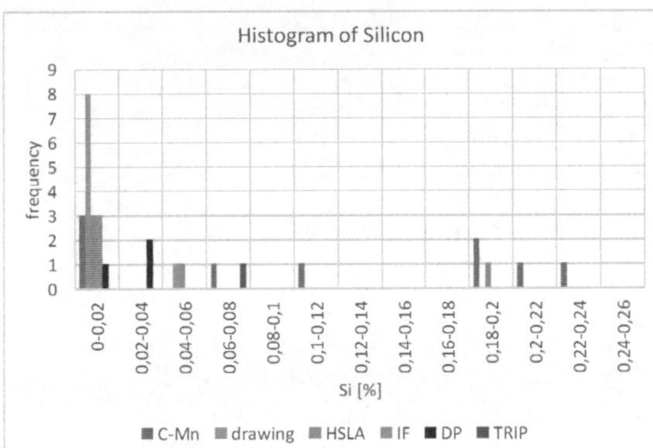

Figure 5. Silicon content in 6 material groups

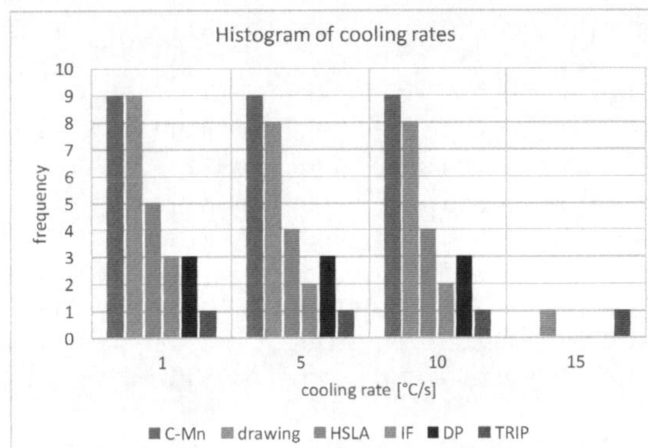

Figure 6. Histogram of cooling rates in 6 groups

4. Results and discussion

The temperature ranges [Ar_3- Ar_1] of $\gamma \rightarrow \alpha$ transformation during cooling stated by dilatometry for the four different cooling rates via the intersectional method are listed in Table 2.

5. Regression Analysis

Using the measured transformation temperatures Ar_3 and Ar_1, the respective regression equations were constructed. Each set of independent variables consists of cooling rate as well as content of 22 available chemical elements. Three types of multi-regression were used: stepwise forward (starting with no variables in the model, testing the addition of each variable using a chosen model fit criterion), stepwise backward (starting with all variables, testing the deletion of each variable using a chosen model fit criterion) and regression with fixed set of variables. P-values less than 0.05 of all components were required for all variables in every equation, and correlation between pairs of components were examined to eliminate mutual dependences between chemical elements. Note that in specimens measured in this investigation, an unexpected dependence among Mn, Ti and Nb was discovered, so that only Mn is present in the final equation. Moreover, the distribution of Al content in the set of examined specimens was revealed to be abnormal, and so Al (Al sol) was also excluded from the regression analysis. Thus, although the above elements are very important from the metallurgical point of view, they had to excluded from the final equations due to statistical reasons.
Final equations are as follows:

$$Ar_3=+914 - 6.85\cdot\text{cooling rate} - 650\cdot C - 134\cdot Mn + 179\cdot Si \qquad R^2_{adj}=0.91 \qquad (1)$$

$$Ar_1=+814 - 9.08\cdot\text{cooling rate} - 532\cdot C - 121\cdot Mn + 165\cdot Si \qquad R^2_{adj}=0.90 \qquad (2)$$

Table 2. Transformation temperatures of particular grades during cooling

ID	Cooling rate 1°C·s^{-1}		Cooling rate 5°C·s^{-1}		Cooling rate 10°C·s^{-1}		Cooling rate 15°C·s^{-1}	
	Ar_3, °C	Ar_1, °C	Ar_3, °C	Ar_1, °C	Ar_3, °C	Ar_1, °C	Ar_3, °C	Ar_1, °C
C-Mn (a)	775	680	725	655	705	635		
C-Mn (b)	650	595	605	520	570	460		
C-Mn (c)	860	735	830	700	660	600		
C-Mn (d)	765	675	780	650	700	625		
C-Mn (e)	815	720	770	675	750	655		
C-Mn (g)	670	630	640	580	620	545		
C-Mn (i)	640	580	605	510	575	475		
C-Mn (j)	650	580	615	530	590	495		
C-Mn (l)	650	595	620	555	600	515		
DP (a)	690	640	660	590	635	555		
DP (b)	710	630	665	595	640	510		
DP (c)	630	560	590	520	570	485		
Draw (a)	845	790	800	740	765	700		
Draw (b)	840	795	815	745	795	720		
Draw (c)	740	620	700	635	685	615		
Draw (d)	835	755						
Draw (e)	835	795	830	750	820	720		
Draw (f)	880	780	870	710	865	690	855	680
Draw (g)	855	800	830	750	805	715		
Draw (h)	840	765	810	710	785	690		
Draw (i)	825	785	780	690	750	670		
HSLA (a)	730	605	660	585	645	520		
HSLA (b)	845	670						
HSLA (c)	775	645	690	630	680	620		
HSLA (d)	725	560	650	595	645	520		
HSLA (e)	770	695	720	610	700	600		
IF (a)	860	835	855	805	845	785		
IF (b)	910	820						
IF (c)	745	715	725	695	720	685		
TRIP	590	550	540	430	460	370	430	370

The comparison of measured temperatures with those estimated via regression equations (1) and/or (2) with an account of chemical composition and cooling rate is shown in Figure 7 (Ar_3) and Figure 8 (Ar_1). Differences between the measured value and that estimated from regression equation are shown in histograms plotted Figure 9 (Ar_3) and Figure 10 (Ar_1).

There are many empirical equations to calculate $\gamma \rightarrow \alpha$ transformation temperatures derived by several authors for special steel grades and special conditions. The most comprehensive list of them is available in [4], other can be found in [1][2], [6][7]. In [9]total, seven equations for Ar_1, as well as 25 equations for Ar_3, are compared with equations (1) and (2). Variables in the mentioned equations are of three types: (a) plain linear variable, (b) combination of two chemical elements (8% of Ar_3 equations, e.g., Si·Ni) and (c) non-linear function (54% of Ar_3 equations - e.g. logarithm, exponent, square root) – see tTable 3. The coefficient of particular variable in regression equation (only those which are used in more than 1/3 of equations) are shown in real scale in

Figure **11** for Ar_3 and in

Figure **12** for Ar_1: red circle denotes value of coefficients derived for equations (1)-(2) in this paper.

The intercept of all Ar_3 equation is between 810°C and 914°C (910°C is known as A3 for ideal equilibrium iron-carbon binary phase diagram), except for that of Yuan [4] for non-deformed austenite (marked as YN in

Figure **11**), which consists mainly from the combination and/or non-linear functions of variables. The intercept used in all Ar_1 – related equations is between 706°C and 814°C. The value of 723°C is referred to as A1 for the ideal equilibrium iron-carbon binary phase diagram.

IOP Conf. Series: Materials Science and Engineering **283** (2017) 012024 doi:10.1088/1757-899X/283/1/012024

It is generally adopted that the cooling rate affects the transformation temperature, while increasing the cooling rate decreases transformation temperatures – in equations (1) and (2) increasing cooling rate by 1°C·s^{-1} causes decrease of Ar_1 by 9°C and/or Ar_3 by 7°C. The cooling rate is included in only approx. 1/7 of equations (probably because transformation temperatures were estimated from the process with a very slow cooling – sometimes they are denoted as Ae1 and/or Ae3, where "e" stands for equilibrium, see [4]).

The coefficients of carbon C, Mn and Ni have the same negative orientation for all equations caused a decrease of transformation temperatures with their increasing contents. The Si coefficients are all positive what is in concordance with expected increasing of transformation temperatures with increasing Si content according to binary Fe-Si diagram. The regression coefficient of chromium (not included in the equations derived in this paper) is mostly negative with 2 exceptions for Ar_3 and positive in all equations for Ar_1.

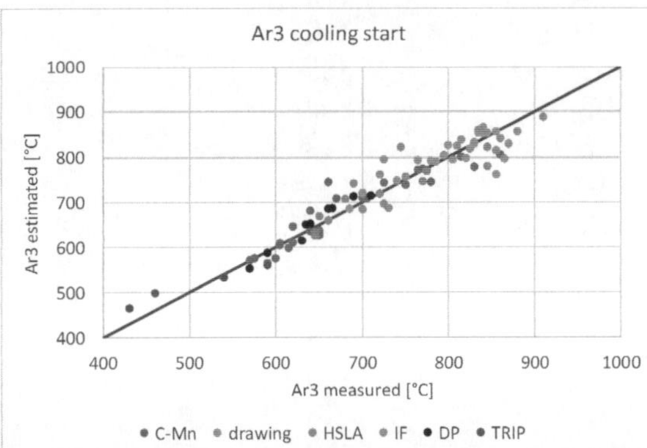

Figure 7. Comparison of measured temperatures Ar_3 with those obtained via equation (1)

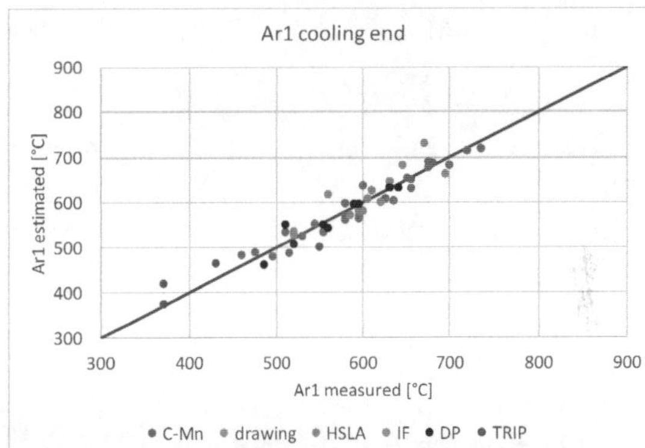

Figure 8. Comparison of measured temperatures Ar_1 with those obtained via equation (2)

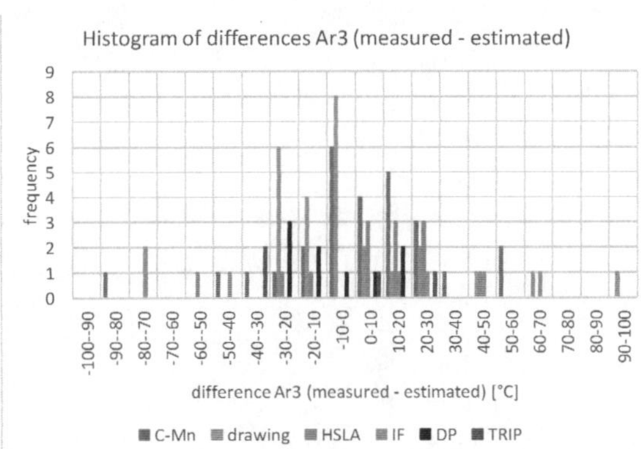

Figure 9. Histogram of differences between measured Ar_3 /Ar_3 and those obtained via regression equation (1)

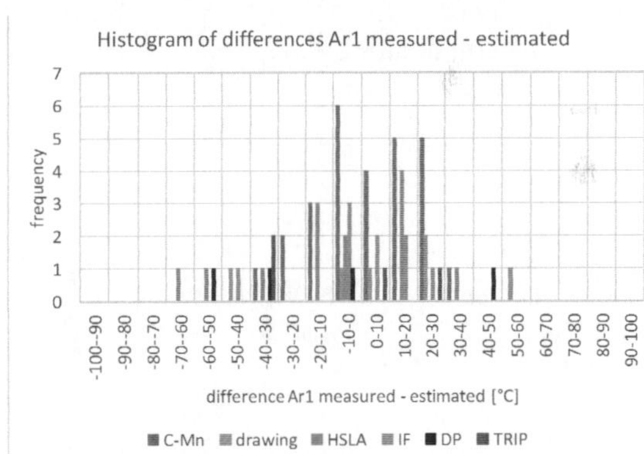

Figure 10. Histogram of differences between measured Ar_3 /Ar_3 and those obtained via regression equation (2)

Given the above discussion, it can be concluded that the obtained results are sufficient for practical purpose of avoiding the crack formation during continuous casting and hot rolling control as well. However, a deeper insight into the embrittlement mechanism requires additional experiments including qualitative and quantitative phase analysis.

Table 3. Number of variables used in the regression by various authors

Variable	Number of equations		Share of equations in the total scope	
	Ar_1	Ar_3	Ar_1 (% from 8)	Ar_3 (% from 26)
Intercept	8	26	100.0	100.0
Cooling rate	1	4	12.5	15.4
C	5	20	62.5	76.9
Mn	8	23	100.0	88.5
Si	7	18	87.5	69.2
Ti		3	0.0	11.5
Nb		7	0.0	26.9
V	1	5	12.5	19.2
Cu		7	0.0	26.9
Al		3	0.0	11.5
Al_sol			0.0	0.0
Ni	5	10	62.5	38.5
Cr	5	13	62.5	50.0
Mo	2	9	25.0	34.6
P		4	0.0	15.4
W	1	2	12.5	7.7
S		2	0.0	7.7
Zr			0.0	0.0
B		1	0.0	3.8
Ca			0.0	0.0
As	1	1	12.5	3.8
Sn			0.0	0.0
Sb			0.0	0.0
N		1	0.0	3.8
Others – linear combination		2	0.0	7.7
Others – non-linear		14	0.0	53.8

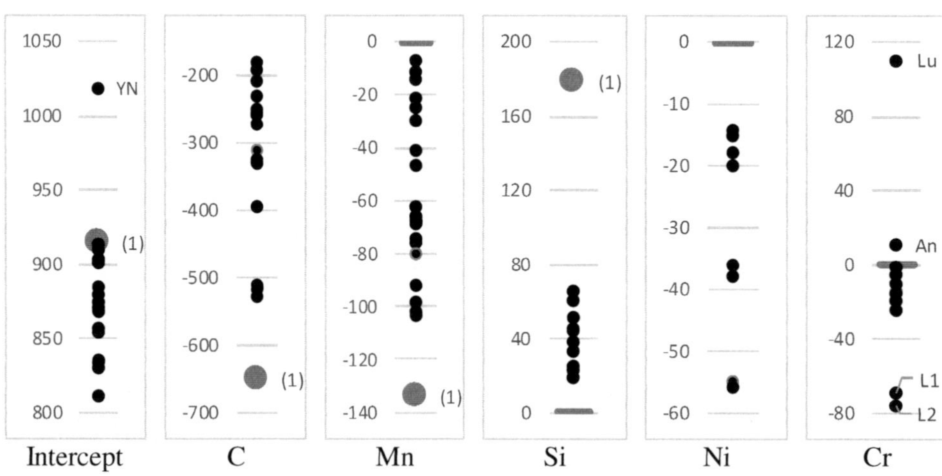

Figure 11. Comparison of regression coefficients of various authors for selected variables in regression equation (1) for Ar_3

In figure 11, "An" stands for Andrews [4], L1 and L2 mean Lotter A and Lotter B in [4], respectively, "Lu" stands for Lutsenko in [4], and "YN" refers to the equation of Yuan for non-deformed austenite in [4].

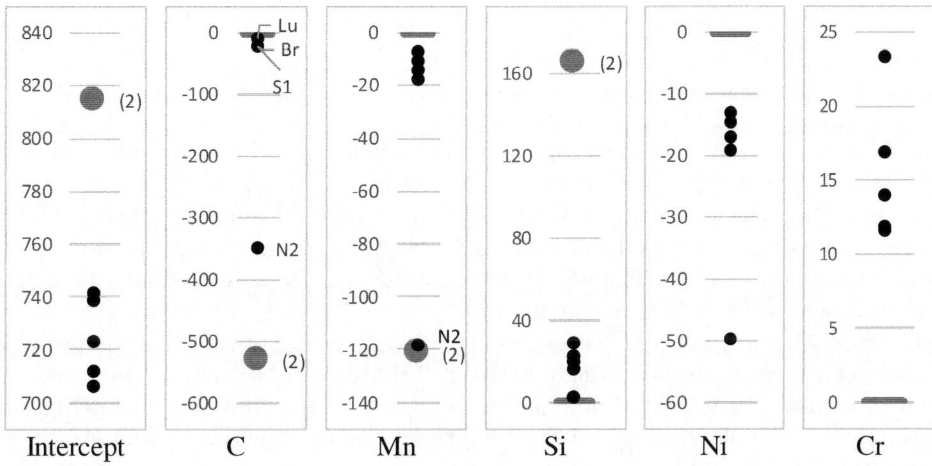

Figure 12. Comparison of regression coefficients of various authors for selected variables in regression equation (2) for Ar1

In f
Figure **12**, "Br"refers to Brandis in [4], Lu to Lutsenko in [4], N2 stands for Proprietary 3 in [4] or Nippon 2 in [9], while S1 stands for Schacht in [4].

6. Conclusions

- The experimental material characteristics can be utilized to avoid crack formation during bending and especially at the unbending section below the caster. This way, the surface quality of the final product can be improved.
- The data and regression equations on the austenite-ferrite transformation temperatures are instrumental for the control of process parameters during hot rolling.

Acknowledgment
This study was prepared and published with the support of the research and innovation operational program for the project "Support to Industrial Research and Development Centre in the field of new materials and technologies of coating and welding" (ITMS NFP313011B635), funded from the budget of the European Fund of Regional Development.

References

[1] Blás J G et al 1989 Influência da composição química e da velocidade de resfriamento sobre o ponto Ar3 em amos de baixo C microligados ao Nb. In: Congresso Anual da Associação Brasileira de Metais, ABM, São Paulo, **1**, pp.11-29.
[2] Choquet P et al 1985 Mathematical Model for Predictions of Austenite and Ferrite Microstructures in Hot Rolling Processes. IRSID Report, St. Germain-en-Laye, p.7
[3] Dossett, J L and Boyer H E 2006 Practical Heat Treating, ASM International, 2nd Edition, p.296.
[4] Gorni A A 2008 Steel Forming and Heat Treatment Handbook, **15**, p.13

[5] Lee Y K, Hong J M, Choi C S and Lee J K 2005 Continuous cooling transformation temperatures and microstructures of niobium bearing microalloyed steels. *Materials Science Forum*, **475-479**, pp.65-68

[6] Ouchi C, Sampei T and Kozasu I 1982 The effect of hot rolling condition and chemical composition on the onset temperature of γ–α transformation after hot rolling. Trans. ISIJ, pp.214-222

[7] Pickering F B 1986 Steels: Metallurgical Principles. In: *Encyclopedia of Materials Science and Engineering*, vol. 6, The MIT Press, Cambridge.

[8] Shiga C et al 1981 Development of Large Diameter High Strength Line Pipes for Low-Temperature Use. Kawasaki Steel Technical Report, Dec. 1981, pp. 97-109

[9] Staňková H, Motyčka P, Meyer L W and Mašek B 2006 Verification of the validity of the common phenomenological models for low-alloyed trip steel. In: Trends in the Development of Machinery and Associated Technology, 10th International Research/Expert Conference, TMT 2006, Barcelona-Lloret de Mar, Spain, pp. 11-15.

[10] Suzuki H G, Nishimura S and Yamaguchi S 1982 Characteristics of hot ductility in steels subjected to the melting and solidification *Trans. ISIJ*, 22, p.48-56.

[11] Vega M I, Medina S F, Chapa M and Quispe A 1999 determination of critical temperatures (t_{nr}, a_{r3}, a_{r1}) in hot rolling of structural steels with different Ti and N contents. *ISIJ International*, **39**, 12, pp.1304-1310

IOP Conf. Series: Materials Science and Engineering **283** (2017) 012025 doi:10.1088/1757-899X/283/1/012025

Abrasive slurry jet cutting model based on fuzzy relations

C H Qiang[1] and C W Guo[1*]

[1]School of Electrical and Power Engineering, China University of Mining and Technology, Xuzhou, China
*E-mail: cwguo@cumt.edu.cn

Abstract. The cutting process of pre-mixed abrasive slurry or suspension jet (ASJ) is a complex process affected by many factors, and there is a highly nonlinear relationship between the cutting parameters and cutting quality. In this paper, guided by fuzzy theory, the fuzzy cutting model of ASJ was developed. In the modeling of surface roughness, the upper surface roughness prediction model and the lower surface roughness prediction model were established respectively. The adaptive fuzzy inference system combines the learning mechanism of neural networks and the linguistic reasoning ability of the fuzzy system, membership functions, and fuzzy rules are obtained by adaptive adjustment. Therefore, the modeling process is fast and effective. In this paper, the ANFIS module of MATLAB fuzzy logic toolbox was used to establish the fuzzy cutting model of ASJ, which is found to be quite instrumental to ASJ cutting applications.

1. Introduction

In recent decades, with the development of industrial processing field, high precision machining methods are becoming more and more mature. High-pressure waterjet cutting is one of the high-precision machining methods [1].

Water jet cutting is considered to be one of the best adaptive cutting methods [2-4]. Not only it can be used for cutting of hard alloy, ceramic, die steel, high silicon cast iron and high hardness material [5], but also it has a good performance for the cutting of viscoelastic materials such as titanium alloy and rubber. The earliest use of water jet is in the breaking of rock, the auxiliary drilling, hydraulic coal mining, and building demolition [6]. The addition of abrasive (slurry) makes the performance of water jet more powerful.

However, the cutting process of the abrasive water jet is a complex stochastic process affected by many parameters, which proper selection is very important. Since the determination of these parameters depend on the experience of operators, so it is difficult to establish an appropriate mechanism model.

In the current process databases, the conditions and data provided are limited and discrete. Therefore, many input data are not applicable for the process database output in practical applications, which brings inconvenience to end users. In order to solve these problems, the concept of "fuzzy logic" is introduced into the database system, so that in any conditions in the effective range of the corresponding process data can be obtained.

Adaptive neural fuzzy inference system (ANFIS) has a strong ability in dealing with the nonlinear, high complexity and discontinuous problems, and it also has many characteristics such as fast convergence speed, good stability, repeatability of the training process and higher precision of the predict. Therefore, it is especially suitable for the problems of engineering prediction. These problems are highly complex because of their characteristics of diversity, spatial and temporal, but that's the good part of ANFIS predicting technology [7, 8].

To solve the problem of constructing an accurate cutting model, a cutting model of ASJ was developed in this study using the fuzzy inference system, which can provide the roughness prediction of the cutting process of abrasive slurry/water jet. In addition to incorporating the discrete values of the parameters into the training sample task, the model can also predict the output of any other value within the effective range via the membership function. Therefore, the fuzzy inference model established in this paper can be used for guiding the cutting process of ASJ, which can be lucrative for the intelligent control system of ASJ cutting.

2. Brief introduction to fuzzy theory

The fuzzy theory is to use fuzzy logic to describe the rank of things in real life, the concept of fuzzy is defined by the form of the fuzzy set, it can quantify the rank of events that belong to this collection then obtain the membership function, the membership degree is used to deal with various problems.

For a classic set A, any of the elements X in the space should correspond to X∈A or X∉A. This feature can be expressed as a function:

$$\begin{aligned} \chi_A(x) &= 1, x \in A \\ \chi_A(x) &= 0, x \notin A \end{aligned} \tag{1}$$

where $\chi_A(x)$ is the characteristic function of set A. If the characteristic function is extended to the fuzzy set, it is found that only 0 and 1 values are applied to the classical set, after being extended to the fuzzy set, the values are changed into the interval [0, 1].

In the domain of U, definition mapping $\mu_{A`}$:

$$\mu_{A`}: U \to [0,1] \qquad u \text{ a } \mu_{A`}(u) \tag{2}$$

It assumed that set U defines a fuzzy subset $\mu_{A`}$, record as A`, called $\mu_{A`}$ is the membership function of A`. The value of $\mu_{A`}$ at $u$ is called the membership degree of u to A`. Record as $\mu_{A`}(u)$ or $A`(u)$.

3. Prediction model of upper roughness

The input parameters of roughness prediction model in this paper are outlet pressure of pipe of the water jet, cutting target distance, cutting speed and abrasive particle size, the output parameters are the roughness of the upper cutting surface. This system adopted in this paper is the Takagi-Sugeno (T-S) fuzzy logic system. As the output of this fuzzy system is single point fuzzy, so the multi-input and the multi-output system can be decomposed into multi-input and single-output system.

The conclusion of the T-S fuzzy model is a linear function of the input parameters, so it has these following advantages: high computational efficiency, be easy to use optimization and adaptive technology, guarantees the continuity of the output space and be suitable for mathematical analysis. The model uses the following "if-then" rule:

$$R^k: \ if \ x_l \ is \ F_l^k, \ldots, x_n \text{ is } F_n^k, then$$

$$y_k = p_0^k + p_0^k x_1 + p_1^k x_2 + \ldots + p_n^k x_n = p_0^k + \sum_{i=1}^{n} p_i^k x_i \tag{3}$$

where F_l^k is the fuzzy set of fuzzy system, p_j^k is the parameter of the fuzzy system, and y^k is the output according to fuzzy rule.

3.1. Model elaboration

The process of fuzzy modeling is a dynamic process that uses fuzzy systems to gradually approach the unknown nonlinear system, to achieve the goal of forcing the whole system. The network structure of the fuzzy system is shown in figure 1:

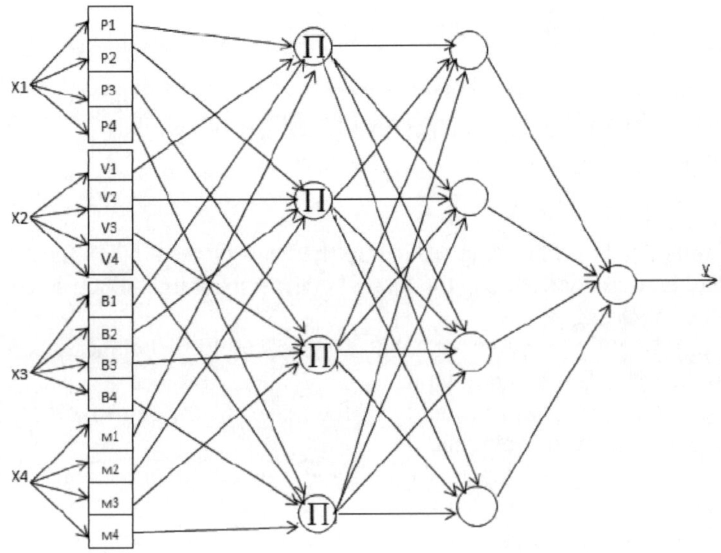

Figure 1. The network architecture of fuzzy systems of upper roughness.

There are four input parameters of the fuzzy system (x1, x2, x3, x4), which are outlet pressure of pipe of the water jet, cutting speed, target distance, and abrasive particle diameter, respectively. After inputting the parameter values, they will be transferred to the first layer of the network. The range of each variable can be specified according to different applications.

The first layer is membership generating layer, the input variables will be fuzzy transformed in this layer, the number of nodes is determined by the number of fuzzy subsets of each input variables, where each node represents a subset (fuzzy linguistic variables). The number of nodes is based on a number of horizontal parameters, so the number of nodes in this layer is 16. The function of this layer is to compute the membership degree of each input component of the parameter to each fuzzy subset (linguistic variable value). The Gauss type membership function is considered here, as shown in equation (4):

$$\mu(x_{ij}) = \exp\left(-\left(\frac{x_j - c_{ij}}{b_{ij}}\right)^2\right) \tag{4}$$

The second layer is the fuzzy inference layer, and each node of the layer represents a possible fuzzy rule. Taking the output of the second layer as the antecedent of this rule, the output of the fuzzy inference layer is the product of the membership of each part of the antecedent, which can be expressed by equation (5):

$$w(i) = m(x_{i1}) * m(x_{i2}) * m(x_{i3}) * m(x_{i4}) \tag{5}$$

The transfer function of each node of the third layer is a linear function, that is, the local linear model, so the output rule of the third layer is:

$$y_i = w(i) * (p_0^i + p_1^i x_1 + p_2^i x_2 + p_3^i x_3 + p_4^i x_4) / \sum_{i=1}^{n} w(i) \tag{6}$$

The set of parameter variables consisting of $\{p_0^i, p_1^i, p_2^i, p_3^i\}$ is called the parametric parameter of the conclusion.

The fourth layer: calculating the total output of the input parameter:

$$y = \sum_{i=1}^{n} \varpi_i y_i = \sum_{i=1}^{n} \omega_i y_i / \sum_{i=1}^{n} \omega_i \tag{7}$$

After determining the central values and widths (c, b) of the membership functions of each node of the second layers and the value of the connection weight (p) of the network post, the above fuzzy system can be determined

3.2. Process of model training

3.2.1. Acquisition of sample data. The sample data are the basis of the ANFIS model. The experiment o obtaining the sample data was carried out for an ASJ cutting machine. When selecting the sample, one needs to ensure the following points:

a. In the training sample, the parameters of each variable should be balanced as much as possible to avoid bias to a certain factor or a horizontal value;

b. In order to ensure the generalization ability of the network, the relevance of the data to be read should be as small as possible, even irrelevant;

c. The input samples and output samples should have a nice relevance, and both these two values can be extracted or controlled;

d. The interference of external causes should be avoided as much as possible in training sample selection.

Usually, the process of determining the capacity of sample data, too much capacity or too little capacity is unreasonable. Too much capacity will increase the difficulty of obtaining sample data, while too little capacity will not reflect the nature of the problem. Based on the cutting experiment of ASJ, 80% of the samples could be used for modeling, and 20% of the sample data would be used as model checking. In this paper, 6 datasets of experimental data were randomly selected as model checking data sets, and the remaining data were used as training samples for training adaptive fuzzy neural networks. Table 1 is part of the experimental data of ASJ cutting, which is used as the sample data of the model.

Table 1. Training data on 1060 aluminum.

No.	Pressure MPa	Velocity mm/s	Distance mm	Material	Upper roughness μm	Lower roughness μm
1	33	1.2	0.5	m1	3.274	4.115
2	33	0.9	0.5	m1	3.0815	3.888
3	30	1.2	1	m1	3.5856	4.562
4	30	0.9	1	m1	3.4465	4.321
5	25	0.6	2	m1	3.187	4.107
6	25	1.2	2	m1	3.54	4.476
7	27	0.6	1.5	m1	3.0522	3.853
8	27	1.2	1.5	m1	3.402	4.218
9	33	1.5	2	m1	3.1071	3.926
10	33	0.6	2	m1	2.6835	3.1325

3.2.2. Normalization of input and output data. Since the function depends on the inner product of the input sample vector, in this experiment, each attribute had a larger range of values, which makes the calculation more complex and the training time longer. In order to avoid the above situation, the parameters should be normalized.

Normalization is to limit the amount of data that need to be processed within a certain range through an algorithm. After normalization, the data will be more convenient to use, and the convergence speed of the program can be accelerated. In MATLAB, there are 3 main methods for normalization.

a. Linear function transformation, the expression is as follows:

$$y = (x - x_{\min}) / (x_{\max} - x_{\min}) \tag{8}$$

where x and y are the values before and after conversion, respectively; x_{max} and x_{min} respectively are the maximum and minimum values of samples, respectively.

b. Logarithmic function transformation, the expression is as follows:

$$y = log10(x) \tag{9}$$

c. Inverse cotangent conversion, the expression is as follows:

$$y = arctan(x) * 2 / \pi \tag{10}$$

By comparison, it can be found that the linear function transformation method has advantages of simpler operation and easier to implement compared with logarithm function transformation method and inverse cotangent transformation method. So linear function transformation method would be used as the normalized method.

The function "mapminmax" can make each row of the matrix normalized to [-1 1], and the syntax of the "mapminmax" statement is: [y1, PS] =mapminmax (x1). Where x1 is the normalized matrix, and y1 is the result of normalization. When it need restore the normalized data, using the following command: x1_again=mapminmax ('reverse', y1, PS)

3.2.3. Determination of the fuzzy system parameters. In the process of fuzzy modeling in the previous section, the membership function of the network structure of the second layer has been determined. Gauss type function is most widely used in MATLAB, and the output of the fuzzy inference system is a linear function of the input parameters, so it would be selected, and the expressions are as follows:

$$\mu(x_{ij}) = \exp\left(-\left(\frac{x_j - c_{ij}}{b_{ij}}\right)^2\right) \tag{11}$$

$$y_i = w(i) * (p_0^i + p_1^i x_1 + p_2^i x_2 + p_3^i x_3 + p_4^i x_4) / \sum_{i=1}^{n} w(i) \tag{12}$$

In the equation, the values of c, b, and p are the parameter values of the system, and the fuzzy system is determined to determine the values of c, b and p of the system.

The program first assigned *t* for a column of 0.3, while the determination of *b* and *c* was realized by the system's random assignment.

When determining the number of fuzzy iterations, iterations of 100, 200, 300, 500, 800 and 1000 times were set. The variance analysis of training data is shown in tables 2 and 3 below:

Table 2. Analysis of different iterations' variance of the upper roughness model.

Times	100	200	300	500	800	1000
Variance	0.01334	0.00643	0.0056	0.00515	0.00371	0.00270

Table 3. Extrapolation analysis of variance of upper roughness model.

Times	100	200	300	500	800	1000

Variance	0.06225	0.05425	0.05278	0.05075	0.05774	0.05942

As can be seen from table 2, the error of the training data decreases gradually with the increased number of iterations,.

In case depicted in table 3, the error decreases first and then increases with the number of iterations. For the number of iterations equal to 500, the error is the smallest.

Through a comprehensive analysis of tables 2 and 3, it was found that with the increase in the number of iterations, the degree of fitting of the training array would gradually improve, but then, due to the array extrapolation overfitting, the error became larger, which makes the model less adaptive to extrapolation arrays. Therefore, the number of iterations of the model was limited to 500.

Then, the system error has to be calculated. According to the calculation error, the system updated the values of membership parameters b and c, and the agjusted weight p of the model.

3.3. Analysis of the model realization results

When executing the prediction program via the model, the predicted and actual values of the 32 sets of training samples were displayed in the MATLAB command window, as well as the respective error estimations.

After the program realization, the actual values of the training data, the ANFIS prediction values, and their deviations/errors appeared in the MATLAB command window, which are partially listed in table 4.

Table 4. Deviation of the results predicted by the model from the actual ones.

Group No.	4	5	6	7	8	9	10	11	12
Roughness deviation	0.0923	0.0266	0.0485	-0.0140	0.0141	-0.0252	0.0162	-0.0463	-0.713

By the error values in table 4, it can be clearly seen that the error is at the 10^{-2} level. Compared with the surface roughness value, the error is very small, so the T-S fuzzy system can be used to predict the cutting surface roughness, and the result is accurate. The reliability of the T-S fuzzy system would be more visually and intuitively judged by comparing the actual values of the cut surface roughness with the predicted values, as shown in figure 2.

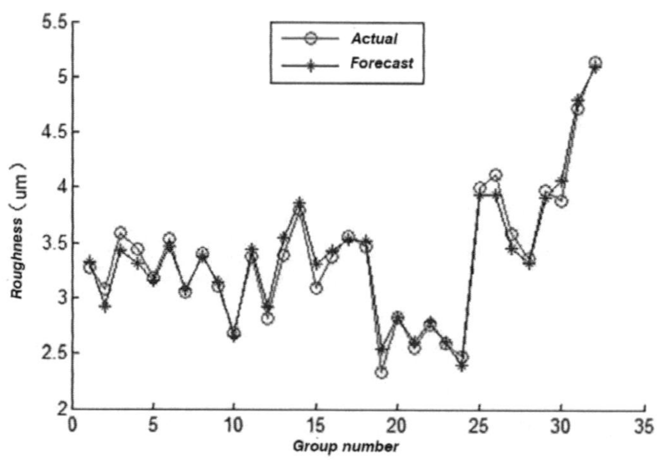

Figure 2. The actual and predicted roughness of training data.

The red line in figure 2 represents the actual result of the roughness of the training sample, and the blue line is the predictive value of the upper roughness model. As can be seen from figure 2, the actual

IOP Conf. Series: Materials Science and Engineering **283** (2017) 012025 doi:10.1088/1757-899X/283/1/012025

and predictive values of the 32 sets of sample data used for training are relatively close, and the error is small, which shows that the model has a good fitting effect, and the system training is very effective.

The model was validated by the additional six sets of data that were not involved in the previous training process. After the input of the above six datasets, the program was run and the outputs were displayed in the MATLAB command window, as shown in table 5:

Table 5. The roughness of the additional data.

Output data	3.342	3.2936	3.7637	3.3751	3.4060	1.9240
Test validation data	3.6762	3.5356	3.9025	3.3087	3.5209	1.7216

The integrated fitting results of the training and validation datasets are depicted in figure 3.

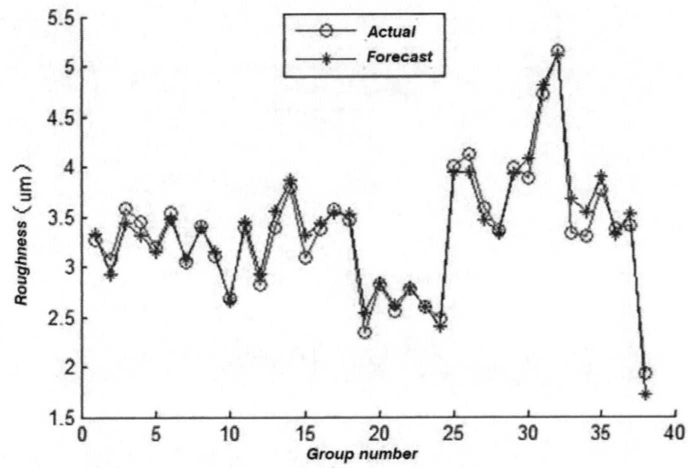

Figure 3. The actual and predicted roughness of training and validation data.

The six datasets on the right side of figure 3 correspond to the validation array. It can be observed that the difference is small in the fitting process with the training array, and the predicted values are close to the actual results. Therefore, the model has better generalization function, and it has certain guiding significance for the cutting process of ASJ. Therefore, the ANFIS can be used to predict the quality of ASJ cutting.

4. Prediction model of the lower roughness

The modeling of lower section roughness and the realization process of the ANFIS program are similar to those of the upper section roughness, with the only exception: in the process of data input, one needs to select the lower roughness for training purposes. Therefore, the description of modeling and training processes are omitted for brevity sake.

When determining the number of fuzzy iterations, their number was set to 100, 200, 300, 500, 800, and 1000. The variance analysis of training data is shown in tables 6 and 7.

Table 6. Analysis of different iterations variance of lower roughness model.

Number of iterations	100	200	300	500	800	1000
Variance	0.01774	0.01300	0.01066	0.00796	0.00692	0.00626

As seen from table 6, the training data error decreases gradually with the number of iterations.

IOP Conf. Series: Materials Science and Engineering **283** (2017) 012025 doi:10.1088/1757-899X/283/1/012025

Table 7. Extrapolation analysis of variance of the lower roughness model.

Times	100	200	300	500	800	1000
Variance	0.09277	0.06841	0.05834	0.04371	0.04576	0.04866

As can be seen from table 7, as the number of iterations increases, the error decreases first and then increases. When the number of iterations is 500, the error is the smallest.

In tables 6 and 7, it is seen that with the increase in the number of iterations, the array fit training is improved, but because of the phenomenon of the array extrapolation overfitting, the error became larger, which makes the model less adaptive to extrapolation arrays. Therefore, the number of iterations of the model was set to 500. After the input of training data (lower roughness), the program was run, the lower roughness values predicted by the model were obtained and compared with the actual values, the comparison results being shown in figure 4.

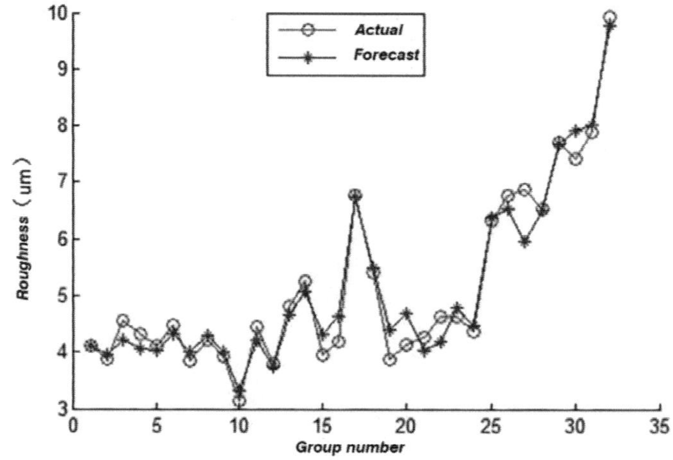

Figure 4. The actual and predicted lower roughness values of training data.

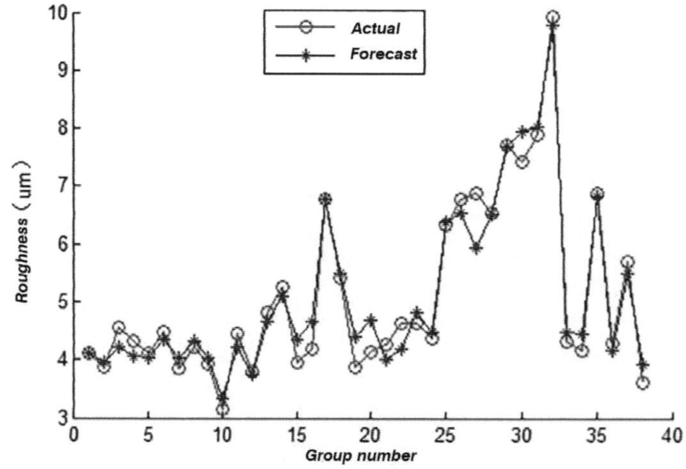

Figure 5. The actual and predicted lower roughness of training and validation data.

Similar to the upper surface roughness prediction model, after the input of six validation datasets not used for training, the program execution resulted in adequate fitting results of the training and validation datasets, as shown in figure 5.

IOP Conf. Series: Materials Science and Engineering **283** (2017) 012025 doi:10.1088/1757-899X/283/1/012025

A joint analysis of figures 4 and 5 has revealed that the lower roughness model can also well predict the plane roughness, and the fitting results on trained and validation datasets meet the basic requirements. Therefore, the model can be incorporated into the ANFIS fuzzy inference system to predict the cross-sectional roughness value of ASJ cutting process.

5. Conclusions

- Based on the theory of fuzzy inference system, combined with the characteristics of ASJ, this article established the roughness prediction model, selected appropriate membership functions, the output of the system as a linear combination of the input.
- Through the comparison between forecast value and actual value of the analysis by using the dates in the real experiment as training samples of the cutting model, it is found that the training effect of the upper and lower roughness prediction model is good. The roughness prediction model has good function of training and generalization, so it has a good guidance to the experiment of ASJ cutting.

Acknowledgments
This work was supported by the National Natural Science Foundation of China (U1510113) and the Jiangsu College Postgraduate Research Innovation Plan Project of 2016 (KYLX16_0522).

References

[1] Chen G M 2007 The feature and application of NC high-pressure water jet cutting technology *Mach Tool Hydraul* **(35)** 8-9
[2] Liu L H, Liu B L, Liu P and Li B K 2011 A review of the fundamental research of premixed abrasive jet *Mech Sci Technol Aerospace Eng* **(3)** 457-62
[3] Hashish M 1991 Characteristics of surface machined with abrasive water jet *Eng. Mater. Technol.* **113** 354-62
[4] Hashish M 1988 Visualization of surfaces machined with abrasive water jet cutting process *Exp Mech* **28(2)** 159-68
[5] Yang Z, Chen S M, Zhang Y J and Li M 2009 The development and application of the high-pressure water jet *Mech Mana Dev* **24(5)** 87-90
[6] Tan Z Y, Li Y, Su R B and Zhang D 2013 Simulated cutting experimental research on heavy concrete from reactor by mixed abrasive water jet *Construct Technol* **42(S1)** 176-8
[7] Zhang Z J and Ding D X 2008 Study on the training parameters setting of ANFIS *J Univ S China (Sci-Tech)* **(22)** 2-3
[8] Wu W and Guo S X 2005 The safe driving speed analytical system on the basis of MATLAB auxiliary fuzzy control *J Shanxi Normal Univ (Nat Sci Ed)* **(S1)** 202-4

IOP Conf. Series: Materials Science and Engineering **283** (2017) 012026 doi:10.1088/1757-899X/283/1/012026

Design of a new abrasive slurry jet generator

F C Wang[1], L L Shi[1] and C W Guo[1*]

[1]School of Electrical and Power Engineering, China University of Mining & Technology, Xuzhou, China
*E-mail: cwguo@cumt.edu.cn

Abstract. With the advantages of a low system working pressure, good jet convergence and high cutting quality, abrasive slurry jet (ASJ) has broad application prospects in material cutting and equipment cleaning. Considering that the generator plays a crucial role in ASJ system, the paper designed a new type ASJ generator using an electric oil pump, a separate plunger cylinder, and a spring energized seal. According to the determining of structure shape, size and seal type, a new ASJ generator has been manufactured out and tested by a series of experiments. The new generator separates the abrasive slurry from the dynamic hydraulic oil, which can improve the service life of the ASJ system. And the new ASJ system can reach 40 MPa and has good performance in jet convergence, which deserves to popularization and application in materials machining.

1. Introduction

As a branch of abrasive water jet machining, abrasive slurry jet machining (ASJM) aims at improving the uniformity of abrasive particles and reducing the nozzle wear. In recent years, the scholars carried out a lot of research works and obtained numerous achievements in ASJ.

Liu [1] used the abrasive slurry jet to incise the carbon steel and aluminum plate under different pressure when the jet contained polyacrylamide. After the experiment, she discovered it has different depths with different concentrations of polyacrylamide jets. Kowsari [2] made an experiment using abrasive slurry jet machining to drill holes on the glass plate, 110 copper and 6061 Aluminum. The jets pressure was 4 MPa to 7 MPa containing 50 weight parts per Million polyethylene oxide. He revealed that elasticity and viscosity of the abrasive slurry jet had a strong effect on the shape, depth, and diameter of drill holes. Qi [3] investigated a machining process on quartz crystals, and the pressure of abrasive slurry jet was 8 MPa to 20 MPa. He found the material removal process, surface roughness and channel depth depend on the characteristics of abrasive slurry jets. Haghbin [4] carried out experiments to compare the abrasive water jet machining (BWJM) and abrasive slurry jet machining (ASJM) in glasses and aluminum cutting. When the pressure was 235 MPa, using 25 μm diameter Al_2O_3 particles, the ASJM was 50% lower centerline waviness and 16% lower centerline roughness than the AWJM. Nouraei [5] compared the characteristics of abrasive slurry jet micro-machining and abrasive air jet micro-machining. Seiji [6] used different types of particles: steel bead, alumina, glass bead and plastic shot to investigate abrasive slurry jets. He found that the larger particles jets are more breakable than small ones. Luo [7] made an experiment under the pressures of 4.5 MPa, nozzle diameter 400 μm to study the jet erosion morphology and material mechanism. With different incident angles having different polishing depths, the 30 deg incident angle ensured the largest depth. Wan [8] found that the pump shutdown resulted in the blocking of ASJ nozzle and pipe. In order to increase the nozzle service life, Katz [9] designed an innovative porous nozzle using

IOP Conf. Series: Materials Science and Engineering **283** (2017) 012026 doi:10.1088/1757-899X/283/1/012026

lubricant oil device with abrasive particles. Liu [10] and Cadavid [11] used electrochemical slurry and helium atmosphere to enhance abrasive slurry jet machining.

In recent years, most scholars focus on the structure of abrasive slurry jet and preparation of slurry. However, there are fewer studies on the abrasive slurry jet generator. The paper presented a novel ASJ generator using a plunger cylinder, which can isolate the abrasive slurry from the dynamic hydraulic oil. The performed experiments verified the feasibility of the new ASJ system.

2. ASJ machining system

As shown in figure 1, the paper presents an ASJ system, which contains the new generator. The electric oil pump was adopted as a power source, which can provide power oil within 63 MPa by adjusting the pressure (1 MPa is the minimum pressure). The system was used by the device of the hydraulic cylinder with the abrasive slurry. The double independent abrasive slurry loop and dynamic oil circulation avoided the impact of the hydraulic oil.

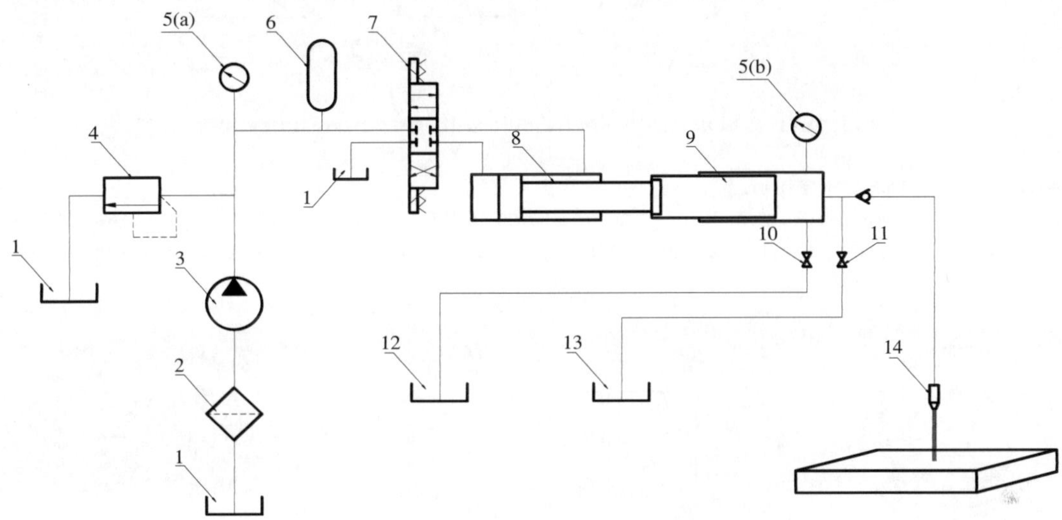

Figure 1. Abrasive slurry jet system diagram:
1. oil tank; 2. filters; 3. oil pump; 4. overflow valve; 5. pressure gauge; 6. energy accumulator; 7. magnetic exchange valve; 8. power cylinder; 9. ASJ generator; 10. globe stop valve; 11. globe stop valve; 12. water tank 13. abrasive slurry mix tank; 14. nozzle.

If the pressure of the hydraulic is higher than the set pressure, the hydraulic oil will through overflow valve flow back into the oil tank. In order to observe the pressure of hydraulic oil and abrasive slurry we designed two pressure gauges (5(a) and 5(b)). The energy accumulator is used to absorb the pressure pulsation of the hydraulic pump to ensure that the system pressure is stable. When open the global stop valve 11 and close valve 10, we can pump water from the water tank to clearing the ASJ generator. Figure 2 shows the flow of abrasive slurry and hydraulic oil for the ASJM feeding and jetting processes.

It is simple, economical and easy to operate for the working principle of the system. It is set as 50 MPa for the design maximum pressure of the system power resource, which can realize abrasive slurry pressure within 1-40 MPa by regulating the infinite pressure and provide support for the abrasive slurry jet under different pressure conditions, polishing, and cutting performance.

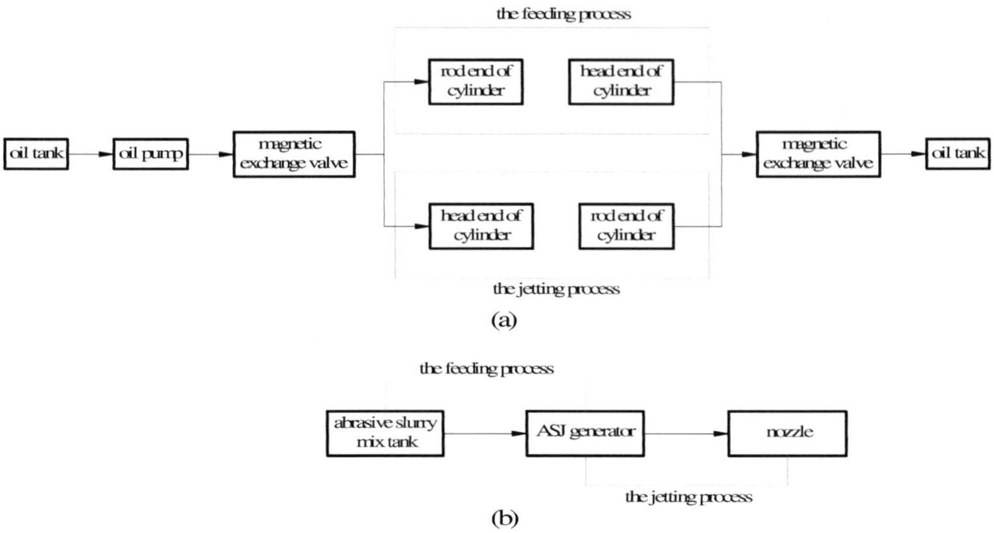

Figure 2. (a) Dynamic oil circulation (b) Abrasive slurry loop.

3. Abrasive slurry generator

Figure 3 shows the design of the slurry generator. The plunger type cylinder of the overall structure is unidirectional, driven by dynamic plunger cylinder to run the suction and compression processes.

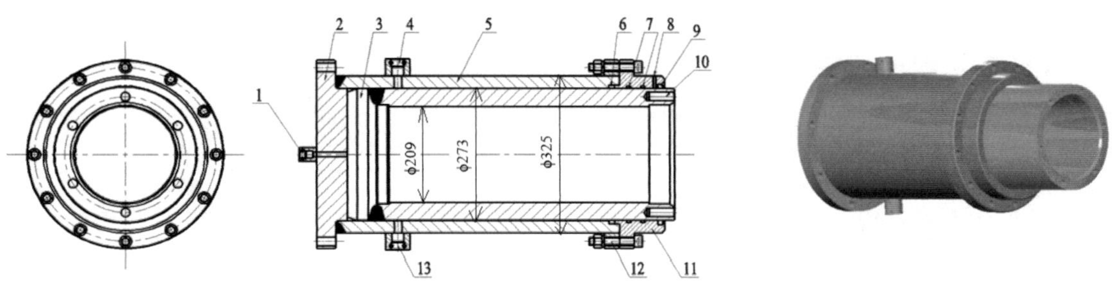

Figure 3. The abrasive slurry jet generator:
1. jet connection joint; 2. generator cylinder bottom; 3. generator plunger; 4. pressure gauge holes; 5. generator cylinder; 6. spring energized seal; 7. O-type ring; 8. oil filter point; 9. scraper seal; 10. flange hole; 11. generator extended period; 12. fixed bolt; 13. abrasive slurry drain hole.

3.1. The operating principle of the generator

As shown in figure 1, the stop valve 10 is turned off and the stop valve 11 is opened, and the abrasive slurry generator is driven by power cylinder. The abrasive slurry is suctioned into the generator when the power cylinder goes back. The power cylinder takes forward movement after the electromagnetic directional valve reverses direction. The abrasive slurry will be turned to the jet stream by the driven generator plunger from the nozzle jet after the increasing pressure. As shown in figure 2, generator extended period plays a guiding role in fixing the spring energized seal. The friction can be reduced by adding the lubrication through the oil injection hole 8, between the plunger and the extended. Jet pressure can be adjusted in a range of 1-40 MPa through regulating the electric pump oil pressure, which is used to study the cutting performance of the abrasive slurry jet in various pressure conditions.

3.2. The design of the main technical performance parameters of the generator

The Bernoulli equation has the following form:

$$p_1 + \frac{\rho v_1^2}{2} + \rho g h_1 = p_2 + \frac{\rho v_2^2}{2} + \rho g h_2 \tag{1}$$

where p_1 is jet pressure, ρ is fluid density, v_1 is jet velocity, p_2 is internal fluid pressure, v_2 is internal fluid velocity, h is the vertical height.
When the generator is placed in the horizontal position, we get $h_1 = h_2$.

$$p_1 + \frac{\rho v_1^2}{2} = p_2 + \frac{\rho v_2^2}{2} \tag{2}$$

The parameter p_1 is reduced to zero after the fluid leaves the nozzle. The level of the fluid speed (v_2) is a few millimeters per second in the generator, while the internal fluid velocity can be neglected, as compared to the jet velocity.

$$\frac{\rho v_1^2}{2} = p_2 \tag{3}$$

The maximum fixed pressure is 40 MPa when designing the inner part of the abrasive slurry generator. Assuming $\rho=1.0*10^3$ kg/m^3, v_1 can be proved to 282.8 m/s, if the fluid is considered as an ideal one. This amounts to about 5-6 minutes when designing the single continuous stream. There are two kinds of nozzle diameter specifications, namely 0.6 mm and 0.8 mm. The inner space capacity in the generator is about 30 liters. The standard seamless steel pipe will be chosen when considering the economy and process. The plunger stroke is designed to 500 mm, the external diameter of the generator plunger is designed to 273 mm, the abrasive slurry injection volume is 29.23 L.

The equation of the hydraulic oil cylinder thickness is as follows:

$$\delta = \delta_0 + C_1 + C_2 \tag{4}$$

$$\delta_0 \geq \frac{P_{max} D}{2.3\sigma_p - 3P_{max}} \tag{5}$$

where C_1 is cylinder outer diameter tolerance allowance, C_2 is corrosion allowance, $Pmax$ is the maximum operating pressure in the cylinder, D is cylinder inner diameter, σ_p is allowable stress.

The rated pressure in the design system is 40 MPa. In other words, P_{max} = 40 MPa, the inner diameter of the generator cylinder is 274.5 mm, but it does not meet the requirements of the material mechanics using the ordinary pipe steel. Alternatively, the 27SiMn seamless steel tube can be selected as cylinder material. Considering the convenience and economy, the generator keeps a steady speed when reaches the rated working pressure that is regarded as working under the static load, so take 4 as the safety coefficient, σ_p = 250 MPa, and $\delta_0 \geq$ 24.13mm. According to the national standard, the model ϕ325*30 seamless tube is chosen. It is flat at the bottom of the cylinder of the generator, the thickness δ_1 can be calculated, according to the formula of the intensity of the embedded disk,

$$\delta_1 \geq 0.433 * D\sqrt{\frac{P}{\sigma_P}} \tag{6}$$

where D is the inner diameter of the generator, $\delta_1 \geq$ 60.1mm. The design thickness is 63 mm, which can be achieved by using the 63 mm-thick 45 carbon steel tube.

Long-term operation will lead to wear and rust accumulation inside the generator, considering the working medium is an abrasive slurry. To improve the hardness and rusting resistance, the inside surfaces of the whole generator are covered with chromium plating. Chromium plating layer has good chemical stability, which can be applied to the different abrasive slurry additives, in order to prevent the corrosion due to acidic and alkaline slurry in the generator and reduce its internal surface wear.

IOP Conf. Series: Materials Science and Engineering **283** (2017) 012026 doi:10.1088/1757-899X/283/1/012026

3.3. The design of the generator seal

The rated internal pressure is 40 MPa inside the generator, which falls into the high-pressure range, the working medium contains garnet abrasive particles, and seal needs to be applied to the high pressure and wear resistant properties, the spring energized seal (figure 4) is selected to use with O-rings, which consists of two parts: sealing and spring. Sealing is filled with PTFE, and spring is stainless steel that is used in heavy-duty conditions, has low friction coefficient, good wear resistance, high-pressure resistance, stable composition, etc. The spring perforce will compress seal lip to ensure the sealing of the plunger when the generator internal pressure is low. The radial force between the liquid and seal lip, which is mainly invoked by the liquid, increases with the liquid pressure in the system. The spring perforce can also be used to prolong the service life by compensating backlash of the seal lip and plunger. The spring-invoked seal designed in this system is used for reciprocating motion whose pressure is below 60 MPa.

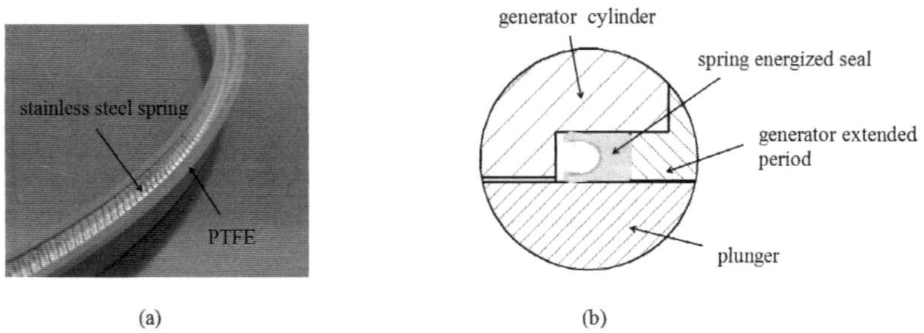

Figure 4. (a) Spring energized seal (b) Installation instruction.

4. Experimental

Figure 5 shows the proposed ASJ machine design, which mainly consists of power system, ASJ generator system, and control system. The water jet, slurry water jet, and abrasive slurry jet tests were conducted, respectively, when the abrasive slurry generator pressure was set as 5 MPa and 40 MPa. The methyl cellulose was used as the thickener, at weight concentration of 2.0 percent. The diameter of garnet sand used was 74 μm, and the volume concentration was 10%. Figure 6 shows the comparison of the experimental jets structure when the pressure is 5 MPa. It can be observed that the aggregation of the slurry jet and abrasive slurry jet outperforms the water jet by its cutting performance.

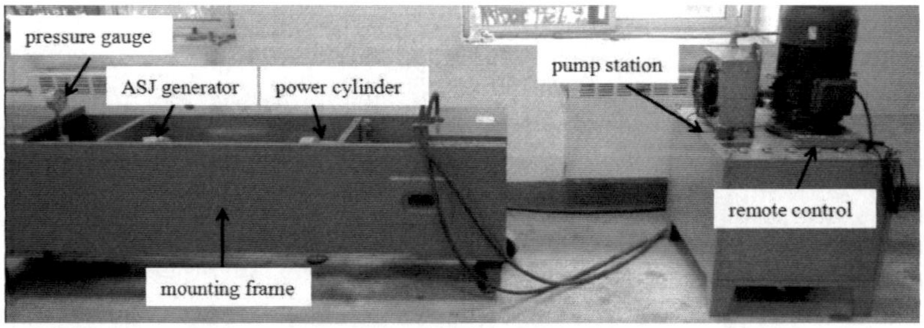

Figure 5. The proposed ASJ machine design.

5. Conclusions

- A simple and efficient design of the abrasive slurry generator is proposed and validated. The double independent system can avoid leakage influence of the system of hydraulic oil and the abrasive slurry. An abrasive slurry jet can be achieved in the pressure range of 10-40 MPa.
- Spring energized seal, which is filled with PTFE, is applicable of the abrasive slurry jet generator at pressures below 40 MPa.

- The particular generator service life can be assessed by additional tests of the ASJ system.

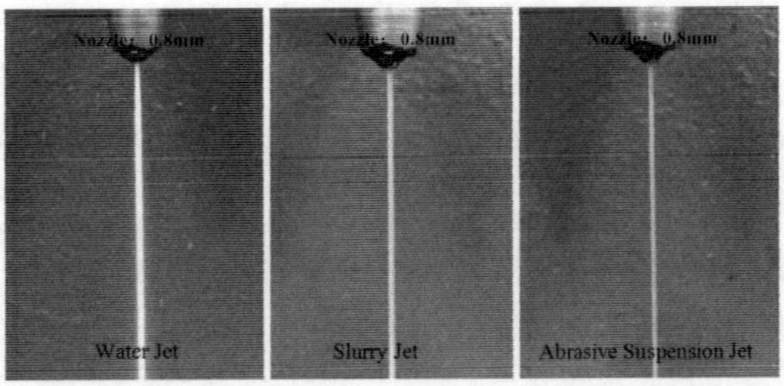

Figure 6. The appearance of water, slurry, and abrasive suspension at pressure of 5 MPa.

Acknowledgments
This work was supported by the National Natural Science Foundation of China (U1510113), and the Jiangsu College Postgraduate Research Innovation Plan Project of 2016 (KYLX16_0522).

References

[1] Liu X J and Yu T 2005 The studies on polyacrylamide applications in abrasive slurry jet cutting *Modern Manufacturing Engineering* **6** 53-5
[2] Kowsari K, Nouraei H, James D F et al 2014 Abrasive slurry jet micro-machining of holes in brittle and ductile materials *Journal of Materials Processing Technology* **214** 1909-20
[3] Qi H, Fan J and Wang J 2014 A study of the micro-machining process on quartz crystals using an abrasive slurry jet *Journal of Engineering Manufacture* **229** 421-34
[4] Haghbin N, Ahmadzadeh F and Spelt J K 2015 High-pressure abrasive slurry jet micro-machining using slurry entrainment *International Journal of Advanced Manufacturing Technology* **84** 1-13
[5] Nouraei H, Wodoslawsky A and Papini M 2013 Characteristics of abrasive slurry jet micro-machining: A comparison with abrasive air jet micro-machining *Journal of Materials Processing Technology* **213** 1711-24
[6] Stinnakre M G, Soulier S and Schibler L 2002 Instantaneous photographic observation of abrasive water suspension jets *JSME International Journal* **45** 830-5
[7] Luo Z, Wang C Y and Wang R J 2013 Research of micro-abrasive suspension jet erosion morphology and material removal mechanism *Advanced Materials Research* **797** 79-84
[8] Wan S and Lim G C 2003 Transient flow in abrasive suspension jet cutting machines on pump shutdown *Journal of Manufacturing Science & Engineering* **125** 172-8
[9] Katz J 2003 Prevention of nozzle wear in abrasive water suspension jets (AWSJ) using porous lubricated nozzles *Journal of Tribology* **125** 168-80
[10] Liu Z, Nouraei H and Papini M 2014 Abrasive enhanced electrochemical slurry jet micro-machining: Comparative experiments and synergistic effects *Journal of Materials Processing Technology* **214** 1886-94
[11] Cadavid R, Wüstenberg D and Louis H 2005 Effect of helium atmospheres on abrasive suspension water jets *International Journal of Advanced Manufacturing Technology* **26** 1246-54

IOP Conf. Series: Materials Science and Engineering **283** (2017) 012027 doi:10.1088/1757-899X/283/1/012027

Experimental study and numerical simulation of the salinity effect on water-freezing point and ice-melting rate

N Qin, Y Wu*, H W Wang and Y Y Wang
College of Mechanical and Electronic Engineering, Qingdao University of Science and Technology, Qingdao 266061, China
Corresponding author: Y Wu, 403751612@qq.com

Abstract. In this paper, based on the background of snowmelt de-icing tools, we studied the effect of salt on freezing point and melting rate of ice through laboratory test and FLUENT numerical simulation analysis. It was confirmed that the freezing point is inversely proportional to the salt solid content, and with the salt solid content increasing, the freezing process of salt water gradually accepts the curing rule of non-crystal solids. At the same temperature, an increase in the salt solid content, the ice melting rate increase by the empirical formula linking the melting time with temperature and salt content. The theoretical aspects of solid/fluid transformation are discussed in detail.

1. Introduction

In cold seasons, snow/ice cover on the roads strongly deteriorate transportation conditions and public safety, which makes the problem of snow cover melting quite topical [1, 2] As the most commonly and largely used snow-melting agent, salt has a great contribution to snow- and ice-melting. However, the utilization of salt causes irreversible pollution to the environment. Thus, it is important to reduce the dosage of salt during the snow-melting process or research environmentally friendly snow-melting agent to replace salt. In the winter of 2008, a one-off snowstorm occurred in south China. The roads were slippery and crowded due to continuous snowstorm and low temperature, numerous areas experienced a blackout/power loss due to high-voltage electric lines were covered by ice and torn. According to statistics, the total direct economic loss caused by the snowstorm amounted to 130 billion yuan (19.5 billion USD). Therefore, the research of transformation about water and ice is emergency due to the theoretical significance and applied value. In publication [3], the author studied the fluid-solid transformation theory based on transformation thermo dynamics and molecular dynamics, and the author deduced the melting speed formula. Authors [4] studied the influence of humidity and temperature on the water-ice fluid-solid transformation and deduced the empirical formula linking the melting time and humidity at different temperatures. In study [5], the influence of pH value and pressure on melting rate was investigated, and the empirical formula linking the pH value, pressure, and melting time was proposed.

The research object of our study is the salinity effect on water freezing point and ice melting rate. The results of this study can provide theoretical substantiation of snow melting process and the respective solid-flow transformation mechanism.

2. Solid-flow transformation mechanism of ice and water

In the field of heat transfer, the solid-flow transformation of ice and water is called Stefan Problem [6]. It bases on thermodynamics and heat transmission science, and the research contains four parts: (a) impact of external environment on the solid-flow transformation process; (b) distribution of temperature field; (c) impact of free convection or forced convection of the solid-flow transformation process; and (d) motion rule of the phase variation interface.

2.1 Differential equation of heat conduction

Practice shows that the quantity of heat transmission per second and per square meter is directly proportional to the temperature alteration ratio in the normal direction x to the cross-section:

$$\frac{\phi}{A} \sim \frac{\partial T}{\partial x} \tag{1}$$

where φ is constant.

Formula (2) is Fourier heat conduction law

$$\phi = -\lambda A \frac{\partial T}{\partial x} \tag{2}$$

where λ is heat conductivity coefficient (where "minus" sign in formula(2) means the direction of heat transfer is opposite to the direction of temperature rise); A is the cross-sectional area; T is the temperature; $\frac{\partial T}{\partial x}$ is the gradient of temperature along the direction of x.

In a rectangular coordinate system, consider a random micro parallelepiped (figure 1) derived from heat conductor, which is assumed to be isotropic.

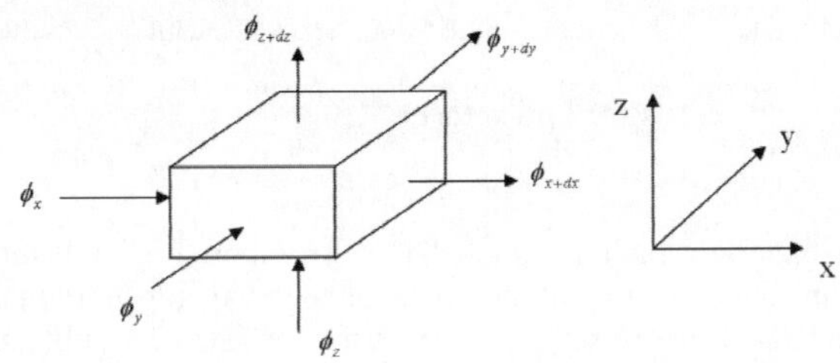

Figure 1. Thermal analysis of the infinitesimal parallelepiped

According to the quantity of heat through following infinitesimal area in figure 1, we can deduce the general type of three-dimensional heat conduction differential equation in the Cartesian coordinate system using the Fourier heat conduction law and the energy conservation law (i.e., the first law of thermodynamics).

$$\rho c \frac{\partial T}{\partial t} = \frac{\partial}{\partial x}(\lambda \frac{\partial T}{\partial x}) + \frac{\partial}{\partial y}(\lambda \frac{\partial T}{\partial y}) + \frac{\partial}{\partial z}(\lambda \frac{\partial T}{\partial z}) + \dot{\phi} \tag{3}$$

In equation (3), parameters ρ, c, φ, and t represent density, specific heat capacity (i.e., heat from internal heat source per second and per square meter), and time, respectively.

2.2 Model of ice melting

Every mathematical description of physical phenomenon needs approximate treatment, the governing equation of phase change problem based on the concept of continuous medium, we suppose that each material is homogeneous and isotropous. The sketch map of the solid-flow two-phase system is shown in figure 2. Here V means a random control body of the space, the value of V is invariable. At the time of t, the control body is divided into two parts by phase interface, which are liquid zone (V_l) and solid zone (V_s). V_n means the speed of liquid arise from density fluctuation or other possible convection reasons. The motion of ice-water interface and the system's temperature field are controlled by the thermal boundary condition on the surface S, the distribution of initial temperature and the thermo-physical property of ice. The subscripts L and S represent the related properties of liquid and solid, respectively.

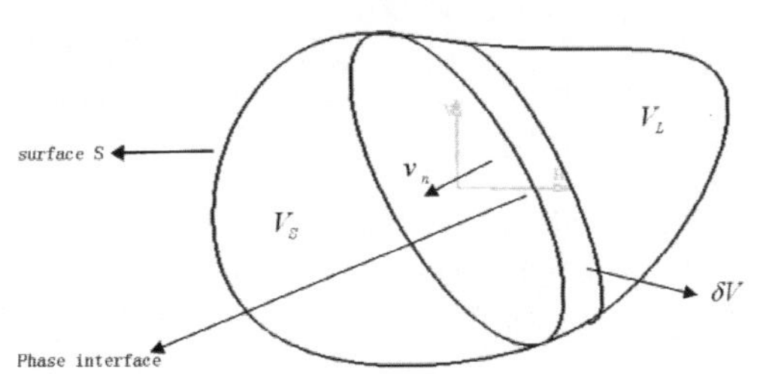

Figure 2. The system of solid phase and liquid phase [5]

The models of heat transfer with phase transformation account can be classified into two types: (i) temperature-based ones, which treat the temperature as a single variable and apply two energy conservation equations to solid and liquid phases; and (ii)) enthalpy-based ones, which regard enthalpy and temperature as two variables and do not use the governing equations. In this study, the former type of (temperature-based) model is used to analyze the ice-melting process and the solid-flow transformation process of ice and water.

According to the energy conservation law, the governing equations of the temperature-based model are as follows:

$$\rho_L c_L\left(\frac{\partial T_L}{\partial t}+v\bullet\nabla T_L\right)=\nabla\bullet(\lambda_L\nabla T_L)+\dot{\phi}_L \tag{4}$$

$$\rho_S c_S\frac{\partial T_S}{\partial t}=\nabla\bullet(\lambda_S\nabla T_S)+\dot{\phi}_S \tag{5}$$

In Eqs. (4) and (5), ρ is density, c is specific heat, T is temperature, ▽ is gradient operator, t is time, υ is velocity vector of liquid phrase, λ is heat conductivity coefficient, and $\dot{\phi}$ is volumetric heat source.

3. Solid-flow transformation test of ice and water

3.1 Test tools and instruments

The used equipment and materials include: an RPH-800 temperature and humidity chamber, the Agilent 34972 data acquisition unit, measuring temperature gun, refrigerator, computer, slot type screwdriver, temperature sensing wires, salt, measuring glass, electric saw, 20-mm cubic molds, vernier caliper, ruler, knife, and paper cups.

3.2 The experimental operation process

3.2.1 Solidification Test of Salt. The following procedure is recommended. Connect Agilent 34972A data acquisition unit to computer and set the temperature inRPH-800 at -20℃, as shown in figure 3.

Figure 3. Laboratory equipment and parameter settings

Firstly, prepare salt water; then measure the water and salt weight with electronic scales.

The solid content is assessed as follows:

$$C=\frac{G_{\text{salt}}}{G_{\text{water}}+G_{\text{salt}}}\times 100\% \tag{6}$$

where C is solid content of salt; G_{water} is weight of water; G_{salt} is weight of additive salt

According to formula (6), prepare salt solution with the solid content of salt at the ratio of 5%, 8%, 10%. The Agilent 34972A data acquisition unit has multi-channels data acquisition [7]. Put three kinds of salt solution into temperature and humidity chamber and collect test data. Ensure the environment of salt solution with different solid content of salt is uniform, which will enhance comparability of the data acquired from the test. Repeat each test three times and average the data. Plot temperature variation curve of salt solution with different solid content of salt in the process of freezing with MATLAB program.

3.2.2 Melting test of brine ice. Firstly, prepare pure water and salt solution with three values of salt solid content, then put them into refrigerator for 24 hours. Due to water evaporation, once the solution is prepared, the pure water and salt solution should be placed into refrigerator immediately to ensure accurate and comparable test results. Secondly, the surface of ice is irregular after freezing, so it is necessary to smooth the surface of the ice before the test and ensure the size of the ice is uniform, which can enhance the comparability of the results. Then, put the ice into thermal/constant humidity chamber to observe the melting, record the temperature and time. In the test, the error is reduced by repeating the ice-melting tests for three times with the same temperature and solid content of salt, and averaging the data.

4. Test data analysis and discussion

4.1 Results and analysis of salt solution freezing tests

The time step of data acquisition is 20 seconds, export the data into computer for analysis. The process of pure water freezing can be divided into three steps; in the first step, the temperature is dropping continuous from initial temperature to freezing point. In the second step, the temperature is stable around the freezing point of salt solution. In the third step, the temperature continuously drop until it is equal to the setting temperature of incubator.

4.1.1 Process of pure water freezing. The temperature variation behavior during the process of pure water freezing is shown in figure 4.

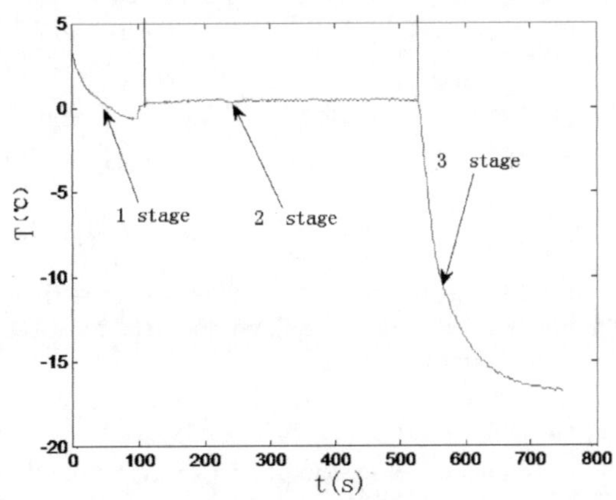

Figure 4. The temperature variation curve for the freezing process of pure water

As shown in figure 4, pure water freezing can be divided into three steps, the first step is temperature drop process before freezing. The second step is temperature-constant process, the average freezing point of 0.44℃ can be obtained by averaging the temperature of three tests. The third step is the temperature drop process after freezing.

There are two kinds of change in the first step, the one is the temperature continuous drop, the other one is the temperature going back to the freezing point and remaining unchanged. The freezing point means that water starts to freeze. By analyzing the first step, we can conclude that the reason why the temperature decreased at first and then increased is that the temperature of environment is below that of pure water, so the heat is transferred from pure water to the environment. At the moment, the internal resistance of water is the intermolecular cohesion, only the molecular force greater than intermolecular cohesion can violate the stable state conditions [8]. This is why the temperature drops to the freezing point at first.

It is known that the higher the temperature, the greater the potential energy, while the potential energy of liquid water exceeds that of ice. Energy release from water to the environment occurs when liquid turns into ice, which results in the temperature rise. When temperature increases to the freezing point, the energy balance occurs: the temperature exhibits a continuous drop after pure water is completely frozen.

4.1.2 Data analysis of salt solution freezing test. As seen from table 1, with salt solid content increase, the freezing point of salt solution is reduced, and since salt solid content is inversely proportional to freezing point, the second step shorter. A probable reason is that pure NaCl is a crystal, but table salt contains some iodine, which makes it non-crystal. With the increase in the solid content of salt, salimity has a stronger impact on the freezing of salt solution. Generally, the freezing curve of salt solution tends to non-crystal behavior.

Table 1. Freezing points of salty water for different solid content values

Solid Content/%	data	T/min	Ice Point/℃
5	55	18	-2.62
8	46	15	-5
10	31	10	-6.32

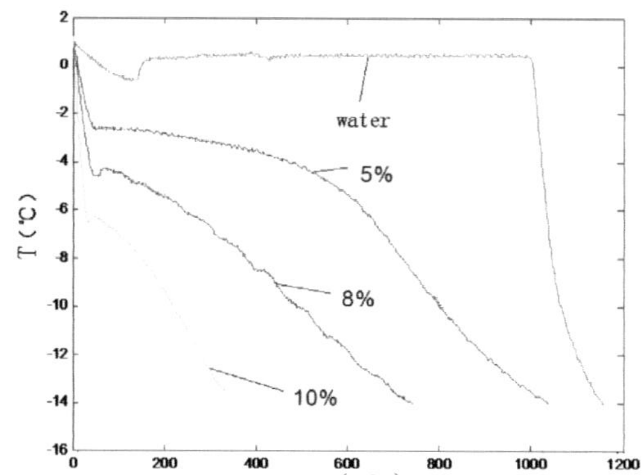

Figure 5. Temperature evolution for freezing of water with four different salinity levels (0, 5, 8, and 10%)

We can see from figure 5 that a higher salinity of water results in a sharper temperature drop and less duration of freezing point attainment. Moreover, there is no distinct transition between the second and third steps, which is characteristic of non-crystal behavior.

4.2 Results and analysis of salt ice melting test

The initial temperature of salt ice taken out of refrigerator is -30.9℃. The melting test data for the ice from pure and salty water are shown in table 2.

Table 2. Summary melting process data (temperature *T*, time *t*, salt solid content *s*)

s / t / T	0%	5%	8%	10%
10	4823	3024	2806	2407
20	3864	2300	1872	1742
30	2652	1470	1374	1231
40	1734	1054	935	864
50	1235	810	769	704
60	864	430	340	286
70	653	319	279	243
80	428	285	237	198

4.2.1 The relationship of pure water ice melting time and temperature

The test data are listed in table 2. They are incorporated into M file, which is imported to Matlab. The curve of pure water ice melting time evolution with temperature is then constructed. By fitting different kinds of function, we find two kinds of functions which can fit the data properly, that is $f(x)=a*e^{bx}$ and $f(x)=a*e^{bx}+c*e^{dx}$. Hereinafter, the two functions are referred to as Power1 and Power2.

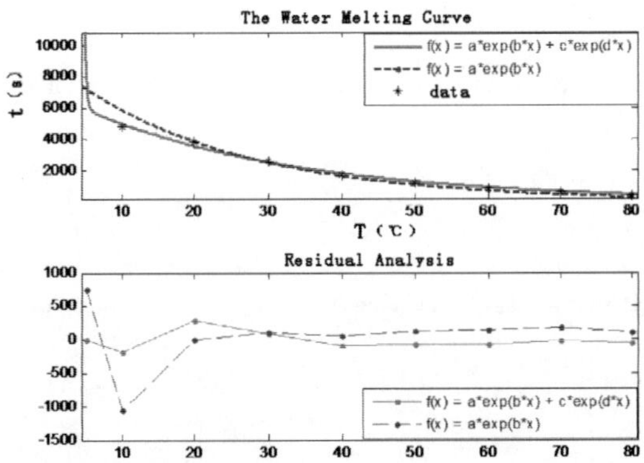

Figure 6. The melting curve of ice from pure water

Power1-fitting formula is:

$$t_1 = 8937 * e^{-0.0419T} \quad (7)$$

Power2-fitting formula is:

$$t_2 = 3.711 * 10^{12} * e^{-4.26T} + 6990 * e^{-0.0335T} \quad (8)$$

Hereinafter, t is melting time, in s; T is temperature, in ℃.

As shown in figure 6, the amplitude of Powerl is higher than that of Power2, which means Power2 is better-fitting than Power1. The fitting degree index of Powerl and Power2 are 0.9643 and 0.9971, which means the fitting degree of Power2 is better than Power1.

The curve of Power2 has a sudden inflexion point, whereas the curve of Powerl has no polar axis around the freezing point, so we choose Power2 as the empirical formula for the melting temperature-time variation of ice from pure water.

4.2.2 Temperature variation of melting time of salty ice with 5%, 8%, 10% solid contents. For the salty ice with three levels of salt solid content, the fitting curves of melting time and temperature are depicted in figures 7, 8, and 9.

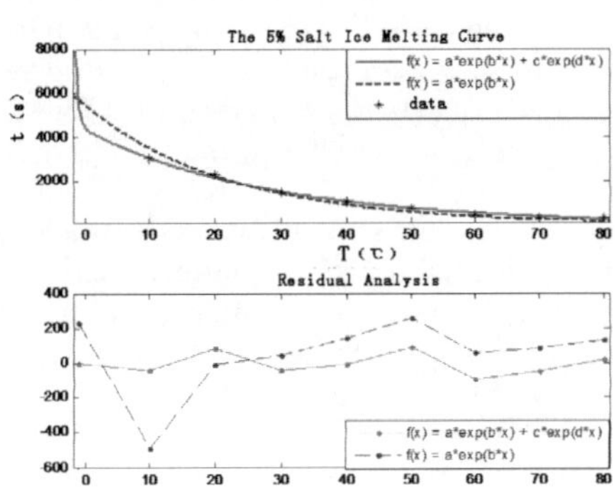

Figure 7. The melting curve of the ice with 5% solid content of salt

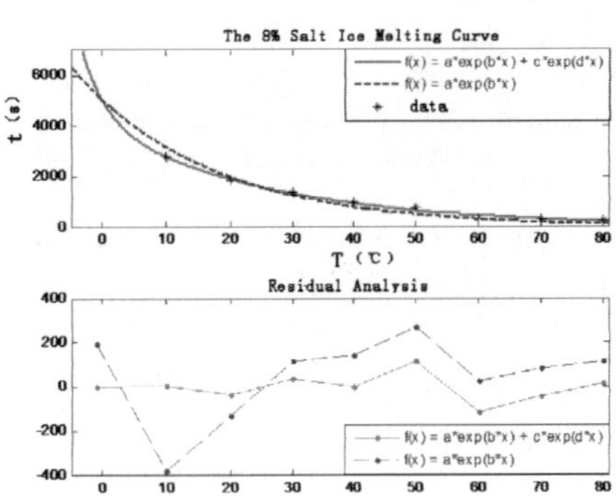

Figure 8. The melting curve of the ice with 8% solid content of salt

When the solid content of salt is 5%, the curves are fitted by the respective test data.
Power1-fitting formula is:

$$t_1 = 5515 * e^{-0.04505T} \quad (9)$$

Power2-fitting formula is:

$$t_2 = 4363 * e^{-0.03535T} + 144.2 * e^{-2.329T} \quad (10)$$

When the solid content of salt is 8%, the curves are fitted by the respective test data.
Power1-fitting formula is:

$$t_1 = 5064 * e^{-0.04637T} \quad (11)$$

Power2-fitting formula is:

$$t_2 = 3883 * e^{-0.0356T} + 1134 * e^{-0.2639T} \quad (12)$$

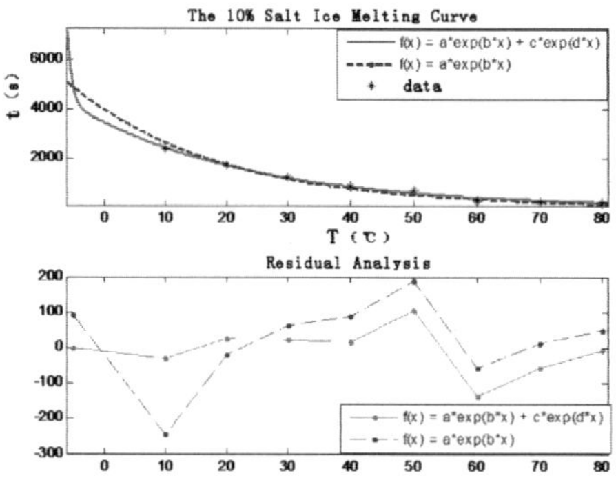

Figure 9. The melting curve of salty ice with 10% solid content of salt

When the solid content of salt is 10%, the curves are fitted by the respective test data.
Power1-fitting formula is:

$$t_1 = 3998 * e^{-0.041T} \quad (13)$$

Power2-fitting formula is:

$$t_2 = 3465 * e^{-0.03514T} + 0.6681 * e^{-1.434T} \quad (14)$$

As shown in the residual analysis of figures 7, 8, and 9, in each couple of melting curves, the amplitude of Powerl is higher than that of Power2, so that Power2 is better than Power1. In each couple of melting curves, the fitting degree index R^2 values in Powerl and Power2 are: 0.9935 and 0.9981; 0.985 and 0.9989; 0.9859 and 0.9986; respectively, which means that fitting degree of Power2 is better than that of Power1. Since there is no polar axis near the freezing point, we selected formulas (10), (12), and (14) as salt's empirical formulas for the 5%, 8%, 10% solid content, which are the formulas for the temperature variation with melting time.

4.2.3 The curve of different solid content of salt. As shown in figure10, for the same temperature, the higher the solid content, the quicker the melting. Analysis based on energy, the reason is that with the increase of solid content of salt, the freezing point of salt ice is decreased. In the melting process under same temperature, the smaller of freezing point, the bigger temperature difference between salt ice and external environment, and the more heat can be absorbed in the same time, which means melting is faster.

Analysis based on enthalpy value, the lower the temperature, the lower the enthalpy value, which means the less energy of specific salt ice has, the less heat can be absorbed during the temperature-rise period. In conclusion, the addition of salt decrease the freezing point, which bring two main advantages in the melting process, (a) the one is increase the temperature difference; (b) the other one is decrease the absorption of heat, that means melting process is accelerated and melting time is shorter.

We can see from Figure10 that four curves tend to straight line y=0, which means the temperature tends to infinity, the melting time tends to 0. On the left of freezing point, the melting time tends to infinity, which confirms the rationality of the fitting curves.

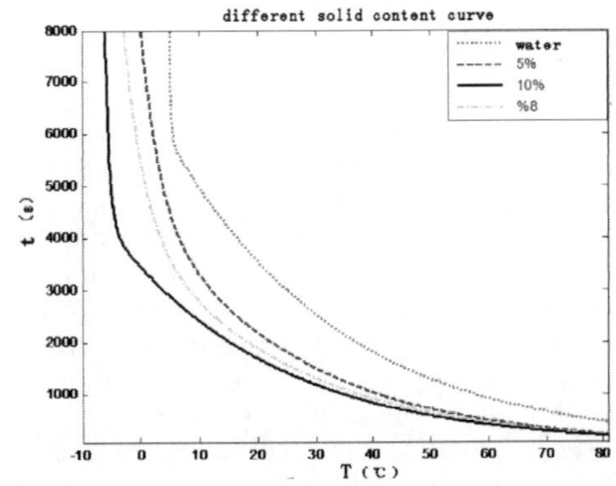

Figure 10. The melting curve of salt ice with different solid contents

5. Numerical simulation of solid-flow transform for ice-water with FLUENT

In order to contract the test melting time with simulation melting time, we choose salt ice with solid content of 5% to simulate the process of melt, using the melt/freezing model in FLUENT [9]. The freezing point of the salt ice is 270.52K and the initial temperature is 243.15K, we suppose the undersurface of the salt ice is insulating. Boundary condition of undersurface is 270.52K,and the environment temperature is 313.15K.

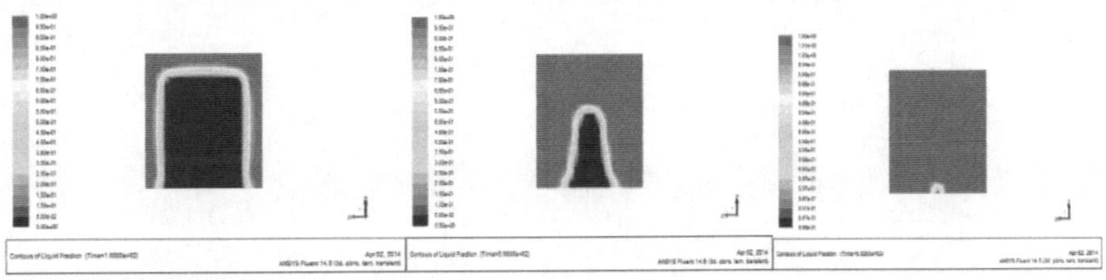

(a) (b) (c)

Figure 11. Melting process of ice with 5% solid content: after 100s (a), 500s (b), and 982s (c).

The melting time from the test is 1024s, whereas the value obtained by the numerical simulation is 982s. The main reasons of the difference can be as follows: (i) the temperature of saline ice will increase during its transfer from refrigerator to incubator; (ii) the temperature in incubator will decrease when the saline ice is placed into incubator; (iii) the enthalpy value changes after the saline ice freezing point drops; and (iv) the heterogeneity of saline ice: the closer to the center of salt ice, the higher the salt content. Thus, the simulation results have a good fit with the experimental data.

6. Conclusions

From the above findings for three different solid contents of salt (5, 8, and 10%), the following conclusions can be drawn:

1. The relationship between freezing point and solid content of salt is inversely proportional, the freezing point decreases with the solid content of salt.
2. At the second step, the duration of freezing point attainment is reduced, and the freezing process of saline solution tends to non-crystal pattern.
3. By analyzing the data from laboratory test for different solid contents of salt, the empirical formulas between ice-melting time and temperature were derived, which provide the theoretical basis for further study.
4. Under the same temperature, with the increase of solid content of salt, the melting time decreases, which implies higher melting rates.

Acknowledgments

This research was financially supported by the National Natural Science Foundation of "Research of theory and technology on rheological, creep and solid-fluid transformation of deep soft rock" (51674149).

References

[1] Zhang H J 2010 Analysis of harm to the environment by using deicing salt to remove snow. *Environmental Protection and Recycling Economy*, **1**: 75-76.
[2] Huang Y, Gao Q, Ma C Q et al 2010 Heat melting characteristic of ice layer in ice-snow melting process on road pavement. *Journal of Jilin University (Engineering and Technology Edition)*, **40** (2): 391-396.
[3] Guo T T 2011 The Theoretical Research on Transformation between Solid and Liquid of Water and Ice. Qingdao: Qingdao University of Science and Technology.
[4] Geng X H 2014 The Experimental Study and Numerical Simulation on the Effect of Humidity and Temperature Between Solid and Liquid of Ice Water. QingDao: Qingdao University of Science and Technology.
[5] Du X P 2014 The Experimental Study of the Impact of pH and Pressure on Ice Melting Rate. QingDao: Qingdao University of Science and Technology.
[6] Bao S H, Shen J M and Wang S F 2011 Agilent 34972A temperature sensor in the system, *Journal of Mechanical and Electrical Engineering*, No.6, http://en.cnki.com.cn/Journal_en/C-C029-JDGC-2011-06.htm
[7] Xu Y M, Wang L P, Qi E Y et al 2011 Study on melting ice of composite CMA environment-friendly snowmelt agent, Dalian Nationalities University, Dalian, China.
[8] Hou S G 2000 Urumqi River winter snow leaching experimental results *Glacial permafrost*. **22** (4): 362-364.
[9] Tang J P 2013 FLUENT14.0 Super study manual. Beijing: Post and Telecom press.

IOP Conf. Series: Materials Science and Engineering **283** (2017) 012028 doi:10.1088/1757-899X/283/1/012028

Development of 3mm three-port Y-junction magnetic optical circulator with two-dimensional magneto-photonic crystals

Y Wang[1], D G Zhang[2,*], S X Xu[2], B G Xu[2] and Z Dong[2]

[1]College of Optoelectronic Engineering, Shenzhen University, Shenzhen 518060, China

[2]College of Electronic Science and Technology, Shenzhen University, Shenzhen 518060, China

[*]Corresponding author: E-mail: dgzhang@szu.edu.cn

Abstract: Two kinds of defect structures of two-dimensional (2D) magnetic-photonic crystal (MPCs) are envisaged to realize circulators in millimeter wave band. The band gaps of the two kinds of photonic crystals are calculated by the plane wave expansion method. The function of the circulator is simulated by finite element method. The calculated excellent external characteristics of the circulator show that MPCs is a promising way for generating an optical circulator.

1. Introduction

In recent decades, photonic crystals (PCs), a new class of artificial materials [1, 2], have attracted a lot of attention, because they can be used for the light flow control. With the special characteristics of photonic localization and band gap (PBG), PCs have found comprehensive applications in sensors [3], mirrors [4], waveguides [5, 6], lasers and cavities [7, 8], filters [10, 11], splitters [12], circulators [13-16] and other important components for optical integrated systems.

Circulators suppress multiple reflections between components and thereby improve tolerance with respect to fabrication imperfections and environmental fluctuations. In this paper, two kinds of defect structures of two-dimensional (2D) magnetic photonic crystals (MPCs) are designed to realize 3mm Y-junction circulators. The MPCs are formed by distributing air holes in bismuth iron garnet (BIG) or introducing gyrotropic ferrite post in the Al_2O_3 rods array. In the millimeter wave region, Faraday rotation effect of the circulator is expressed as tensor permeability. The band gaps of the PCs are calculated by the plane wave expansion method (PWEM). The function is simulated by the FEM. The calculated excellent external characteristics of the circulator show that MPCs is a promising way for generating an optical circulator.

2. Air hole Y-shaped MPCs circulator

Several kinds of 2D MPCs design schemes have been envisaged to realize three-port circulators in [11, 13, 14]. A windmill-typed circulator formed by air holes array is shown in figure 1(a). An air hole Y-shaped circulator is shown in figure 1(b). A T-typed circulator composed by dielectric posts is shown in Figure1(c).

The first two 2D MPCs defect structures shown in figure 1(a) and figure 1(b) are composed of triangular lattice 2D MPCs, which is made of air holes array distributed in BIG. A point defect is introduced in the central position of the defect structures and a micro resonant cavity is constituted by the adjacent six holes around the central defect. The three line defects composed of air holes with different radius form three optical waveguides labeled Port A, B, and C. Under external DC magnetic field, light wave launched from input port spread into micro resonator by the coupling between waveguide and micro-cavity. Duo to Faraday rotation effect, the mode generates frequency division in the micro-cavity, and the wave front occur an angle deflection.

Finally, light wave go output to the corresponding waveguide and achieve the desired ring function.

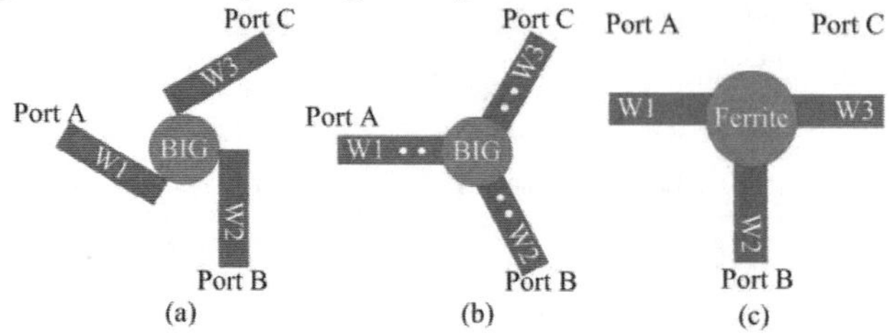

Figure 1. Three kinds of circulators: (a) A windmill-shaped circulator; (b) Three ports Y-shaped circulator; (3) Three ports T-shaped circulator.

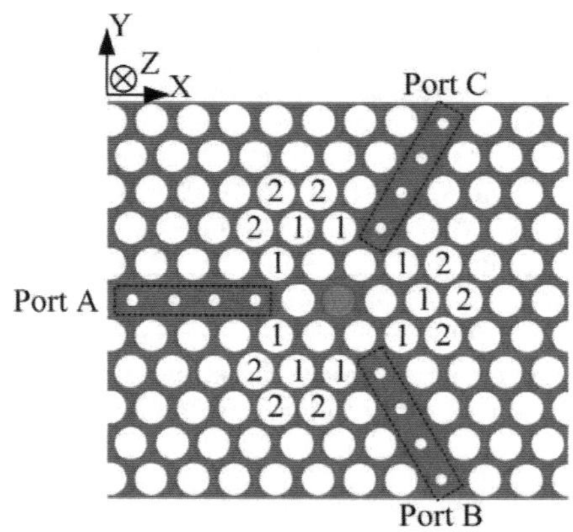

Figure 2. Triangular lattice 2D MPCs with air hole array.

2.1. Defect structure with air holes

Due to the scaling theory of Maxwell's equations, 2D MPCs defect structure of air holes with optical band in [14] will be scaled to the size of millimeter-wave band. Actually, the structure shown in figure 2 is used to achieve a 3mm three-port Y-junction magnetic optical circulator.

As shown in figure 2, a 2D MPCs defect structure is composed of 2D triangular lattice air holes array distributed in BIG, where the white circles (radius is r_0) represent air holes and the three rows of air holes with smaller radius ($r_4 = 0.4r_0$) form three line defect waveguides marked as Port A, B, and C. A bigger point defect with radius of $r_3 = 1.39r_0$ is lead into the central location of the designed structure and a micro resonator is taken shape at the central area. The micro-cavity can be treated as a cylindrical cavity approximatively and the radius is $R = a - r_0$. In Figure2, the lattice constant a is 1mm, which is amplified from the corresponding lattice constant 1 μm in [14]. When the circular area (red) is biased DC magnetic field, the Y-typed structure composed of air holes array has excellent external characteristics as a single circulator. As compared to the structure of [12], this design seems more simple and compact.

2.2. Photonic band gap and mode field analysis of point defect

As shown in Figure3, the band gap of TM mode for 2D triangular lattice PCs is calculated by PWEM without external DC magnetic field. It is expressed as a normalized frequency range $0.2903(2\pi c/a)$ ~ $0.3728(2\pi c/a)$, where c is speed of light.

In order to obtain a circulator with the specified operating frequency ($\lambda = 3\text{mm}$), the radius of resonant cavity in figure 2 will be precisely adjusted. The radius of micro-resonator is determined by the lattice constants a and the air hole radius r_0. When the lattice constant $a = 1\text{mm}$ is fixed, the center frequency of resonant cavity versus with the radius r_0 of air holes is shown in figure 4. As the radius of air pore increasing, the corresponding wavelength progressively decreases. Exactly, when the radius of the air holes is $r_0 = 0.321689\,\mathrm{mm}$, the wavelength of circulator is $\lambda = 3\mathrm{mm}$.

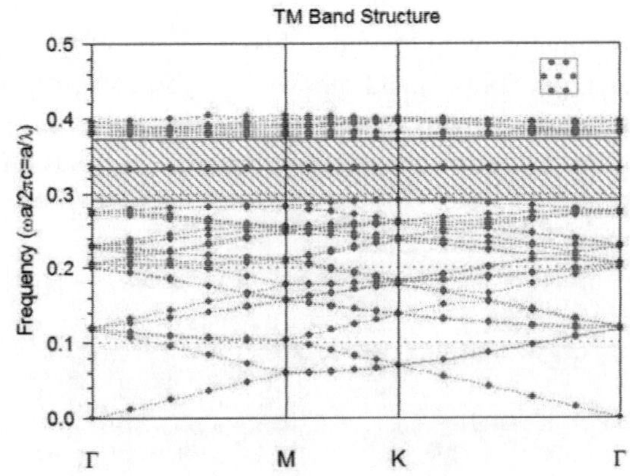

Figure 3. PBG for triangular lattice PCs.

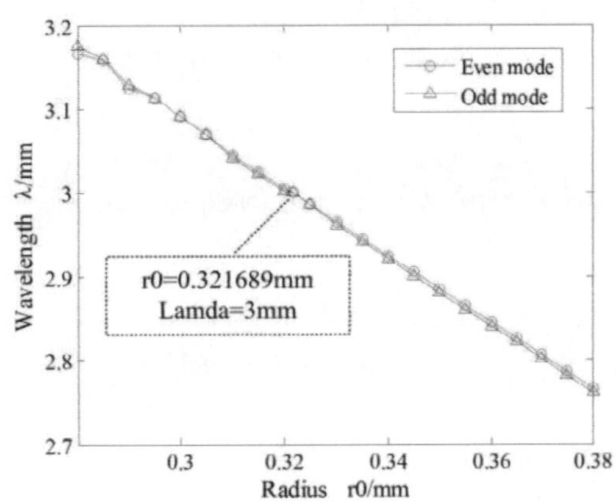

Figure 4. The relationship between the radius of air hole and the center frequency of micro-cavity.

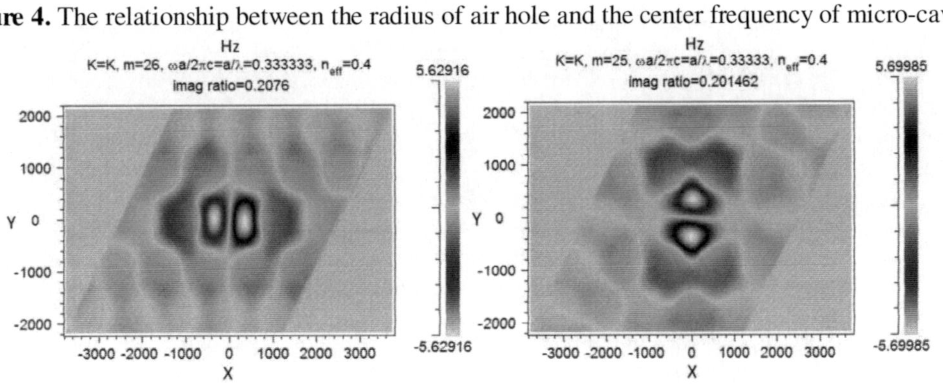

Figure 5. Hz field distribution of modes in micro resonator cavity: (a) odd mode; (b) even mode.

Numerical results show that the cylindrical resonant cavity can support a pair of dipole degenerate modes at the central normalized frequency of $0.333333(2\pi c/a)$, which are represented respectively as even

IOP Conf. Series: Materials Science and Engineering **283** (2017) 012028 doi:10.1088/1757-899X/283/1/012028

mode $|e\rangle$ and odd mode $|o\rangle$. The distribution of magnetic field intensity Hz is shown in figures 5(a) and 5(b).

2.3 2D MPCs circulator with air holes array

When external magnetic field is applied along z-direction in the central position, the electromagnetic field propagation in the circulator is numerically simulated by FEM at the operating frequency of $0.333333(2\pi c/a)$, as shown in figure 6. Figure 6(a) shows that the light wave launched from the input port (A) is almost totally and averagely transmitted to the output port (B) and the isolated port (C) without any external DC magnetic field ($\varepsilon_a = 0$). In the case of external DC magnetic field ($\varepsilon_a = 0.6$), figure 6(b) exhibits that the doubly degenerate modes take place coupling in the central position and the wave front occur a 120° Faraday deflection angle, so light wave launched from Port A is transmitted to Port B and isolating to Port C, analogously also shown in figures 6(c) and 6(d).

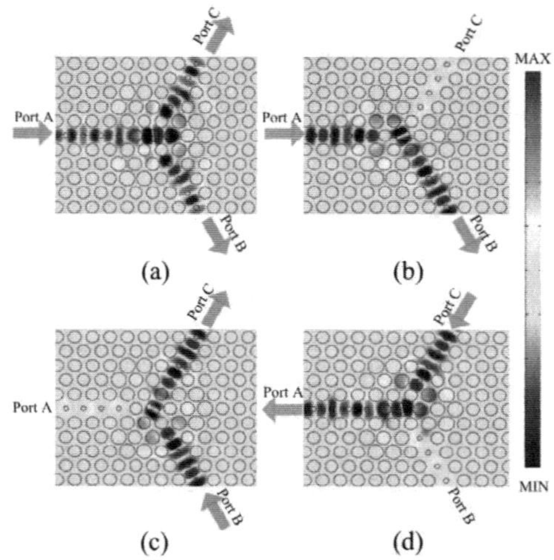

Figure 6. Light wave transmission in 2D MPCs circulator:(a) Light wave transmission without any external DC magnetic field; (b)(c)(d) Light wave transmission with external DC magnetic fields.

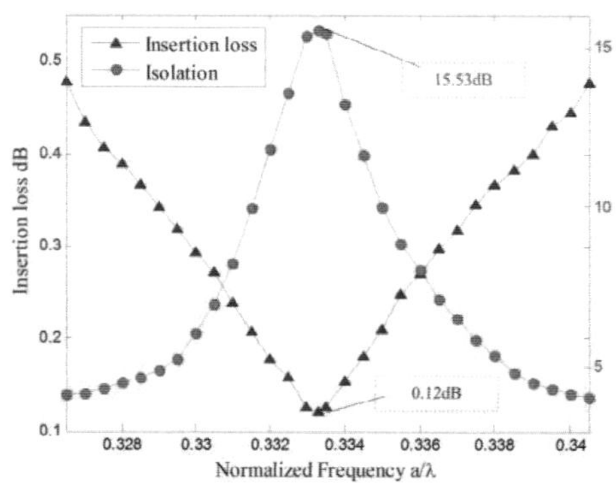

Figure 7. Numerical insertion loss and isolation

The insertion loss and isolation of the circulator are then studied by changing the frequency of incident light wave. When light wave is launched from one of the three ports, the power of light wave at the other two ports is measured. Isolation and insertion loss of the circulator changing with frequency are shown in figure 7.

Numerical results show that at the center frequency of $\lambda = 3$mm or the normalized frequency is $0.333333(2\pi c/a)$, circulator has highest isolation of 15.53dB and lowest insertion loss of 0.12dB. When the frequency of the input light wave diverges from the central frequency, transmission efficiency of the circulator gradually decreases. With the increasing frequency offset, the transmission characteristic of the circulator sharply deteriorate for the constantly reducing isolation and the increasing insertion loss.

3. Dielectric cylinder Y-shaped MPCs circulator

3.1 Defect structure with dielectric rods

As shown in figure 8(a), a T-shaped three-port [13] circulator has been designed based on square lattice photonic crystals (SLPCs) with posts array. The numerical isolation of this design reaches 25dB, but the bandwidth is still narrow. Based on SLPCs, a four-port cross-typed circulator is designed in [16].

Summarizing the former several designs, a novel simpler 2D MPCs defect structure is envisaged and a more compact Y-typed circulator is numerically studied. By introducing a ferrite post ($\varepsilon_r = 12.9$) in the central position of the triangular lattice PCs with dielectric rods array, the circulator is shown in figure 8(b). The ferrite post with bigger radius plays a role of resonator. When the central ferrite post is longitudinal magnetization, the light wave takes place a 120° Faraday rotation in the central area.

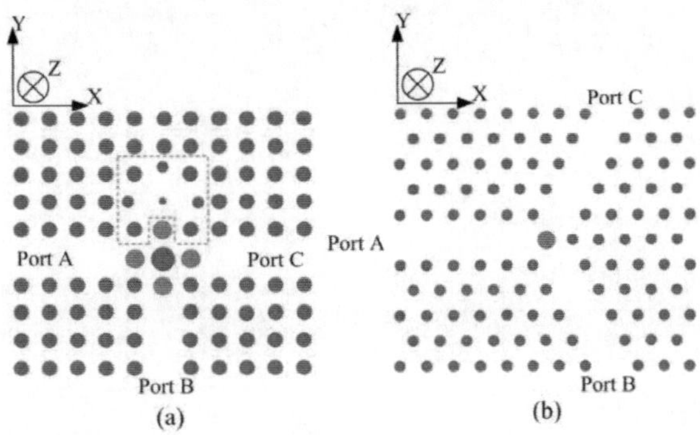

Figure 8. Rods array 2D-MPCs circulator: (a) T-shaped; (b) Y-shaped

3.2 Low-loss 2D MPCs circulator

Photonic band gap structure for TE mode and point defect micro-cavity of two-dimensional triangular lattice rods array have been analyzed in [15]. Similar to the previous section, the structural parameters of circulator will be set in the millimeter wave band and lattice constants take $a = 1.125$mm and the radius of dielectric post is $r_0 = 0.18a$. This time the dielectric post is chosen low-loss ceramics material with refractive index $n_0 = 3.4$ [17]. The PBG of TE mode for 2D triangular lattice PCs is also calculated by PWEM. The normalized frequency range of the PBG is $0.2385 \sim 0.4216(2\pi c/a)$. By changing the radius r_1 of the central ferrite post to $0.27a$, the operating wavelength of circulator can achieve 3mm, at this moment, the normalized frequency is $0.375(2\pi c/a)$.

When the external DC magnetic field is $H_0 = 2.85\times10^5$ A/m , and loss coefficient is $\alpha = 3\times10^{-5}$, the elements of tensor permeability are represented by following equations [16]:

$$\mu_r = 1 + \omega_m(\omega_0 + i\alpha\omega)\Big/\left[(\omega_0 + i\alpha\omega)^2 - \omega^2\right] \tag{1}$$

$$\kappa = \omega_m\omega\Big/\left[(\omega_0 + i\alpha\omega)^2 - \omega^2\right] \tag{2}$$

$$\omega_0 = \mu_0\gamma H_0 \tag{3}$$

$$\omega_m = \mu_0\gamma M_s \tag{4}$$

where $\gamma = |e/m| = 1.759\times10^{11}$ C/kg is gyromagnetic ratio and saturation magnetization is $M_s = 2.39\times\ 10^5$ A/m . In different situations, the electromagnetic field propagations in circulator are simulated at the central frequency of $0.375(2\pi c/a)$ by FEM, as shown in figure 9.

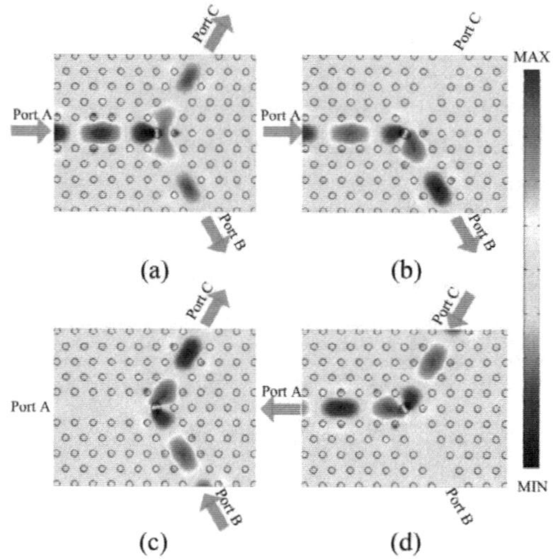

Figure 9. Light wave transmission in 2D MPCs circulator,(a) Light wave transmission without any external magnetic field; (b)(c)(d) Light wave transmission with external magnetic field.

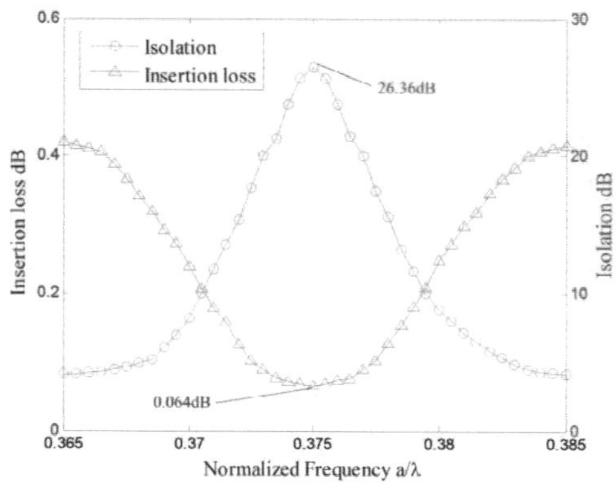

Figure 10. Isolation and insertion loss versus frequency

The numerical results show that the variations trend with frequency for isolation and insert loss in this paper perfectly conform to those numerical results of [13], as shown in figure 10. The maximum isolation is by nearly 26.36dB higher than the optimal result in [13]. Furthermore, the lowest insertion loss of 0.064dB has a significant improvement. In [13], four ferrite rods were used as the central defect, which undoubtedly will bring more energy loss.

4. Conclusions

Two kinds of defect structures of 2D MPCs are envisaged to realize circulators in the millimeter wave band. The band gaps of the two kinds of PCs are calculated by PWEM. The function of the circulator is simulated by FEM. The excellent performance of insertion loss 0.064dB and isolation 26.36dB for the circulator show that MPCs is a promising way for generating an optical circulator.

Acknowledgments
We thank the support from the National Natural Science Foundation of China (NSFC; grant 61171006).

References

[1] Yablonovitch E 1987 Inhibited spontaneous emission in solid state physics and electronics *Phys. Rev Let* **58**(20) 2059-2062

[2] John S 1987 Strong localization of photons in certain disordered dielectric super lattices *Phys. Rev Let* **23** 2486-2489

[3] Zhang D G, Yung E K N and Ding H Y 1995 On the operating modes of a millimeter-wave H-plane partial height ferrite post waveguide circulator *Microwave and Optical Tech. Letters* **10**(4) 230-235

[4] Inoue M, Arai K, Fujii T and Abe M 1999 One dimensional magneto-photonic crystal J*ournal of Applied Physics* **85**(8) 5768-5770

[5] Kato H, and Inoue M 2002 Reflection-mode operation of one-dimensional magneto-photonic crystals for use in film-based magneto-optical isolator devices *Journal of Applied Physics* **91**(10) 7017-7019

[6] Sakaguchi S and Sugimoto N 1999 Transmission properties of multilayer films composed of magneto-optical and dielectric materials *Journal of Lightwave Technology* **17** (6) 1087- 1092

[7] Zhang D G, Li J, Ni N and Ouyang Z B 2004 Electromagnetic field analysis and numerical calculation of one-dimensional magneto-photonic crystal *Acta Photonica Sinica* **33**(8) 1007-1010

[8] Zhou H, Zhang D G, Ouyang Z B and J. Z. Li 2008 Structure Research on One-dimensional Magneto-photonic Crystals Used in Optical Isolator *Acta Photonica Sinica* **37**(1) 106-109

[9] Wang Z and Fan S H 2005 Magneto-optical defects in two-dimensional photonic crystals *Applied Physics B* 81 369-375

[10] Jalali A A and Friberg A T 2005 Faraday rotation in a two dimensional photonic crystal with a magneto-optic defect *Optics Letter* **30**(10) 1213-1215

[11] Wang Z and Fan S H 2005 Optical circulators in two dimensional magneto-optical photonic crystals," *Optics Letters* **15**(30) 1989-1991

[12] Fan S H and Wang Z 2006 An ultra-compact circulator using two-dimensional magneto-optical photonic crystals *Journal of Physical Society of Japan* **30**(6) 641-645

[13] Wang Q, Ouyang Z B, Tao K, Lin M and Ruan S 2012 T-shaped optical circulator based on coupled magneto-optical rods and a side-coupled cavity in a square-lattice photonic crystal *Physics Letters A* **376**(4) 646-649

[14] Wang Q, Ouyang Z B and Liu Q 2011 Multiport photonic crystal circulators created by cascading magneto-optical cavities *Journal of the Optical Society of America* B **28**(4) 703-708

[15] Wang Y, Zhang D G, Xu S X, Ouyang Z B and Li J Z 2016 Low-loss Y-junction two-dimensional magneto-potonic crystals circulator using a ferrite cylinder *Optics Communications* 369 1-6

[16] Wang Y, Zhang D G, Ouyang Z B and Li J Z 2014 Four-Port Cross-Shaped Circulator Based on Two-Dimensional Magneto-Photonic Crystals *Acta Optica Sinica* **34**(10) 1023001

[17] Sebastian M T and Jantunen H 2008 Low loss dielectric materials for LTCC applications: a review *International Materials Reviews* **53**(2) 57-90

IOP Conf. Series: Materials Science and Engineering **283** (2017) 012029 doi:10.1088/1757-899X/283/1/012029

Study on performance of waterborne anticorrosive coatings on steel rebars

S N Ramaswamy[1 *], R Varalakshmi[1] and R Selvaraj[2]

[1]Department of Civil Engineering, Kalasalingam University, Tamilnadu, India
[2]CSIR-CECRI, Karaikudi, Tamil Nadu, India.
[*]Corresponding author: E-mail s.n.ramasamy@klu.ac.in

Abstract. Durability of reinforced cement concrete structures is mainly affected by corrosion of steel reinforcements. In order to protect the reinforcing bars from corrosion and to enhance the lifetime of reinforced cement concrete structural members, anticorrosive treatment to steel is of prime importance. Conventional coatings are solvent based. In this study, water based Latex was used to formulate anticorrosive coating. Latex is applied to steel specimen substrates such as plates and rods and their mechanical properties such as flexibility, abrasion, bendability, adhesive strength, impact resistance, etc. were studied. It was inferred that coating containing latex, micro silica, zinc phosphate, ferric oxide, aluminum oxide, titanium oxide and silica fume was found to possess more corrosion resistance under marine exposure conditions.

1. Introduction

Concrete is the most widely used construction material having very good compressive strength but very low tensile strength. To impart its tensile strength, steel reinforcements are normally provided in reinforced cement concrete (RCC). Reinforcing steel bars are normally well protected against corrosion by embedment in high-alkaline environment of concrete. Corrosion of embedded reinforcing steel in RCC construction is a world-wide problem affecting durability of concrete which can cause minor disfigurement to catastrophic collapse of the whole structure. Examples of corrosion of reinforcement include buildings and bridges which results in increased volume of steel, cracking, delamination and spalling of cover concrete. Also, structures under marine exposure are suffering severe damages due to penetration of chloride ions causing corrosion of reinforcing steel. When a small but uniform amount of moisture is present in the atmosphere, concrete will be carbonated to the depth of the steel reinforcement and the steel is likely to corrode uniformly. Alternatively, corrosion may occur at local points within the length of rebars, which may form severe notching or thinning whilst the remainder of the bar lengths is left un-corroded. Because of varying conditions that exist within structures, corrosion proceeds at different rates at different locations. In order to protect reinforcing steel form corrosion, universally adopted method is to provide anti-corrosive treatment to steel reinforcement before embedding in concrete. Different types of treatments used in practice are galvanizing zinc–aluminum coating, fusion bonded epoxy coating, cement polymer composite coating, epoxy phenolic rebar coating, epoxy-based coating, etc., majority of the conventional coatings are solvent borne. At normal temperature and pressure, these solvents emit volatile organic compounds (VOCs) and hazardous air pollutants (HAPs), which are dangerous and cause risk to human health. Most of the studies were conducted using solvent-based anticorrosive coatings. Various studies were carried out on corrosion-induced cracking of reinforced concrete and the performance of anticorrosive coatings.

Gao *et al* [1] conducted experimental and statistical studies on the effects of the load-induced cracks for predicting time needed for the initiation of corrosion and the critical time needed to maintain the integrity of the structure against failure. Gao *et al* [2] also conducted experimental studies under accelerated corrosion on reinforced concrete (RC) slabs and inferred that pre-cracked structure has noticeable influence on the corrosion damage under sustained loading. Authors [2] also conducted experimental studies on the loss of mass of reinforcing steel due to corrosion and the effect of inclusion of polypropylene fibers in concrete under accelerated corrosion. They observed that maximum mass loss of steel was occurred at the center of the corrosion region for the specimens corroded with pre-existing cracks. Also, addition of polypropylene was found effective in reducing the growth of crack speed without any reduction of the ultimate capacity of the specimen [3]. Subsequently, researchers [4] conducted statistical studies on the ranking of corrosion affected concrete bridges using multiple criteria decision making system. The same group of authors performed experimental and statistical studies on the effects of the load-induced cracks for predicting time needed for the initiation of corrosion and the critical time so as to maintain structural integrity [5].

Almusallam *et al* [6] conducted studies on the effectiveness of surface coatings in improving the durability of concrete and inferred that epoxy and polyurethane coatings perform better than acrylic, polymer and chlorinated rubber coatings. Cheng *et al* studied the effect of coatings to rebars on corrosion resistance and bond strength of reinforced concrete. They observed that the corrosion rate of zinc is higher than carbon steel in pH12 solution but lower in 3.5% NaCl solution. The bond strength decreases with an increase in corrosion rate for uncoated rebar and zinc-coated rebar. After 14 days accelerating corrosion process, the reduction ratio of bond strength for zinc-coated rebar is less than that of uncoated rebar [7].

Saravanan *et al* evaluated the performance of polyaniline pigmented epoxy coating for corrosion protection of steel in concrete environment. They stated that the corrosion resistance of polyaniline pigmented epoxy coating decreased initially and then increased due to a passivating ability of polyaniline pigment in electrochemical impedance study [8]. Baldissera *et al* carried out studies on performance of epoxy resin containing polyaniline for corrosion protection of mild steel bars. The corrosion behavior of coated mild steel samples was investigated in 3.5% NaCl solution by electrochemical impedance spectroscopy and concluded that coating containing SPAN was found to have the best performance in corrosion protection [9].

To the best knowledge of the authors, very limited studies were carried out using water as the solvent to disperse the resin for providing corrosion protection coating to reinforcing bars. There are certain specific advantages of using waterborne coating systems for corrosion protection of rebars such as reduced emission of volatile organic compounds and hazardous air pollutants, use of conventional application process, reduced toxicity and odor, good storage life, easy to clean, elimination of the problem of disposal of hazardous waste, etc. The objectives of this research work are: (i) to formulate various composition of water based anticorrosive coatings to steel rebars using latex as the binder and study their performance on abrasion, adhesion and impact resistance; and (ii) to identify the coating, which is the most suitable for corrosion protection of reinforcing steel under marine exposure.

2. Experimental studies

2.1. Materials and methods

2.1.1 Materials used. For undertaking this study, water roofing Latex purchased from Madurai was used as binders. Fly ash, ordinary Portland cement, feldspar, silica fume, china clay, kaolin, micro silica, clay were purchased in Madurai and used as fillers. Chemicals and minerals such as zinc phosphate ($Zn_3(PO_4)_2$), aluminum oxide (Al_2O_3), titanium dioxide (TiO_2), and ferric oxide (Fe_2O_3) were purchased from Hamada Company and utilized as inhibitors, hiding agent, and coloring agent, respectively.

IOP Conf. Series: Materials Science and Engineering **283** (2017) 012029 doi:10.1088/1757-899X/283/1/012029

2.1.2. XRF analysis. The chemical composition of the fillers, pigments and additives were determined using X-ray fluorescence spectrometry (XRF). The summary of XRF results of OPC, flash, feldspar, clay, kaolin, china clay, silica fume, micro silica, ferric oxide are presented in Table 1.

Table 1. Chemical composition of materials used

Material composition	OPC	Fly ash	Micro silica	China clay	Kaolin	Silica fume	Clay	Feld spar	Iron oxide
SiO_2	6.514	43.538	89.484	13.303	73.042	95.265	37.822	49.464	2.829
Fe_2O_3	11.157	29.041	0.037	7.030	3.333	1.231	39.707	1.085	94.98
SO_3	2.044	1.179	1.371	0.874	1.524	3.193	1.522	1.472	0.646
CaO	78.754	3.987	0.353	73.138	1.732	0.306	13.636	-	-
Al_2O_3	-	10.501	-		12.828	-	-	-	1.360
Others	0.432 K_2O	7.472 TiO_2	7.298 Na_2O	4.369 BaO	5.764 K_2O	-	5.402 TiO_2	46.66 K_2O	-

2.1.3. Preparation of surface of steel samples. Steel plates of size 100mm x 150mm x 0.8mm, 100mm x 100mm x 0.8mm, 100mm x 75mm x 0.2mm, 100mm x 20mm x 0.8mm and steel reinforcing bars of 300mm length with varying diameters such as 8mm, 10mm and 16mm were used as samples, to which the formulated coatings were applied. Initially, the steel plates and rebars were cleaned using pickling solution and then washed twice in tap water, dried in desiccators and kept ready for coating. Figure 1 shows the collection of materials, cleaning, and drying in desiccators before applying the formulated coatings.

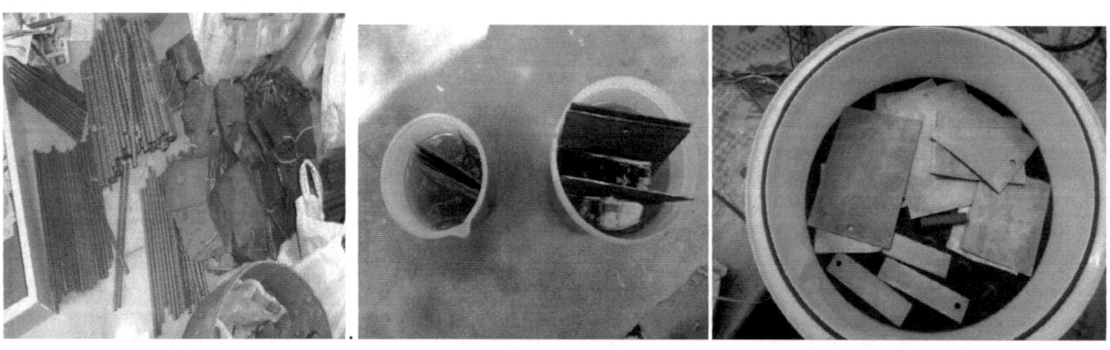

(a) (b) (c)

Figure 1. Collection of materials (a), cleaning (b) and drying (c) process

2.1.4. Formulation of latex coating systems. In the present research, 18 coating composition systems using latex were formulated as given in figure 2. These coatings consist of waterproofing latex as the binder with various compositions of fillers, inhibitors, coloring agents, and hiding agents. After the surface preparation, the coatings were deposited on the specimen using the brush (in two layers). Figure 3 shows the composition of various latex-based coatings.

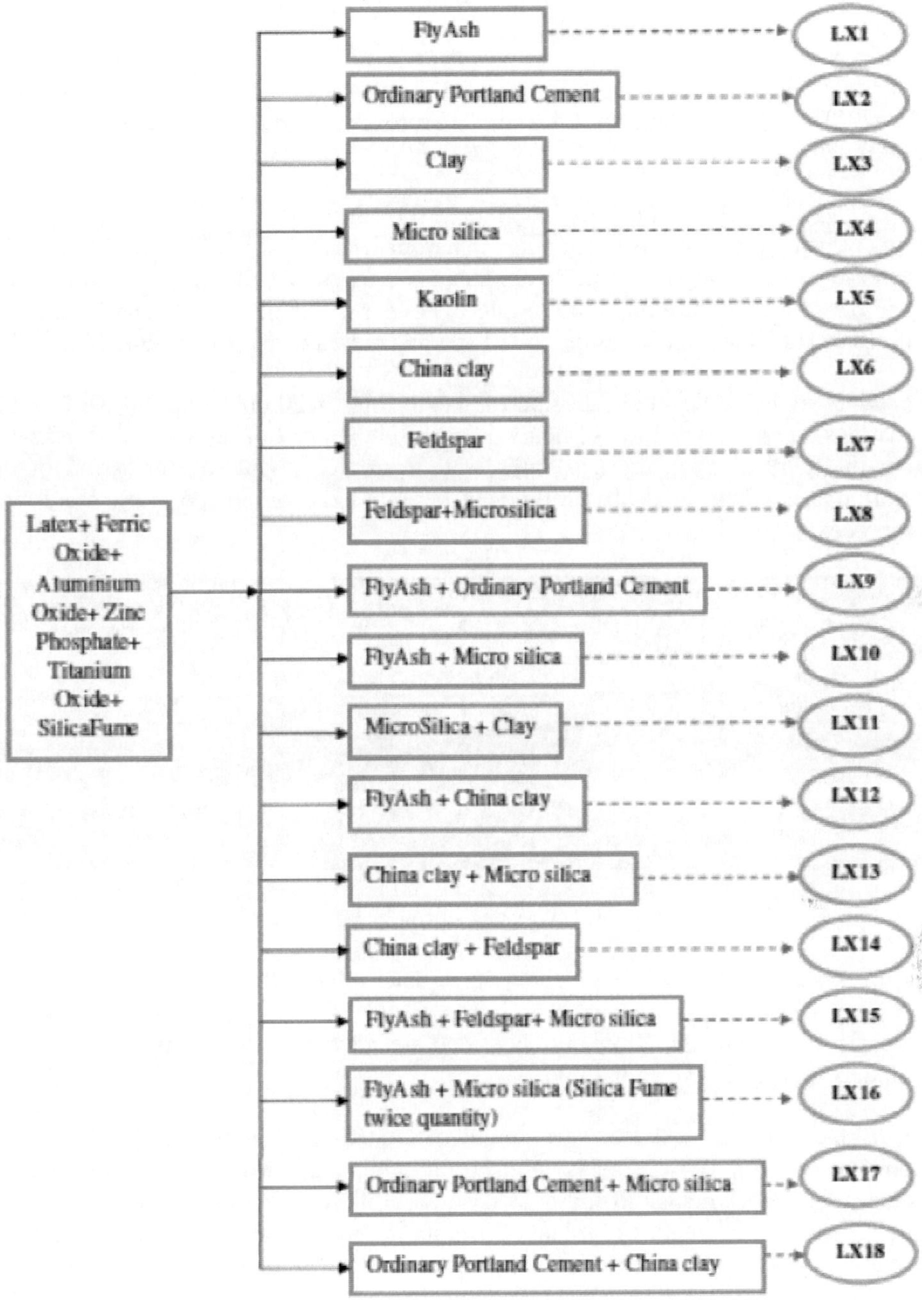

Figure 2. Formulation of latex (LX) coating systems

2.2. Experimental studies

Latex-based coatings are red in color, can be easily deposited, and have brushing consistency. All the above coatings were formulated and applied to the steel rods and substrates for evaluating the performance of coatings such as dry film thickness, bendability, hardness, flexibility, abrasion, film adhesive test, etc. All steel samples with 18 latex-based coatings were examined for their mechanical properties as per relevant ASTM standards.

IOP Conf. Series: Materials Science and Engineering **283** (2017) 012029 doi:10.1088/1757-899X/283/1/012029

2.2.1. Dry film thickness test. Dry film thickness test was conducted in accordance with ASTMD 1186, [10] using the instrument known as Elcometer. The surface coatings were applied on the steel plates of size 100mm x 100mm x 75mm. The coated plates were kept at 60°C in an oven for 3 hours. Using Elcometer, the thickness of coatings was measured in various 6 places on the plate both at front and back faces. The average value of thickness was taken as dry film thickness and the results are presented in table 2.

2.2.2. Hardness test. Resistance to indentation/scratching is termed as hardness. Film hardness measurement using pencil 6 B to 6 H (softer to harder) is a well-known inexpensive and rapid test. All the 18 latex-based coating steel plates were subjected to the pencil hardness as per ASTM D3363-05 [11] using 6B to 6 H by scribing at 45° angle to have a stroke of 6.5mm length. It was found that all the 18 types of latex-based coatings exhibited adequate resistance against scratching.

2.2.3. Bendability test. The steel substrate of size 100mm x 20mm x 0.8mm were coated with latex-based coatings and were dried for 48 hours at room temperature. Then they were bent to 90° and 180° as shown in figure 3. Then the substrates were inspected about the formation of crack or stress concentration area if any. In all the 18 types of latex-based coatings, no crack was found during both bending process.

(a) (b)

Figure 3. Bendability test: specimens bent to 180° (a) and 90° (b)

2.2.4. Abrasion test. Latex-based coated dry steel plates of size100mm x 100mm x 0.8mm were coated with the 18 types of greener coatings. The initial weight of the coated plates was taken first. Then the plates were scribed with E4320 emery sheet and final weight after abrasion was taken. The loss in weight due to abrasion is shown in table 2.

2.2.5. Film adhesion test. Adhesive strength of coated film on the steel substrate was measured using tensometer. [12]. Tensometer is an electrically operated instrument and application of tensional force should be perpendicular to the coated surface of the specimen which is held vertically. The coating film adhesive strength of 18 latex coatings is given in table 2. It is evident from table 2 that LX17 coating gives better adhesion to the metal substrate compared to all other coatings. LX18 coating shows low adhesive strength.

2.2.6. Cracking resistance test. In accordance with ASTM D522 [13]. Steel plates were bent into conical shape using mandrel bend test equipment as shown in figure 4. The latex-coated steel plates were placed in the equipment so as to experience flexibility. In this test, the steel plates were bent in such a way that one end having 3mm diameter and gradually increasing to 38mm diameter at the other end within the length of 200mm. This is for determination of the resistance to cracking of attached green coatings on steel substrates. Latex-coated thin plates were placed in the equipment and were bent up to 180° in a single stroke. Formation of cracks (if any) was observed from 3mm die end to the other end. It was observed that all coatings exhibited very good flexibility characteristics without any cracking.

Table 2. Results of dry film thickness, abrasion and adhesive strength

Designation of coating	Dry film thickness (µm)	Abrasion test				Adhesive strength (N/mm^2)
		Initial weight	Final weight	Loss of weight	% Loss of weight	
LX1	93	84.44	84.43	10	0.012	8.16
LX2	109	86.94	86.92	20	0.023	7.72
LX3	135	87.35	87.34	10	0.011	8.17
LX4	165	89.03	89.01	20	0.022	8.94
LX5	97	84.69	84.67	20	0.024	8.02
LX6	95	84.11	84.10	10	0.012	8.10
LX7	88	85.95	85.94	10	0.012	8.13
LX8	89	86.59	86.57	20	0.023	7.83
LX9	103	86.01	86.00	10	0.012	7.91
LX10	121	83.90	83.89	10	0.012	7.69
LX11	152	89.26	89.23	30	0.034	8.21
LX12	127	86.31	86.30	10	0.012	7.96
LX13	95	86.31	86.29	20	0.023	8.21
LX14	99	84.65	84.63	20	0.024	7.81
LX15	85	83.88	83.87	10	0.012	7.92
LX16	93	83.59	83.58	10	0.012	7.63
LX17	134	87.38	87.35	30	0.034	8.98
LX18	99	84.34	84.32	20	0.024	7.62

(a) (b)

Figure 4 .Conical mandrel test equipment (a) and bent plates (b)

2.2.7. Environmental exposure study. The rate of corrosion on coated surface area with and without diagonal scribes was determined by the environmental exposure study. For this study, one side surface of coated plate was scribed while the other side was left unscribed. All the edges of plates were dipped in wax so as to seal the edges from corrosion. All 18 coated specimens were exposed to a salty environment by spraying sodium chloride solution three times per day as shown in figure 5 for duration of 400 hours [14]. This was done to create severe environment (seashore) as shown in figure 6.

Table 3 represents rating specifications as per ASTM-D1654 [14]. Using carbide tipped pencil type scribing tool, the coated plates were scribed to obtain a uniform V-cut through the coating by holding the tool approximately at 45° angle to the surface. Cutting tool was positioned such that the carbide tip alone should be in contact with the surface. The creep value in was measured in millimeters and rating was given according to the recommendations of ASTM-D1654 [14]. The unscribed surface of the latex coating plates was also observed and percentage of failed area was noted. The comparative values are presented in table 4.

It can be seen from table 4 that LX4 and LX17 were having less creep at scribed surface and also less percentage of failure area on the unscribed surface, compared to other coatings. The comparative results are presented in figure 6.

(a) (b)

Figure 5. Environmental exposure study: (a) spraying of NaCl solution and (b) drying of specimen

IOP Conf. Series: Materials Science and Engineering **283** (2017) 012029 doi:10.1088/1757-899X/283/1/012029

Table 3. Rating specification as per ASTM-D1654

Scribed plate creepage (mm)	Rating number	Unscribed plate Failed %	Rating number
Zero	10	No fail	10
Over 0 to 0.5	9	0 to 1	9
Over 0.5 to 1	8	2 to 3	8
Over 1 to 2	7	4 to 6	7
Over 2 to 3	6	7 to 10	6
Over 3 to 5	5	11 to 20	5
Over 5 to 7	4	21 to 30	4
Over 7 to 10	3	31 to 40	3
Over 10 to 13	2	41 to 55	2
Over 13 to 16	1	56 to 75	1
Over 16 to more	0	Over 75	0

Table 4. Results on the rating number for environmental exposure test

Designation of coating	Scribed plate		Coating ID	Unscribed plate	
	Creepage (mm)	Rating number		Failed %	Rating number
LX1	3	6	LX1	30	4
LX2	3	6	LX2	30	4
LX3	2	7	LX3	35	3
LX4	**1**	**8**	LX4	**6**	**7**
LX5	3	6	LX5	35	3
LX6	4	5	LX6	20	5
LX7	6	4	LX7	35	3
LX8	4	5	LX8	15	5
LX9	4.5	5	LX9	20	5
LX10	6	4	LX10	30	4
LX11	3	6	LX11	20	5
LX12	2	7	LX12	60	1
LX13	3	6	LX13	30	4
LX14	2	7	LX14	85	0
LX15	2.5	6	LX15	40	3
LX16	3	6	LX16	90	0
LX17	0.5	8	LX17	2	8
LX18	2	7	LX18	25	4

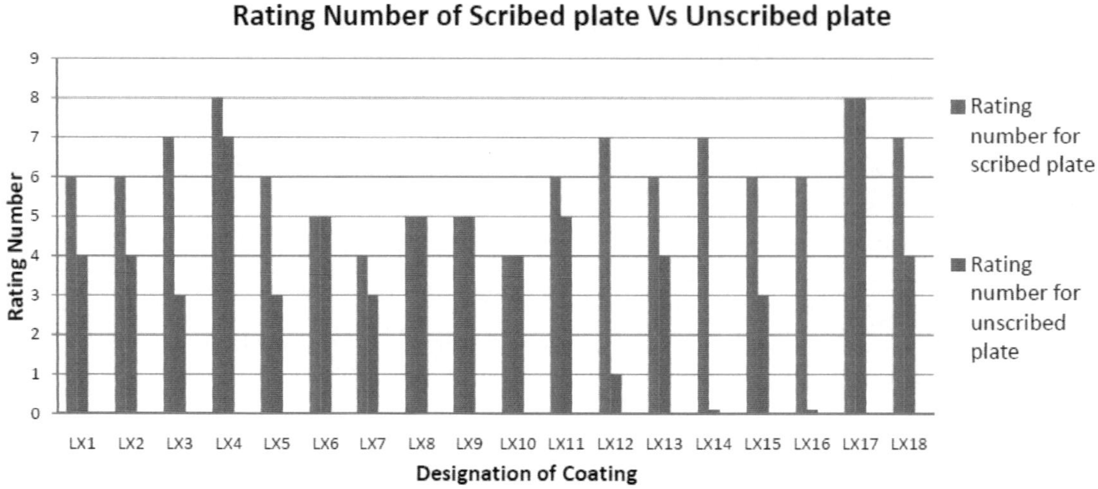

Figure 6. Comparison of rating of scribed and unscribed plates

2.2.8. Impact test: The resistance of latex-based coating against impact was determined as per ASTM D2794-93 [15] and the instrument is shown in figure 7. A standard weight of 1.8kg was dropped to strike an indenter which deforms the coating on the substrate and failure of a coating film was observed using the magnifier. Defected coating was marked as failed. It was observed that no cracking or loss of bond occurred at the impact area. All 18 types of coating passed the impact test.

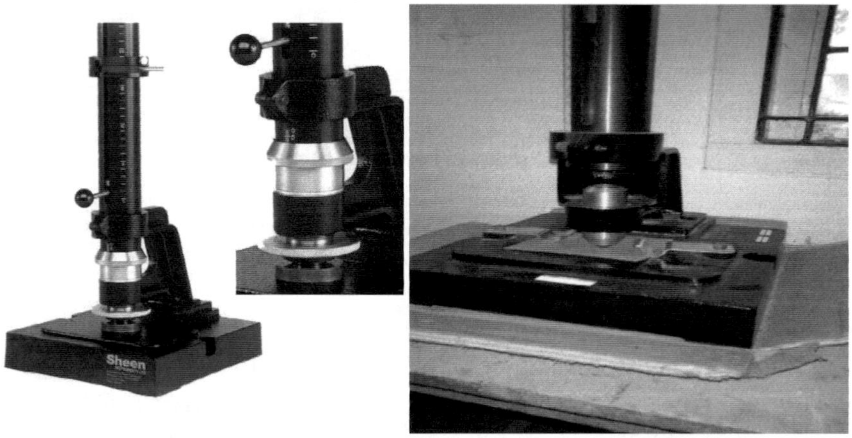

Figure 7 (a) Impact test equipment (b) Impact test set up

3. Results and discussion

The formulated water borne latex coatings exhibited Dry film thickness ranging from 85 microns to 165 microns. Adhesive strength of coatings varied from 7.62 to 8.98 N/mm^2: LX 4 coating and LX17 coating exhibited good adhesive strength. Regarding hardness test, all the 18 types of latex-based coatings exhibited adequate resistance against scratching. Also, all latex-based coatings were found to perform satisfactorily without the formation of any cracks or scratch under Bendability test. Environment exposure study of coatings revealed that LX4 (latex + micro silica + zinc phosphate + ferric oxide + aluminum oxide + titanium oxide+ silica fume) and lx17 (latex + ordinary Portland cement + micro silica + zinc phosphate + ferric oxide + aluminum oxide+ titanium oxide+ silica fume) was found to have less creep at scribed surface and also less percentage of failure area on the unscribed surface. These coatings can be considered as the better coating.

4. Conclusions

The latex- based coating is a waterborne greener manufacturing eco-friendly coating. From the total 18 coatings, only two (namely, LX4 and LX17) exhibited a good corrosion resistance and were found to be the most effective by the performed environmental exposure experimental studies. To ascertain the durability of coatings, further corrosion evaluation studies, such as electro-chemical impedance, anodic polarization studies, Taber abrasion, automatic scratch test, and impressed voltage test have been carried out on the above coatings.

References

[1] Gao Z, Liang R Y and Patnaik A K, 2017 Probabilistic lifetime performance and structural capacity analysis of continuous reinforced concrete slab bridges, *Int. Journal of Advanced Structural Engineering* 9 231.

[2] Gao Z H, Patnaik A K and Liang R Y 2016 Effects of sustained loading and pre-existing cracks on corrosion behaviour of reinforced concrete slabs *Construction and Building Materials* 124 776–85.

[3] Gao Z H, Liang R Y, Payer J and Patnaik A K 2016 Mass Loss and Capacity Loss of Reinforced Concrete Slabs with and without Polypropylene Fibers Corroded with Pre-existing Cracks and Sustained Loading *Cor. Risk Mgmt Conf.*

[4] Gao Z H, Patnaik A K and Liang R Y 2016 VIKOR method for ranking concrete bridge repair projects with target-based criteria *Cor. Risk Mgmt Conf.*

[5] Gao Z H, Patnaik A K and Liang R Y 2016 Effects of cracking & sustained loading n the strength of corroded steel reinforced RC slabs *Cor. Risk Mgmt Conf.*

[6] Almusallam A A, Khan F M, Dulaijan S U et al 2003 Effectiveness of surface coatings in improving concrete durability', *Cement & Concrete Composites* 25 473-81.

[7] Cheng A, Huang R, Wu J K and Chen C H 2005 Effect of rebar coating on corrosion resistance and bond strength of reinforced concrete *Construction & Building Materials* 19 404-12.

[8] Saravanan K, Sathiyanarayanan S, Muralidharan S et al 2007 Performance evaluation of polyaniline pigmented epoxy coating for Corrosion protection of steel in concrete environment *Progress in Organic Coatings* 160-67.

[9] Baldissera A F and Ferreira C A 2012 Coatings based on electronic conducting polymers for corrosion protection of metals *Progress in Organic Coatings* 241–47.

[10] ASTM D1186-01 Standard test methods for non-destructive measurement of dry film thickness of nonmagnetic coatings applied to a ferrous base.

[11] ASTM D3363-05 Standard test method for film hardness by pencil test.

[12] ASTM D2197-86 Standard test method for adhesion of organic coatings by scrape adhesion.

[13] ASTM D522-93a Standard test methods for mandrel bend test of attached organic coatings.

[14] ASTM D1654-08 Standard test method for evaluation of painted or coated specimens subjected to corrosive environments.

[15] ASTM D2794-93 Standard test method for resistance of organic coatings to the effects of rapid deformation (impact).

IOP Conf. Series: Materials Science and Engineering **283** (2017) 012030 doi:10.1088/1757-899X/283/1/012030

Realization of point contact for stacks Al_2O_3/SiN_x rear surface passivated solar cells

S J Wang[1,2], G J Jin[1], L L Wang[1], Z Zuo[3], Y L Meng[1*], Y Zhao[4] and S Z Jin[1]

[1]College of Optics and Electronic Technology, Chia Jiliang University, Hangzhou, China
[2]Laboratoire de Nanotechnologie et d'Instrumentation Optique (LNIO), University of Technology of Troyes, Troyes, France
[3]Department of Physics, University of Toronto, Toronto, Ontario, M5S1A7, Canada
[4] State Key Laboratory on Integrated Optoelectronics, Jilin University, Changchun, China
[*]Corresponding author: myl@cjlu.edu.cn

Abstract. A simple locally-sintered process for realizing rear point contacts of the Passivated emitter and rear cells, of which rear surface is passivated by Al_2O_3/SiN_x double layers, is demonstrated in this paper. As compared to conventional cells passivated by aluminium back surface field, the cells show higher open circuit voltage and short circuit current. The sintering effect on the final passivation quality of stacks Al_2O_3/SiN_x during the sintering process is investigated and discussed in detail.

1. Introduction

Since it has been proven that Al_2O_3 is capable of passivating the surface of silicon, its application becomes more and more attractive for high-efficiency industrial crystalline silicon solar cells, especially for passivated emitter and rear cells (PERC) [1-3]. It is well known that the full area passivation can provide a higher effective lifetime than conventional structure of industrial silicon solar cells with aluminum back surface field (Al-BSF) [4]. Meanwhile, the stack films of $Al_2O_3/SiN_x/Al$ on the rear side of cell also provide an excellent reflector for near-bandgap photons, which can improve the light trapping properties and the short-circuit current of the cell [5].

As the development of equipment for deposition of Al_2O_3 [6,7], the application of Al_2O_3 in high-efficiency PERC cells is generalized rapidly, which also drives the development of other techniques for improving the fabrication processes of PERC cells. Point contacts on rear side of PERC cells play a very important role for the performance of PERC cells, because the status of their contact with the silicon base area directly affects the series resistance and fill factors.

Hence, the process for realizing rear point contacts is a key factor to obtain PERC cells with high efficiency successfully. Up to now, many techniques are introduced to realize point contacts, such as laser ablation [8], wet chemical etch [5], etc. But most of these techniques are not suitable to mass production because of expensive cost and complicated processes. In order to comply with the processes utilized now in solar cells factories, screen-printing point contact is explored in this study. The influence of locally-sintered point contact on the passivation quality of Al_2O_3 is also studied.

2. Experimental

2.1. Fabrication of locally-sintered processing point contacts

Locally-sintered point contacts were fabricated by two steps in which different types of aluminum paste were used to realize different functional layers. The aluminum paste with glass frit was screen printed onto rear side of PERC cells firstly for the purpose of forming point electrodes. Then the locally-sintered point contact was obtained via a high temperature sintering processing. After the formation of point contacts, another screen printing processing or Al evaporation onto those point contacts was carried out to form rear electrode. In order to gain the optimized sinter temperature and time for the formation of locally-sintered point contacts, the full area Al was deposited onto surface of a Cz monocrystalline silicon wafer. Before Al paste with glass frit was printed on the other side, stacked 20nm-thick Al_2O_3 and 50nm-thick SiN_x layers were grown on the silicon as passivation layers sequentially. The Al_2O_3 film was deposited in commercial atomic layer deposition reactor Beneq TFS200. The source of reaction used in ALD was traditional metal-organic compound tri-methyl-aluminum [Al $(CH)_3$]. SiN_x was grown in Plasma-Therm 790+. A reference sample with Al/p-Si/Al structure was also fabricated. The reference sample was fired firstly at 600℃ for 5min to form ohm contact between Al and Si.

2.2. Solar cells fabrication

Figure 1 shows the whole fabrication processes flow of PERC solar cells based on p-type multicrystalline silicon. The size of multicrystalline silicon was 40 mmx40 mm. The thickness was 200 um, and the resistivity was 1.5~3 Ω·cm. After etching the surface damage, an acid-etched texturing was made on both sides. Subsequently, wet chemical cleaning and phosphorus diffusion were performed. The single-step phosphorus emitter was diffused from a $POCl_3$ source, resulting in an n^+-emitter with a sheet resistance of 60 Ohm/square.

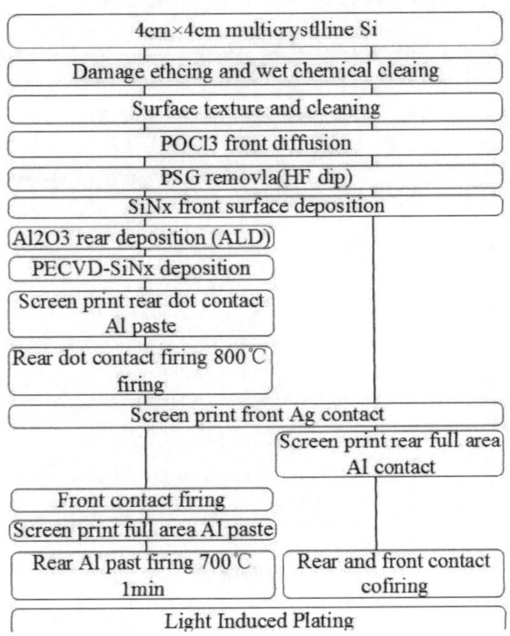

Figure 1. Processing flowchart for the PERC-type solar cells and conventional cells fabricated in this study

The phosphorus glass was removed by a short HF dip before deposition of front SiN_x. Then 80-nm-thick SiN_x was deposited on samples, after that the cells were divided into two groups. Each group had four samples. The rear surfaces of the two groups cells were passivated in different ways. One group were passivated by Al-BSF, while the other group were passivated by stack layers which consist of a 20nm Al_2O_3 and 50nm SiN_x. Afterwards, Ag front grid and a full area Al electrode were screen printed on front sides of the cells in the first group. The cofiring was performed after front contact and Al-BSF generation. As to the second group, the point contacts were screen printed on rears of the cells after the Al_2O_3/SiN_x stacks layers were deposited. The contact size was 500 um×500 um. The contact pitch was 2mm. Before Ag grid on the front surfaces was screen printed, 800 ℃ sintering was performed to form the point contact. After front Ag contact firing, rears full area Al paste was screen printed too. 1 minute firing at 750 ℃ was performed to ensure the contact between full area Al paste and point contact screen printed. At last, light induced 10 minutes plating was adopted to increase the thickness of front Ag grid. All cells were measured under standard testing conditions (25℃, 100mW/cm2, AM1.5G).

3. Results and discussion

Figure 2 shows the I-V characteristics of samples with different contacts which were realized via various sinter conditions. In order to verify the formation of good contact between Al paste and p-Si under different conditions, we made three samples with different structures and sinter temperature. All sintering time is 1 minute. As displayed in figure 2, the Ohm contact between Al paste and p-Si can be observed even at 600 ℃. The current changes linearly as bias changes from -0.7 V to 0.3 V. It indicates that there is no rectifying effect between p-Si and the two electrodes. However, there is no current in the sample with structure Al/p-Si/Al_2O_3/SiN_x/Al paste even the sinter temperature reaches 750 ℃. Hence, we assume that the Al_2O_3/SiNx double films cannot be fired through by such paste at this temperature. It results in the failure of forming contact between Al paste and p-Si layers. When the temperature is increased up to 800℃, the sample with structure Al/p-Si/Al_2O_3/SiN_x/Al paste shows a current comparable to that without Al_2O_3/SiN_x double films. The result indicates that the Al_2O_3/SiN_x stacks layers have been fired through, with generation of the alloy junction between p-Si and Al paste.

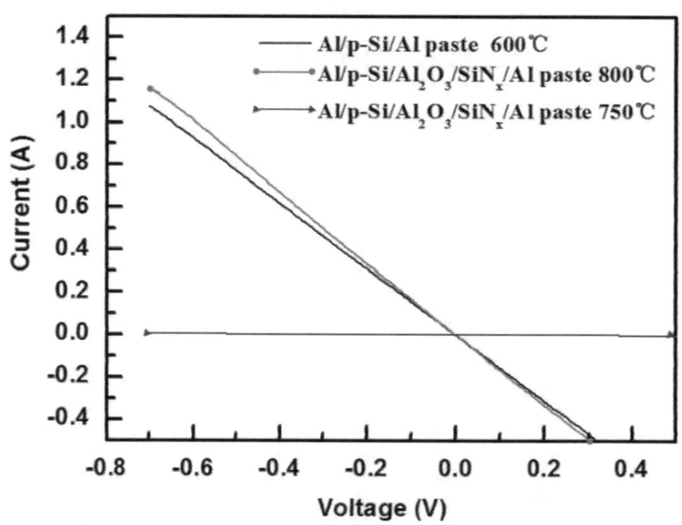

Figure 2. Comparison of I-V characteristics of samples with different firing conditions.

The SEM images show the cross-sections of samples after sintering at different temperatures. Figure 3(a) shows the cross-section of sample sintered at 750℃. It is obvious that there still is a clear boundary between Al and Al_2O_3/SiN_x layers. Since both Al_2O_3 and SiN_x are dielectric materials, it is hard to distinguish the boundary of Al_2O_3 and SiNx in SEM image. The SEM image in figure 3(a) reveals that the pasted Al does not penetrate into Al_2O_3/SiN_x double layers and form an alloy junction with p-Si as well. Figure 3(b) shows the cross-section of the sample sintered at 800 ℃, in which the boundary between Al particles and Al_2O_3/SiN_x vanishes. Additionally, the Al particles ablate with p-Si at some position as shown in the enlarged images of figure 3(b). These images also support the deduction obtained from the I-V comparison. According to the experimental results, the rear point contacts of PERC cells can be realized via sintering.

a) b)

Figure 3. SEM micrograph of a cross-section: for a full area screen-printed Al paste after 1 minute firing at 750 ℃ (a) and point contact Al paste after 1minute firing at 800 ℃ (b).

One-sun parameters of cells in different groups are summarized in table 1. The average open-circuit voltage of cells in the second group with Al_2O_3 rear surface passivation is slight higher than that of cells passivated by Al-BSF. The higher V_{oc} is attributing to the better passivation ability of Al_2O_3 for rear surface [5]. On the other hand, the small point contacts area results a higher contact resistance which is one of parts of series resistance of cells, and that leads to a low fill factor. Though Al_2O_3/SiN_x passivated cells have enhanced V_{oc} and I_{sc}, the average conversion efficiency of 9.20% is slightly lower than that of cells keeping Al-BSF. The lower fill factor also affects the efficiency, which results in an efficiency of 9.20%.

Table 1. Average One-sun parameters measured under standard testing conditions of two batches with different (Al-BSF and Al_2O_3/SiN_x) rear surface passivations

Rear side	V_{oc} (mV)	J_{sc} (mA/cm^2)	FF (%)	Eff (%)	R_s (Ω)
Al-BSF	577.1	27.7	58.2	9.25	0.30
Al_2O_3/SiN_x	582.9	28.3	55.6	9.20	0.32

Note: All cells were fabricated on 1.5~3Ωcm multicrystalline p-Si. The aperture cell area is 16cm^2.

In order to confirm the good passivation performance of Al_2O_3/SiN_x on rear surface after firing at 800℃, the internal quantum efficiency (IQE) of the best cell in each group is compared in Figure 4. It is obvious that IQE value of cell with Al_2O_3/SiN_x passivation is higher than that of cell with Al-BSF in the long-wavelength range. The higher IQE value in long-wavelength range due to excellent rear surface passivation of Al_2O_3/SiN_x stacks layers. The results also indicate that the passivating quality of Al_2O_3 is still good even though firing at 800℃. Combined with the results shown in table 1, the comparison of IQE proves that the point contacts formed via screen printing and different firing processes is feasible. Our research also provides a new cost-effective way to fabricate PERC cells.

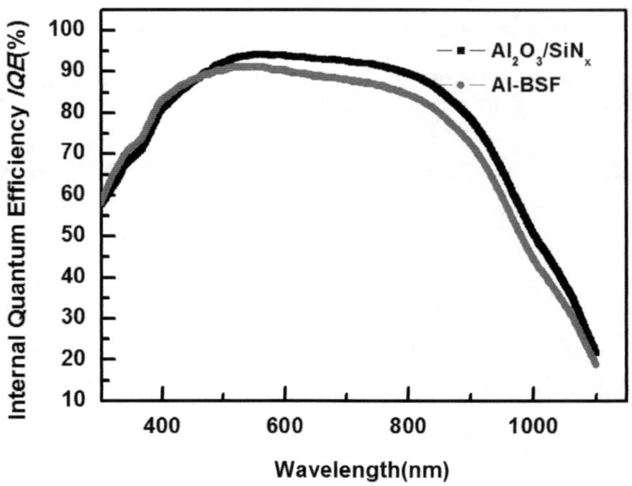

Figure 4. Measured internal quantum efficiency as a function of wavelength for p-type solar cells with different (Al-BSF and Al_2O_3/SiN_x) rear passivating procedures

4. Conclusions

In this study, a co-firing is proven to be a feasible way to form point contacts on rear side of PERC cells of which emitter and rear are passivated by Al_2O_3. Though three steps of firing were utilized in this study to form different contacts, cells with point contacts also exhibit a better V_{oc} and I_{sc} than that of cells with Al-BSF. It is also expected that the printed aluminum paste screen printed can be substituted by fritless Al paste to realize front and rear contact co-firing. The fabrication process to form point contacts on rear side of PERC cells proposed in this study may simplify the whole fabrication process of current industrial solar cells based on bulk silicon.

Acknowledgments

This work was supported by the Open Fund of the State Key Laboratory on Integrated Optoelectronics (No.IOSKL2015KF28), the fund of Zhejiang Provincial Natural Science Foundation (No.LQ15F040004 and No.LQ15F050004), and the National Natural Science Foundation of China (No.61505192).

References

[1] Dingemans G 2010 *Proceedings of the 25th EU PVSEC*. Spain.
[2] Saint-Cast P, Benick J, Kania D, Weiss L, Hofmann M, Rentsch J, Preu R, and Glunz Stefan W 2010 *IEEE Electron Device Letters* 31 695.
[3] Li T T, A C, Tan J, Samundsett C, Saynova D and Geerligs B 2010 *Proceedings of the Conference on Optoelectronic and Microelectronic Materials and Devices*. Australia.
[4] Lauermann T, T L, Scholz S, Raabe B, Hahn G and Terheiden B 2010 *Proceedings of the 35th IEEE Photovolatic Specialists Conference*. Honolulu.
[5] Schmidt J, Merkle A, Hoex B, van de Sanden M C, Kessels W.M.M and Brende R 2008 *Proceedings of the 33rd IEEE Photovolatic Specialists Conference.*
[6] Schmidt J, Bock R, Altermatt P P and Cuevas A 2008 *Proceedings of the 23rd EU PVSEC.*
[7] Sperlich H P, Saint-Cast P, Erben E and Peters L 201 *Proceedings of the 25th EU PVSEC.*
[8] Schneiderlochner E, Preu R, Ludemann R, and Glunz S W 2002 *Progress in photovoltaics: Research and Applications* 10 2.

IOP Conf. Series: Materials Science and Engineering **283** (2017) 012031 doi:10.1088/1757-899X/283/1/012031

Retardation effect of different alcohols on the cement coagulation in polycarboxylate- and naphthalene-based cement admixtures

S M Huang* **and F L Zhou**

Sanmenxia Polytechnic, Sanmenxia 472000, Henan, China

* Corresponding author: hsm362826@163.com

Abstract. Alcohol has great potential to delay the coagulation of cement. The effects of alcohol on paste fluidity and normal consistency coagulation time have been studied for polycarboxylate superplasticizer and naphthene cement admixture. Seven alcohols were combined with polycarboxylate superplasticizer and naphthene at a concentration of 0.01-0.09%, respectively, including n-propanol, methanol, sorbitol, ethylene glycol, glycerol, ethanol, and mannitol. The fluidity and normal consistency coagulation time of each cement admixture were measured. The performance of both polycarboxylate superplasticizer and naphthene cement admixtures were compared to develop cement admixture with delayed coagulation.

1. Introduction

Alcohols and their derivatives are commonly used as cement retarders due to their unique carbon chain structure. With the rapid development of water-reducing agent, in particular, the extensive application of polycarboxylate superplasticizer and naphthalene, it is of paramount importance to investigate the effects of alcohols and their derivatives on water reducing for use in combination with cement admixture. In this study, the effects of alcohols combined with polycarboxylate superplasticizer and naphthalene complex on delayed coagulation and dispersion properties of cement were analyzed and compared.

2. Materials and methods

2.1. Materials

The following alcohols were purchased and used in this study: n-propanol (purity: ≥99.8%, origin: Tianjin), methanol (purity: ≥99.5%, origin: Beijing), sorbitol (purity: ≥ 97.5%, origin: Tianjin), ethylene glycol (purity: 97%, Tianjin), glycerol (purity: ≥98%, origin: Yantai), dehydrated ethanol (purity: ≥99.5%, origin: Tianjin), and mannitol, purity: ≥99.5%, origin: Tianjin). Polycarboxylate superplasticizer (solid content: 34%) was purchased from Henan Province, naphthalene from Beijing, and ordinary Portland cement (strength grade: 42.5R) from Henan Province.

2.2. Methods

The fluidity, normal consistency, and setting time of cement paste were determined using the test method specified in GB/T 8077-2000 [1] and GB/T 1346-2001 [2]. The amount of alcohol was 0%, 0.01 %, 0.03%, 0.05%, 0.07% and 0.09%, respectively. The polycarboxylate superplasticizer was 0.6% in liquid. The naphthalene content was 0.6% in solid. The water to cement ratio was 0.29.

3. Results

IOP Conf. Series: Materials Science and Engineering **283** (2017) 012031 doi:10.1088/1757-899X/283/1/012031

3.1. The effect of n-propanol on fluidity and coagulation time of cement paste

Initially, the n-propanol was added to polycarboxylate superplasticizer or naphthalene at a percentage of 0-0.09%. The effects of n-propanol addition to polycarboxylate superplasticizer and naphthalene cement admixture, respectively, on the fluidity and normal consistency and setting time of the cement paste are shown in figures 1 and 2, respectively. One can see that n-propanol had a feeble effect on the fluidity of polycarboxylate superplasticizer (figure 1a) or naphthalene cement admixture (figure 2a). In fact, when the content of n-propanol was more than 0.05%, the addition of n-propanol to polycarboxylate superplasticizer cement admixture showed delayed coagulation (figure 1b) while no significant effect on naphthalene cement admixture (figure 2b). Therefore, the polycarboxylate superplasticizer in combination with n-propanol (>0.05%) was better than that of naphthalene.

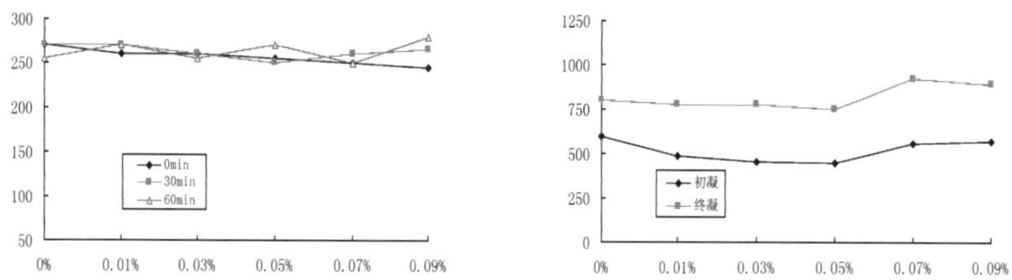

(a) (b)

Figure 1. Dosage curve of n-propanol in polycarboxylate superplasticizer cement admixture: (a) fluidity; (b) normal consistency and setting time.

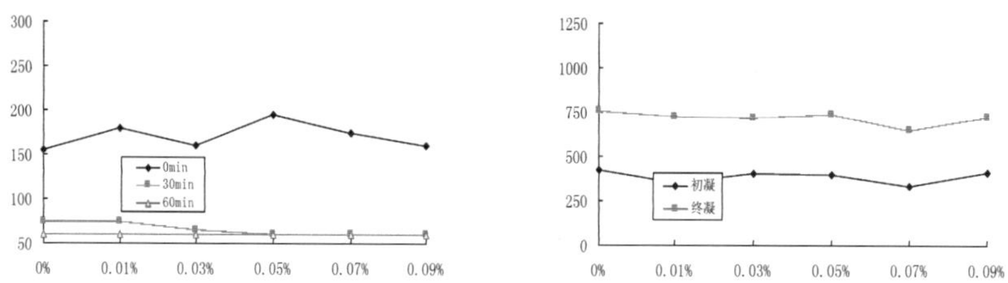

(a) (b)

Figure 2. Dosage curve of n-propanol in naphthalene cement admixture: (a) fluidity; (b) normal consistency and setting time.

3.2. The effect of methanol on fluidity and coagulation time of cement paste

Next, methanol was added to polycarboxylate superplasticizer or naphthalene at a percentage of 0-0.09%. The effect of methanol addition to polycarboxylate superplasticizer and naphthalene cement admixture on the fluidity and normal consistency and setting time of the cement paste fs shown in figures 3 and 4, respectively. It is obvious that methanol has little effect on the fluidity of polycarboxylate superplasticizer cement admixture (figure 3a) while the result for naphthalene cement admixture was not stable, decreasing with time (figure 4a). As far as the delayed coagulation is concerned, when the content of methanol was more than 0.01%, the addition of methanol to polycarboxylate superplasticizer cement admixture showed reduced initial setting time and increased final setting time (figure 3b). There was no significant effect on naphthalene cement admixture (figure 4b). Therefore, neither the polycarboxylate superplasticizer nor naphthalene was suitable to be used in combination with methanol.

IOP Conf. Series: Materials Science and Engineering **283** (2017) 012031 doi:10.1088/1757-899X/283/1/012031

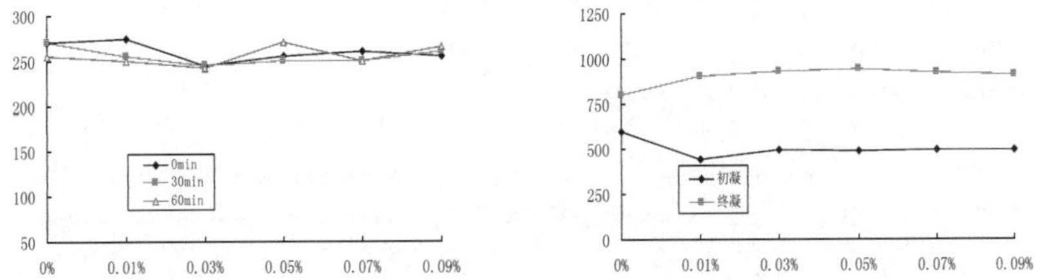

(a) (b)

Figure 3. Dosage curve of methanol in polycarboxylate superplasticizer cement admixture: (a) fluidity; (b) normal consistency and setting time.

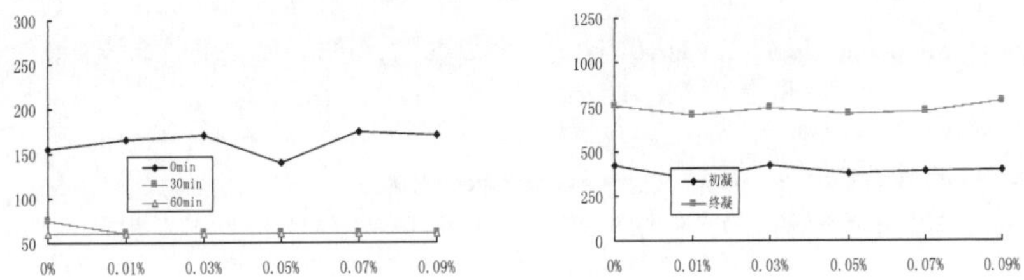

(a) (b)

Figure 4. Dosage curve of methanol in naphthalene cement admixture: (a) fluidity; (b) normal consistency and setting time.

3.3. The effect of sorbitol on fluidity and coagulation time of cement paste

Then sorbitol was added to polycarboxylate superplasticizer or naphthalene at a percentage of 0-0.09%. The effects of sorbitol addition to polycarboxylate superplasticizer and naphthalene cement admixture on the fluidity and normal consistency and setting time of the cement paste are shown in figures 5 and 6.

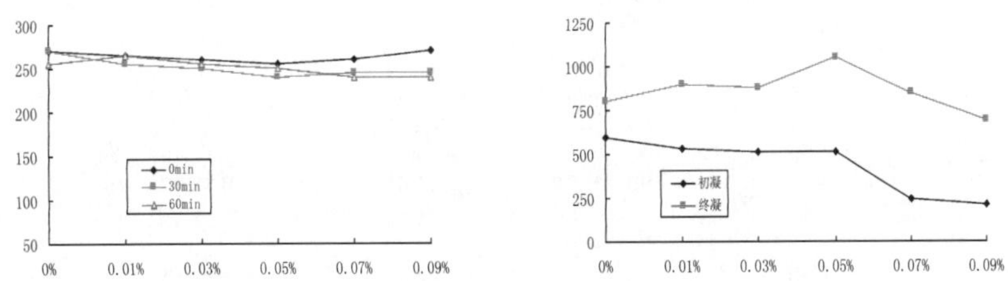

(a) (b)

Figure 5. Dosage curve of sorbitol in polycarboxylate superplasticizer cement admixture: (a) fluidity; (b) normal consistency and setting time.

It is seen that sorbitol had little effect on the fluidity of polycarboxylate superplasticizer cement admixture (figure 5a). When the content of sorbitol was more than 0.05%, the addition of sorbitol to polycarboxylate superplasticizer cement admixture showed shortened normal consistency and setting time (figure 5b). As far as the naphthalene cement admixture is concerned, when the sorbitol was more than 0.03%, the fluidity of the cement paste got worse (figure 6a). The sorbitol had no significant effect on the normal consistency and setting time of the naphthalene cement admixture (figure 6b). Therefore, at low dosage sorbitol and naphthalene cement admixture was slightly better than that of polycarboxylate superplasticizer. However, neither the polycarboxylate superplasticizer nor naphthalene was suitable to be used in combination with sorbitol at a dosage higher than 0.03%.

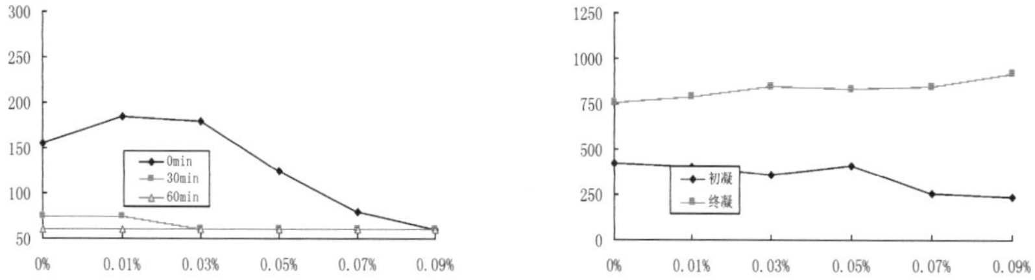

(a) (b)

Figure 6. Dosage curve of sorbitol in naphthalene cement admixture: (a) fluidity; (b) normal consistency and setting time.

3.4. The effect of ethylene glycol on fluidity and coagulation time of cement paste

Next, the ethylene glycol was added to polycarboxylate superplasticizer or naphthalene at a percentage of 0-0.09%. The effects of ethylene glycol addition to polycarboxylate superplasticizer and naphthalene cement admixture, respectively, on the fluidity and normal consistency and setting time of the cement paste are shown in figures 7 and 8, respectively. It is obvious that ethylene glycol had little effect on both the fluidity and normal consistency and setting time of polycarboxylate superplasticizer (figure 7) or naphthalene cement admixture (figure 8). There was no obvious advantage and disadvantage between them.

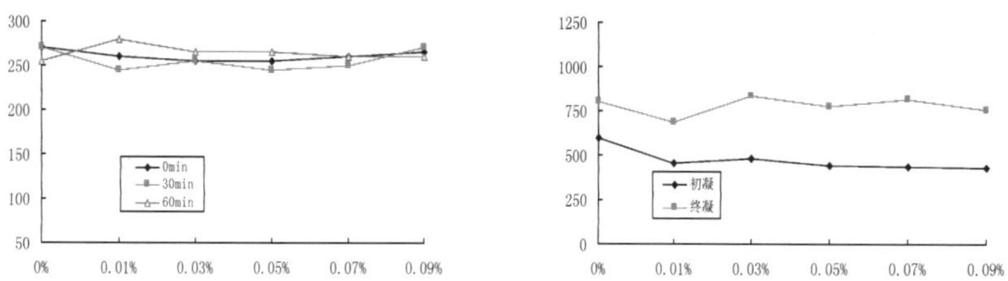

(a) (b)

Figure 7. Dosage curve of ethylene glycol in polycarboxylate superplasticizer cement admixture: (a) fluidity; (b) normal consistency and setting time.

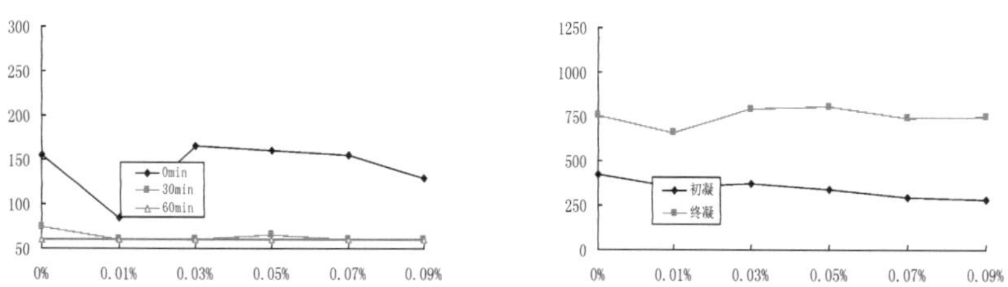

(a) (b)

Figure 8. Dosage curve of ethylene glycol in naphthalene cement admixture: (a) fluidity; (b) normal consistency and setting time.

3.5. The effect of glycerol on fluidity and coagulation time of cement paste

Subsequently, glycerol was added to polycarboxylate superplasticizer or naphthalene at a percentage of 0-0.09%. The effects of glycerol addition to polycarboxylate superplasticizer and naphthalene cement admixture, respectively, on the fluidity and normal consistency and setting time of the cement paste are shown in figures 9 and 10, respectively. It is obvious that glycerol had little effect on both the fluidity and normal consistency and setting time of polycarboxylate superplasticizer of cement admixture (figure 9a and b). Glycerol had impacted the fluidity of naphthalene admixture (figure 10a) while it had little effect on its normal consistency and setting time (figure 10b). When the content of glycerol was lower than 0.03%, it tended to increase the fluidity of naphthalene admixture (figure 10a). When the content of glycerol was over 0.03%, the fluidity could not be improved (figure 10a). It indicated that glycerol could be in combination with naphthalene at low dosage. Therefore, combined with a low dosage of glycerol naphthalene is better than polycarboxylate superplasticizer.

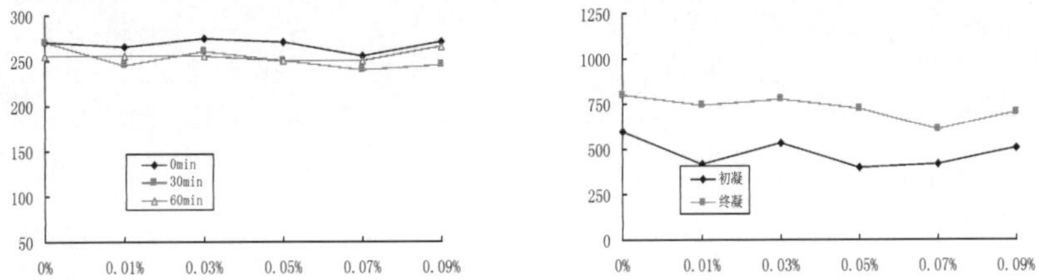

(a) (b)

Figure 9. Dosage curve of glycerol in polycarboxylate superplasticizer cement admixture: (a) fluidity; (b) normal consistency and setting time.

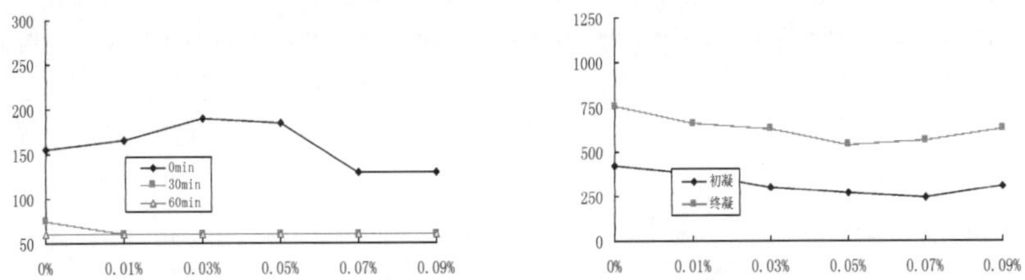

(a) (b)

Figure 10. Dosage curve of glycerol in naphthalene cement admixture: (a) fluidity; (b) normal consistency and setting time.

3.6. The effect of anhydrous ethanol on the fluidity and coagulation time of cement paste

Also, anhydrous ethanol was added to polycarboxylate superplasticizer or naphthalene at a percentage of 0-0.09%. The effects of anhydrous ethanol addition to polycarboxylate superplasticizer and naphthalene cement admixture on the fluidity and normal consistency and setting time of the cement paste are shown in figures 11 and 12, respectively. It is obvious that anhydrous ethanol has little effect on the fluidity of polycarboxylate superplasticizer cement admixture (figure 11a) as well as the normal consistency and setting time (figure 11b), while the result for naphthalene cement admixture was not stable (figure 12a and b). Therefore, neither the polycarboxylate superplasticizer nor naphthalene cement admixture was applicable in combination with anhydrous ethanol alone.

IOP Conf. Series: Materials Science and Engineering **283** (2017) 012031 doi:10.1088/1757-899X/283/1/012031

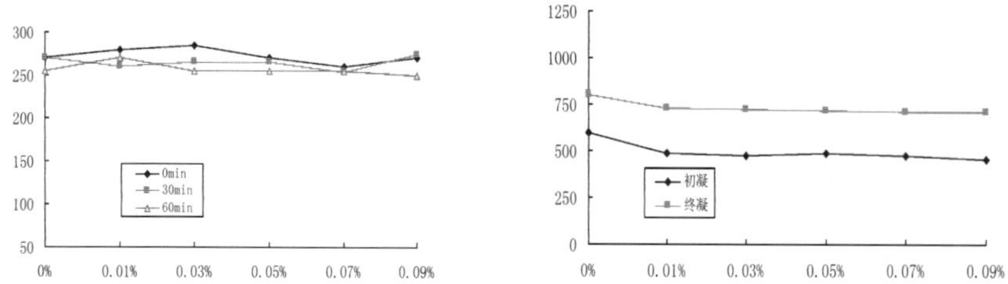

(a) (b)

Figure 11. Dosage curve of anhydrous ethanol in polycarboxylate superplasticizer cement admixture: (a) fluidity; (b) normal consistency and setting time.

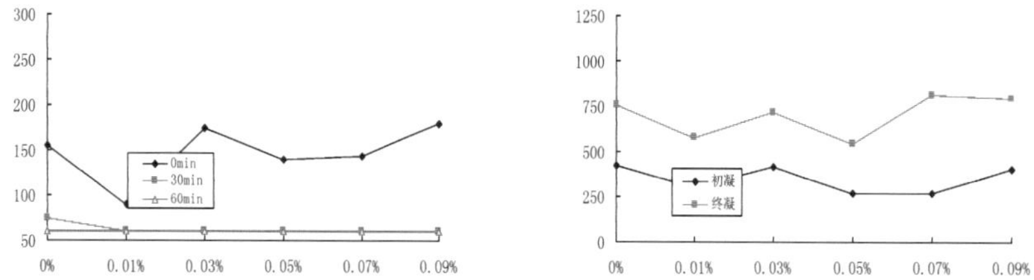

(a) (b)

Figure 12. Dosage curve of anhydrous ethanol in naphthalene cement admixture: (a) fluidity; (b) normal consistency and setting time.

3.7. The effect of mannitol on fluidity and coagulation time of cement paste

Finally, mannitol was added to polycarboxylate superplasticizer or naphthalene at a percentage of 0-0.09%. The effects of mannitol addition to polycarboxylate superplasticizer and naphthalene cement admixture on the fluidity and normal consistency and setting time of the cement paste are shown in figures 13 and 14, respectively.

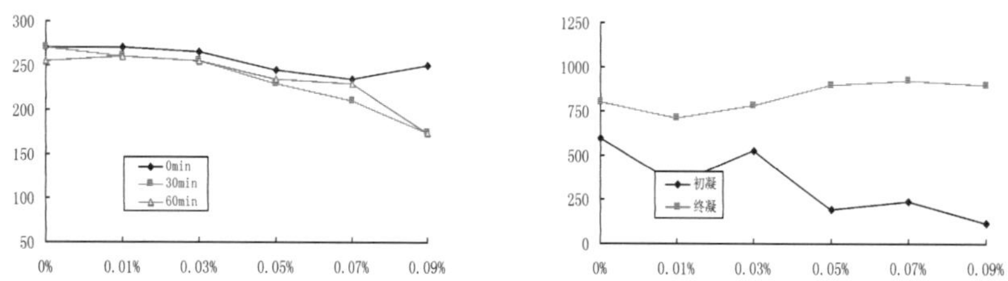

(a) (b)

Figure 13. Dosage curve of mannitol in polycarboxylate superplasticizer cement admixture: (a) fluidity; (b) normal consistency and setting time.

It is obvious that mannitol tended to decrease the fluidity of the polycarboxylate superplasticizer cement admixture (figure 13a). However, with the increase of dosage, mannitol could shorten the initial setting time and increase the final setting time (figure 13b). It indicated that mannitol could be combined with the polycarboxylate superplasticizer cement admixture but need control the dosage. As far as the naphthalene cement admixture is concerned, when mannitol was lower than 0.03%, the fluidity tended to increase; when the sorbitol was more than 0.03%, the fluidity of the cement paste got worsen (figure 14a). Meanwhile, mannitol could shorten the normal consistency and set time of the naphthalene cement admixture (figure 14b). Therefore, a low dosage of mannitol can be combined with naphthalene cement admixture. At low dosage mannitol combined with the naphthalene, cement admixture is better than with polycarboxylate superplasticizer.

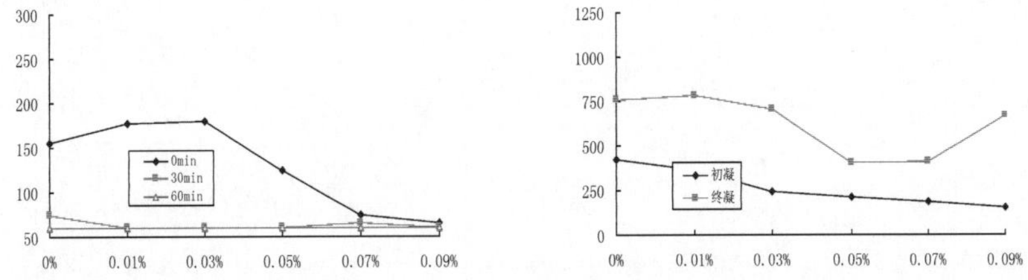

(a) (b)

Figure 14. Dosage curve of mannitol in naphthalene cement admixture: (a) fluidity; (b) normal consistency and setting time.

4. Conclusions

We compared the effect of commonly used alcohols (e.g., n-propanol, methanol, sorbitol, ethylene glycol, glycerol, ethanol, mannitol) on the fluidity and normal consistency and setting time of cement paste in combination with polycarboxylate superplasticizer and naphthene. We found out that the combination of a low dosage of sorbitol, glycerol, and mannitol with naphthalene is better than that of the polycarboxylate superplasticizer. Furthermore, the combination of high dosage of n-propanol with polycarboxylate superplasticizer is better than that of the naphthalene complex. Also, methanol, ethylene glycol, and anhydrous ethanol are not suitable for combining with either polycarboxylate superplasticizer or naphthalene complex.

References

[1] Methods for testing uniformity of concrete admixture 2000 GB/T 8077-2000.
[2] Test method for water consumption, coagulation time, consistency of cement standard consistency 2001 GB/T 1346-2001.

Institute of Physics
Dirac House, Temple Back
Bristol BS1 6BE UK

ISSN: 1757-8981
ISBN 978-1-5108-5500-7

CURRAN ASSOCIATES INC.
proceedings.com

9 781510 855007